Differential Geometry of Submanifolds and its Related Topics

Differential Geometry of Submanifolds and its Related Topics

Proceedings of the International Workshop
in Honor of S. Maeda's 60th Birthday

Saga University, Saga, Japan, 4 – 6 August 2012

Editors

Sadahiro Maeda (Saga University, Japan)

Yoshihiro Ohnita (Osaka City University, Japan)

Qing-Ming Cheng (Fukuoka University, Japan)

 **World Scientific**

NEW JERSEY · LONDON · SINGAPORE · BEIJING · SHANGHAI · HONG KONG · TAIPEI · CHENNAI

Published by

World Scientific Publishing Co. Pte. Ltd.

5 Toh Tuck Link, Singapore 596224

USA office: 27 Warren Street, Suite 401-402, Hackensack, NJ 07601

UK office: 57 Shelton Street, Covent Garden, London WC2H 9HE

British Library Cataloguing-in-Publication Data
A catalogue record for this book is available from the British Library.

DIFFERENTIAL GEOMETRY OF SUBMANIFOLDS AND ITS RELATED TOPICS
Proceedings of the International Workshop in Honor of S. Maeda's 60th Birthday

ISBN 978-981-4566-27-8

Printed in Singapore by B & Jo Enterprise Pte Ltd

FOREWORD

Professor Sadahiro MAEDA
and his friends in Mathematics

It is a great honor for me to have this opportunity to write this biographical profile on Professor Dr. Sadahiro MAEDA. On behalf of his friends, students and colleagues, I will give you a little history of this man and his achievements and describe his particular brand of mathematics. He completed his undergraduate studies in mathematics at Saga University in March, 1975 and went on to obtain his MA from Kumamoto University in March, 1978. During his master's studies, he met Professor Koichi Ogiue of Tokyo Metropolitan University, who came to Kumamoto University to give a series of lectures on differential geometry of submanifolds in complex space form. This was the key turning point of Sadahiro MAEDA's career. He immediately decided to apply for admission to enter the doctoral course at Tokyo Metropolitan University under Professor Ogiue. He was able to pass the entrance examination successfully, and began to study the geometry of submanifolds there. Among the many topics he was interested in at the time, he was particularly fascinated with real hypersurfaces in complex projective spaces and extrinsic appearances of geodesics of the submanifolds. In 1989, he published two joint papers. The first, entitled "On real hypersurfaces of a complex projective space", Math. Z. 202 (1989), 299-311, written with Professor Makoto Kimura. The second, entitled "Helical geodesic immersions into complex space forms", Geom. Dedicata 30 (1989), 93-114, written with Professor Yoshihiro Ohnita. MAEDA successfully defended his thesis to obtain his doctoral degree of mathematics from Tokyo Metropolitan University in May, 1989. He was offered his first position as a lecturer at Kumamoto Institute of Technology in April, 1985 and collaborated with several colleagues in mathematics, Professors Kazumi Tsukada, Yoshihiro Ohnita, Makoto Kimura, Toshikai Adachi and myself on a variety of projects. This marked the starting point of MAEDA's greatest achievements. It was then that he began studying extrinsic shapes of submanifolds from a geometrical perspective, and he subsequently conducted several intriguing and innovative geometrical studies. His papers are a plethora of geometrical ideas coming from his geometrically refined sense of taste. With these achievements in hand, he was offered a position as an associate professor in 1991 at Nagoya Institute of Technology where Professor T. Adachi was working. The paper "Circles in a complex projective space", Osaka J. Math. 32 (1995), 709-719, coauthored with Toshiaki Adachi and I, marked the beginning of a fruitful collaboration with Toshiaki Adachi. In 1995, MAEDA was awarded a professorship of mathematics at Shimane University. After some time, Makoto Kimura joined him there as a colleague in the mathematics department. By then, Professor Sadahiro MAEDA had become a mature geometer. In 2002, he wrote a paper with Professors T. Adachi and M. Yamagishi entitled "Length spectrum of geodesic spheres in a non-flat complex space form", J. Math. Soc. Japan 54 (2002). In that study, he discovered that the estimation on the number of the congruency classes of geodesics in geodesic spheres was determined by the length of geodesics. He has also collaborated with Professor K. Ogiue, Professor Ryoichi Takagi, Professor Bang-Yen Chen and others on many more studies. MAEDA has been a professor of Mathematics at Saga University since April, 2007. No less than 120 academic papers have been published since his graduation from Saga University, and Professor Sadahiro MAEDA is now recognized as a world-famous geometer comparable to his respected teacher Professor Koichi Ogiue. Not only known as a researcher but also as an accomplished educator MAEDA has sent several fine geometers into mathematics world. Nobutaka Boumuki, Hiromasa Tanabe and Kazuhiro Okumura were supervised by Professor Sadahiro MAEDA when they were predoctoral students.

This volume was planned to be published on the occasion of his 60th birthday.

In closing, Professor Sadahiro MAEDA expresses his heartfelt appreciation to all of his teachers, friends, students, the contributors of the present volume and especially to his wife, his son and two daughters for their kind support over the years.

Seiichi UDAGAWA

PREFACE

The workshop on Differential Geometry of Submanifolds and its Related Topics was held at Saga University, Saga City, Japan, during the period of 4-6 August, 2012 in honor of Professor Sadahiro Maeda (Saga University) for his 60th birthday, which was supported by the warm hospitality of Professor Q. M. Cheng (Fukuoka University).

This international workshop provided opportunities for geometers in Japan, China and Korea to exchange ideas and discuss geometry quite freely and in a homely atmosphere. This volume is concerned with minimal surfaces, real hypersurfaces of a nonflat complex space form, submanifolds of symmetric spaces and curve theory. Each participant shows his new results or gives a brief survey in these areas.

The publication of this volume is financially supported by the official fund of the Japanese government, which is entitled "Research on Submanifold Geometry and Harmonic Map Theory in Symmetric Spaces" with JSPS Grant-in-Aid for Scientific Research (C) No. 24540090, (2012-2014) whose principal investigator is Professor Y. Ohnita (Osaka City University and OCAMI). The purpose of this academic program is to study the submanifold geometry and harmonic map theory in symmetric spaces from viewpoints of the geometric variational problems, integrable systems and Lie theory of finite and infinite dimensions. This volume is published to realize such a purpose. We gratefully acknowledge Professor T. Hamada (Fukuoka University) for his devotion as the TeX Editor of our volume. The organizers would like to dedicate the present volume to Professor Sadahiro Maeda.

The Organizing Committee
May 5, 2013

ORGANIZING COMMITTEE

T. Adachi
Nagoya Institute of Technology, Japan

N. Boumuki
Tokyo University of Science, Japan

K. Okumura
Asahikawa National College of Technology, Japan

T. Shoda
Saga University, Japan

SCIENTIFIC ADVISORY COMMITTEE

S. Udagawa
Nihon University, Japan

T. Hamada
Fukuoka University, Japan

Saga University, Japan, August 2012

PRESENTATIONS

1. **Kim Young Ho** (Kyungpook National University, Korea),
   Archimedean Theorems and W-curves

2. **Toshihiro Shoda** (Saga University),
   Morse index of the classical periodic minimal surfaces

3. **Lingzhong Zeng** (Saga University),
   The spectrum of the poly-Laplacian with arbitrary order

4. **Toshiaki Adachi** (Nagoya Institute of Technology),
   Trajectories for Sasakian magnetic fields on homogeneous Hopf hypersurfaces in a complex hyperbolic space

5. **Yoshihiro Ohnita** (Osaka City University and OCAMI),
   Differential geometry of submanifolds in symmetric spaces

6. **Naoyuki Koike** (Tokyo University of Science),
   The classifications of certain kind of isoparametric submanifolds in symmetric spaces of noncompact type

7. **Seiichi Udagawa** (Nihon University),
   Real hypersurfaces with constant principal curvatures in a complex projective space

8. **Hiroshi Tamaru** (Hiroshima University),
   Some topics on homogeneous submanifolds in the complex hyperbolic space

9. **Nobutaka Boumuki** (Tokyo University of Science),
   Totally geodesic Kaehler immersions into complex space forms and a nonexistence theorem for Hessian metrics pf positive constant Hessian sectional curvature

10. **Sadahiro Maeda** (Saga University),
    Homogeneous submanifolds and homogeneous curves in space forms

11. **Makoto Kimura** (Ibaraki University),
    Austere hypersurfaces in odd dimensional sphere and real hypersurfaces in a complex projective space

12. **Kazuhiro Okumura** (Asahikawa National College of Technology),
    η-Einstein real hypersurfaces in a nonflat complex space form

13. **Makiko Sumi Tanaka** (Tokyo University of Science),
    Antipodal sets of compact symmetric spaces and the intersection of submanifolds

CONTENTS

Proceedings of the Workshop on
Differential Geometry of Submanifolds
and its Related Topics
Saga, August 4-6, 2012

HOMOGENEOUS SUBMANIFOLDS AND HOMOGENEOUS CURVES IN SPACE FORMS

Sadahiro MAEDA

Department of Mathematics, Saga University,
1 Honzyo, Saga 840-8502 Japan
E-mail: smaeda@ms.saga-u.ac.jp

We study homogeneous submanifolds and homogeneous curves in either a complex projective space, a complex hyperbolic space or a Euclidean sphere, namely they are orbits of some subgroups of the full isomery groups of these ambient spaces. The main purpose of this expository paper is to survey some results of the author on this topic without proof.

1. Homogeneous real hypersurfaces in a nonflat complex space form

All homogeneous real hypersurfaces in $\mathbb{C}P^n(c)$ are as follows (R. Takagi, On homogeneous real hypersurfaces in a complex projective space, Osaka J. Math. 10 (1973), 495-506).

Theorem 1.1. *In $\mathbb{C}P^n(c)$ $(n \geqq 2)$, a homogeneous real hypersurface is locally congruent to one of the following Hopf hypersurfaces all of whose principal curvatures are constant:*

(A$_1$) *A geodesic sphere $G(r)$ of radius r, where $0 < r < \pi/\sqrt{c}$;*
(A$_2$) *A tube of radius r around a totally geodesic $\mathbb{C}P^\ell(c)$ $(1 \leqq \ell \leqq n - 2)$, where $0 < r < \pi/\sqrt{c}$;*
(B) *A tube of radius r around a complex hyperquadric $\mathbb{C}Q^{n-1}$, where $0 < r < \pi/(2\sqrt{c}\,)$;*
(C) *A tube of radius r around the Segre embedding of $\mathbb{C}P^1(c) \times \mathbb{C}P^{(n-1)/2}(c)$, where $0 < r < \pi/(2\sqrt{c}\,)$ and n $(\geqq 5)$ is odd;*

(D) *A tube of radius r around the Plücker embedding of a complex Grass-mannian* $\mathbb{C}G_{2,5}$, *where* $0 < r < \pi/(2\sqrt{c}\,)$ *and* $n = 9$;

(E) *A tube of radius r around a Hermitian symmetric space* SO(10)/U(5), *where* $0 < r < \pi/(2\sqrt{c}\,)$ *and* $n = 15$.

The following theorem gives a geometric meaning of homogeneous real hypersurfaces in $\mathbb{C}P^n(c)$ (M. Kimura, Real hypersurfaces and complex sub-manifolds in complex projective space, Trans. Amer. Math. Soc. 296 (1986), 137-149).

Theorem 1.2. *A connected real hypersurface* M^{2n-1} *in* $\mathbb{C}P^n(c)$ $(n \geq 2)$ *is locally congruent to a homogeneous real hypersurface if and only if M is a Hopf hypersurface all of whose principal curvatures are constant in this ambient space.*

Motivated by Theorem 1.2, we obtain the following classification theorem of Hopf hypersurfaces all of whose principal curvatures are constant in $\mathbb{C}H^n(c)$ (J. Berndt, Real hypersurfaces with constant principal curvatures in complex hyperbolic space, J. Reine Angew. Math. 395 (1989), 132-141).

Theorem 1.3. *Let M be a connected Hopf hypersurface all of whose princi-pal curvatures are constant in* $\mathbb{C}H^n(c)$ $(n \geq 2)$. *Then M is locally congruent to one of the following homogeneous real hypersurfaces*:

(A_0) *A horosphere in* $\mathbb{C}H^n(c)$;

($A_{1,0}$) *A geodesic sphere* $G(r)$ *of radius r, where* $0 < r < \infty$;

($A_{1,1}$) *A tube of radius r around a totally geodesic* $\mathbb{C}H^{n-1}(c)$, *where* $0 < r < \infty$;

(A_2) *A tube of radius r around a totally geodesic* $\mathbb{C}H^\ell(c)$ $(1 \leq \ell \leq n - 2)$, *where* $0 < r < \infty$;

(B) *A tube of radius r around a totally real totally geodesic* $\mathbb{R}H^n(c/4)$, *where* $0 < r < \infty$.

Furthermore, we have the following classification theorem of all homogeneous real hypersurfaces in $\mathbb{C}H^n(c)$. The following theorem shows that there exist many non-Hopf homogeneous real hypersurfaces in $\mathbb{C}H^n(c)$ (J. Berndt and H. Tamaru, Cohomogeneity one actions on noncompact symmetric spaces of rank one, Trans. Amer. Math. Soc. 359 (2007), 3425-3438).

Theorem 1.4. *Let M be a homogeneous real hypersurface in* $\mathbb{C}H^n(c), n \geq 2$. *Then M is locally congruent to one of the following*:

(1) *A tube of radius $r \in (0, \infty)$ around a totally geodesic $\mathbb{C}H^k(c)$ in $\mathbb{C}H^n(c)$ for some $k \in \{0, \ldots, n-1\}$;*
(2) *A tube of radius $r \in (0, \infty)$ around a totally real totally geodesic $\mathbb{R}H^n(c/4)$ in $\mathbb{C}H^n(c)$;*
(3) *A horosphere in $\mathbb{C}H^n(c)$;*
(4) *The minimal ruled real hypersurface S determined by a horocycle in a totally real totally geodesic $\mathbb{R}H^2(c/4)$ in $\mathbb{C}H^n(c)$, or an equidistant hypersurface from S;*
(5) *A tube of radius $r \in (0, \infty)$ around a normally homogeneous submanifold F_k of $\mathbb{C}H^n(c)$ with real normal bundle of rank $k \in \{2, \ldots, n-1\}$;*
(6) *A tube of radius $r \in (0, \infty)$ around a normally homogeneous submanifold $F_{k,\varphi}$ of $\mathbb{C}H^n(c)$ with normal bundle of rank $2k \in \{2, \ldots, 2[(n-1)/2]\}$ and constant Kähler angle $\varphi \in (0, \pi/2)$.*

In Theorem 1.4, real hypersurfaces (1), (2), (3) are Hopf hypersurfaces, but real hypersurfaces (4), (5), (6) are non-Hopf hypersurfaces.

We shall characterize all homogeneous Hopf hypersurfaces of a nonflat complex space form $\widetilde{M}_n(c)(= \mathbb{C}P^n(c)$ or $\mathbb{C}H^n(c))$ from the viewpoint different from that of Theorem 1.2.

We first recall that $\widetilde{M}_n(c)$ admits no totally umbilic real hypersurfaces. This implies that $\widetilde{M}_n(c)$ does not have real hypersurfaces M^{2n-1} all of whose geodesics are mapped to circles in the ambient space $\widetilde{M}_n(c)$. We here weaken this geometric condition on geodesics of M.

Theorem 1.5 ([4, 20]). *Let M^{2n-1} be a connected real hypersurface in a nonflat complex space form $\widetilde{M}_n(c), n \geqq 2$. Then M is locally congruent to a homogeneous Hopf hypersurface if and only if there exist orthonormal vectors $v_1, \ldots, v_{2n-2}$ orthogonal to ξ at each point p of M such that all geodesics $\gamma_i = \gamma_i(s)$ $(1 \leqq i \leqq 2n-2)$ on M with $\gamma_i(0) = p$ and $\dot{\gamma}_i(0) = v_i$ are mapped to circles of positive curvature in $\widetilde{M}_n(c)$.*

We next weaken the fact that the holomorphic distribution $T^0M = \{X \in TM | X \perp \xi\}$ is *not* integrable for each Hopf hypersurface M of $\widetilde{M}_n(c)$.

Theorem 1.6 ([20]). *For a connected real hypersurface M^{2n-1} in a nonflat complex space form $\widetilde{M}_n(c), n \geqq 2$, the following two conditions are equivalent.*

(1) *M is locally congruent to a homogeneous Hopf hypersurface in $\widetilde{M}_n(c)$.*
(2) *The holomorphic distribution T^0M on M is decomposed as the direct sum of restricted principal distributions $V_\lambda^0 = \{X \in TM | AX = \ $*

$\lambda X, X \perp \xi\}$. *Moreover such every* V_λ^0 *satisfies one of the following two properties*:

2_a) V_λ^0 *integrable and each of its leaves is totally geodesic in our real hypersurface* M^{2n-1};

2_b) $V_\lambda^0 \oplus \{\xi\}_\mathbb{R}$ *is integrable.*

Remark 1.1. Theorems 1.5 and 1.6 characterize all homogeneous real hypersurfaces in $\mathbb{C}P^n(c)$.

In $\mathbb{C}H^n(c)$, we here consider a horosphere and the ruled homogeneous real hypersurface which are typical examples of homogeneous real hypersurfaces. Note that the former case is a Hopf hypersurface and the latter case is a non-Hopf hypersurface. We characterize individually these two homogeneous real hypersurfaces.

Proposition 1.1 ([32]). *For a connected real hypersurface M of $\mathbb{C}H^n(c)$, $n \geq 2$, the following three conditions are equivalent.*

(1) *M is locally congruent to a horosphere of $\mathbb{C}H^n(c)$.*
(2) *The exterior derivative $d\eta$ of the contact form η of M satisfies the following equation on $M : d\eta(X,Y) = (\sqrt{|c|}/2)g(X, \phi Y)$ for $\forall X, Y \in TM$ or $d\eta(X,Y) = -(\sqrt{|c|}/2)g(X, \phi Y)$ for $\forall X, Y \in TM$.*
(3) *At any point $p \in M$, there exist orthonormal vectors $v_1, \ldots, v_{2n-2}$ orthogonal to the characteristic vector ξ_p, which satisfies the property that all geodesics $\gamma_i = \gamma_i(s)$ $(1 \leq i \leq 2n-2)$ on M with $\gamma_i(0) = p, \dot{\gamma}_i(0) = v_i$ are mapped to a circle of the same curvature $\sqrt{|c|}/2$ in $\mathbb{C}H^n(c)$.*

Remark 1.2.

(1) In Proposition 1.1, Condition (3) gives a geometric meaning of the equation in Condition (2).
(2) For every real hypersurface M^{2n-1} of a nonflat complex space form $\widetilde{M}_n(c), n \geq 2$, the contact form η of M is not closed (i.e., $d\eta \neq 0$).

Proposition 1.2 ([2]). *For a connected ruled real hypersurface M of $\mathbb{C}H^n(c), n \geq 2$, the following two conditions are equivalent.*

(1) *M is homogeneous in the ambient space $\mathbb{C}H^n(c)$.*
(2) *Every integral curve γ of ξ lies on a totally real totally geodesic $\mathbb{R}H^2(c/4)$. Furthermore, the curvature function of γ does not depend on the choice of γ. That is, for arbitrary two points $x, y \in M$, taking two integral curves γ_x, γ_y of ξ with $\gamma_x(0) = x, \gamma_y(0) = y$, we can*

see that their curvature functions κ_x, κ_y satisfy $\kappa_x(s) = \kappa_y(s + s_0)$ for $\forall s \in (-\infty, \infty), \exists s_0 \in (-\infty, \infty)$.

Remark 1.3. $\mathbb{C}P^n(c)$ does not admit a homogeneous ruled real hypersurface. On the contrary, $\mathbb{C}H^n(c)$ has just one homogeneous ruled real hypersurface with respect to the action of its isometry group. Moreover, this homogeneous ruled real hypersurface is *minimal* in $\mathbb{C}H^n(c)$.

We shall construct and classify all minimal ruled real hypersurfaces in $\widetilde{M}_n(c)$ by using totally real circles (i.e., circles on a totally real totally geodesic $\mathbb{R}M^2(c/4)$ of constant sectional curvature $c/4$) in $\widetilde{M}_n(c)$.

The shape operator A of every minimal ruled real hypersurface of $\widetilde{M}_n(c)$ is expressed as:

$$A\xi = \nu U, \ AU = \nu\xi, \ AX = 0 \quad \text{for } X(\perp \xi, U) \in TM.$$

Here, $\nu = \|A\xi\|$ and U is a unit smooth vector field on an open dense subset $M_* = \{p \in M | \nu(p) \neq 0\}$. Denoting by $\widetilde{\nabla}$ the Riemannian connection of $\widetilde{M}_n(c)$, we see that every integral curve of the characteristic vector field ξ on an arbitrary *minimal* ruled real hypersurface M in $\widetilde{M}_n(c)$ satisfies the following ordinary differential equations (see [11]):

$$\widetilde{\nabla}_\xi \xi = \nu \phi U, \ \widetilde{\nabla}_\xi (\phi U) = -\nu\xi, \ \xi\nu = 0.$$

Hence we have the following

Proposition 1.3 ([11]). *Every integral curve γ of the characteristic vector field ξ on an arbitrary minimal ruled real hypersurface in $\widetilde{M}_n(c)$ is a totally real circle. In general, the curvature $k_\gamma (= |\nu|)$ of the curve γ depends on the choice of γ.*

Using this proposition and investigating the congruence relation between integral curves of ξ with respect to the isometry group $I(\widetilde{M}_n(c))$ of the ambient space $\widetilde{M}_n(c)$, we obtain the following two theorems.

Theorem 1.7 ([1]). *In $\mathbb{C}P^n(c), n \geq 2$, there exists the unique minimal ruled real hypersurface with respect to the action of its isometry group. Moreover, it is not complete.*

Theorem 1.8 ([1]). *In $\mathbb{C}H^n(c), n \geqq 2$, there exist three congruence classes of minimal ruled real hypersurfaces with respect to the action of its isometry group. They are the classes of elliptic type, parabolic type and axial type. Minimal ruled real hypersurfaces of parabolic type and axial type are complete. In these three real hypersurfaces, every integral curve γ of ξ is one of the following totally real circles.*

(1) *elliptic type*: γ *is a circle of curvature* $k_\gamma > \sqrt{|c|}/2$ *on* $\mathbb{R}H^2(c/4)$.
(2) *parabolic type*: γ *is a circle of curvature* $k_\gamma = \sqrt{|c|}/2$ *on* $\mathbb{R}H^2(c/4)$.
(3) *axial type*: γ *is a circle of curvature* $k_\gamma < \sqrt{|c|}/2$ *on* $\mathbb{R}H^2(c/4)$.

In the following three figures, dotted lines show leaves $\mathbb{C}H^{n-1}(c)$ of minimal ruled real hypersurfaces and lines inside of the balls show totally real circles on a totally real totally geodesic $\mathbb{R}H^2(c/4)$, which are denoted by γ in the above. In Fig. 2, two lines show horocyclic totally real circles having the same points at infinity. We may easily guess that the only parabolic one is homogeneous. In this case, our minimal ruled real hypersurface is expressed as an orbit under the action of the direct product of the isometry group $\mathrm{I}(\mathbb{C}H^{n-1}(c))$ of a totally geodesic $\mathbb{C}H^{n-1}(c)$ and a one-parameter subgroup $\{\varphi_s\}$ generating a circle of curvature $\sqrt{|c|}/2$ on $\mathbb{R}H^2(c/4)$.

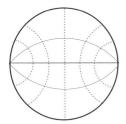 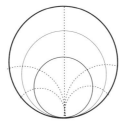 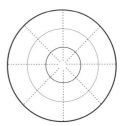

Fig. 1.   axial type   Fig. 2.   parabolic type   Fig. 3.   elliptic type

2. Characterizations of isoparametric hypersurfaces in a sphere

In this section, we study isoparametric hypersurfaces M^n (i.e., hypersurfaces M^n such that all the principal curvatures of M^n are constant) in Euclidean sphere $S^{n+1}(c)$.

The classification problem of isoparametric hypersurfaces M^n in $S^{n+1}(c)$ is still open. However we know that the numbers g of distinct principal curvatures of such hypersurfaces M^n are either $g = 1$ (i.e., M^n is totally umbilic), 2 (i.e., M^n is a Clifford hypersurface), 3 (i.e., M^n is a Cartan hypersurface), 4 or 6.

Moreover, the following are well-known.

(1) The shape operator A of an isoparametric hypersurface M^n in $S^{n+1}(c)$ is parallel if and only if $g = 1, 2$.
(2) When $g = 1$, all geodesics on M^n are mapped to circles (i.e., small circles or great circles) on the ambient sphere $S^{n+1}(c)$.

Weakening these geometric properties of isoparametric hypersurfaces with $g = 1, 2$, we can characterize all isoparametric hypersurfaces in a sphere.

Theorem 2.1. *For a connected hypersurface M^n of $S^{n+1}(c)$ ($n \geq 2$), the following three conditions are equivalent.*

(1) *M is locally congruent to an isoparametric hypersurface in $S^{n+1}(c)$.*

(2) *The tangent bundle TM of M is decomposed as the direct sum of the principal distributions $V_{\lambda_i} = \{X \in TM | AX = \lambda_i X\}$ such that the covariant derivative of the shape operator A of M in $S^{n+1}(c)$ satisfies $(\nabla_X A)Y = 0$ for all $X, Y \in V_{\lambda_i}$ for every principal curvature λ_i, where ∇ denotes the Riemannian connection of M.*

(3) *At each point $p \in M$, there exist an orthonormal basis $\{v_1, \ldots, v_{m_p}\}$ of the orthogonal complement of $\ker A_p$ in T_pM with $m_p = \operatorname{rank} A_p$ such that every geodesic of M through p with initial vector v_i ($1 \leq i \leq m_p$) is mapped to a small circle of positive curvature in $S^{n+1}(c)$.*

By virtue of Conditions (1) and (3) in Theorem 2.1 we have

Theorem 2.2 ([24]). *Let M^n be a connected hypersurface of $S^{n+1}(c)$ ($n \geq 2$). Then M^n is locally congruent to an isoparametric hypersurface without zero principal curvature in $S^{n+1}(c)$ if and only if at each point $p \in M$ there exists an orthonormal basis $\{v_1, \ldots, v_n\}$ of T_pM satisfying that all geodesics $\gamma_i = \gamma_i(s)$ ($1 \leq i \leq n$) on M^n with $\gamma_i(0) = p$ and $\dot{\gamma}_i(0) = v_i$ are mapped to small circles of positive curvature in the ambient sphere $S^{n+1}(c)$.*

Isoparametric hypersurfaces having two distinct constant principal curvatures in $S^{n+1}(c)$ are usually called *Clifford hypersurfaces*. For each pair of positive constants c_1, c_2 with $1/c_1 + 1/c_2 = 1/c$ and each natural number r with $1 \leq r \leq n-1$, we consider a Clifford hypersurface $M_{r,n-r} = M_{r,n-r}(c_1, c_2) := S^r(c_1) \times S^{n-r}(c_2)$. We know that $M_{r,n-r}$ has two distinct constant principal curvatures $\lambda_1 = c_1/\sqrt{c_1 + c_2}$ with multiplicity r and $\lambda_2 = -c_2/\sqrt{c_1 + c_2}$ with multiplicity $n - r$. Hence the tangent bundle $TM_{r,n-r}$ is decomposed as $TM_{r,n-r} = V_{\lambda_1} \oplus V_{\lambda_2}$. Clifford hypersurfaces have the following properties on their geodesics.

Proposition 2.1. *Let γ be a geodesic on a Clifford hypersurface $M_{r,n-r}(c_1, c_2)$ with $1/c_1 + 1/c_2 = 1/c$ and $r \in \{1, \ldots, n-1\}$ in $S^{n+1}(c)$. Then we have the following.*

(1) *γ is mapped to a great circle on $S^{n+1}(c)$ if and only if the initial vector $\dot{\gamma}(0)$ of the geodesic γ is expressed as $\dot{\gamma}(0) = (\sqrt{c_2}\, w_1 +$*

$\sqrt{c_1}\, w_2)/\sqrt{c_1 + c_2}$. Here, w_i is a unit vector of V_{λ_i} $(i = 1, 2)$.

(2) When $\dot{\gamma}(0)$ is neither a principal curvature vector nor the vector in Condition (1), the curve γ is not mapped to a circle on $S^{n+1}(c)$.

Considering the converse of Proposition 2.1, we obtain the following

Proposition 2.2. *A connected hypersurface M^n in $S^{n+1}(c)$ is locally congruent to a Clifford hypersurface $M_{r,n-r}(c_1, c_2)$ with $1/c_1 + 1/c_2 = 1/c$ and $r \in \{1, \ldots, n-1\}$ if and only if there exist a function $d : M \to \{1, 2, \ldots, n-1\}$, a constant α $(0 < \alpha < 1)$ and an orthonormal basis $\{v_1, \ldots, v_n\}$ of $T_p M$ at each point $p \in M$ satisfying the following two conditions:*

(1) All geodesics $\gamma_i = \gamma_i(s)$ $(1 \leqq i \leqq n)$ on M^n with $\gamma_i(0) = p$ and $\dot{\gamma}_i(0) = v_i$ are mapped to small circles of positive curvature on $S^{n+1}(c)$;

(2) All geodesics $\gamma_{ij} = \gamma_{ij}(s)$ $(1 \leqq i \leqq d(p) < j \leqq n)$ on M^n with $\gamma_{ij}(0) = p$ and $\dot{\gamma}_{ij}(0) = \alpha v_i + \sqrt{1 - \alpha^2}\, v_j$ are mapped to great circles on $S^{n+1}(c)$.

In this case, d is automatically $d \equiv r$ and our hypersurface M is expressed locally as a Clifford hypersurface $M_{n,n-r}(c/\alpha^2, c/(1 - \alpha^2))$.

At the end of this section we consider minimal isoparametric hypersurfaces with three distinct constant principal curvatures in $S^{n+1}(c)$. Isoparametric hypersurfaces with three distinct constant principal curvatures in a sphere are usually called *Cartan hypersurfaces*. If we denote by m_i the multiplicity of a principal curvature λ_i of a Cartan hypersurface, then we find that these three principal curvatures have the same multiplicity (i.e., $m_1 = m_2 = m_3$) (H. Münzner, Isoparametrische hyperflären I, II, Math. Ann. **251** (1980), 57-71; Math. Ann. **256** (1981), 215-232).

When a Cartan hypersurface is minimal, it is congruent to one of the following minimal hypersurfaces:

$$M^3 = O(3)/(O(1) \times O(1) \times O(1)) \to S^4(c),$$
$$M^6 = U(3)/(U(1) \times U(1) \times U(1)) \to S^7(c),$$
$$M^{12} = Sp(3)/(Sp(1) \times Sp(1) \times Sp(1)) \to S^{13}(c),$$
$$M^{24} = F_4/\mathrm{Spin}(8) \to S^{25}(c).$$

Principal curvatures of a Cartan minimal hypersurface are $\sqrt{3c}$, 0, $-\sqrt{3c}$ (É. Cartan, Familles de surfaces isoparamétriques dans les espaces à courbure constante, Ann. Mat. Pura Appl **17** (1938), 177-191; Sur de familles remarquables d'hypersurfaces isoparamétriques dans les espaces sphériques, Math. Z. **45** (1939), 335-367). Inspired by this fact, we obtain the following

Theorem 2.3. *Let M be a connected hypersurface of $S^{n+1}(c)$. Suppose that at each point $p \in M$ there exists an orthonormal basis $\{v_1, \ldots, v_{m_p}\}$ of the orthogonal complement of $\ker A_p$ with $m_p = \operatorname{rank} A_p$ such that*

(1) *all geodesics through p with initial vector v_i $(1 \leqq i \leqq m_p)$ are mapped to small circles of positive curvature in $S^{n+1}(c)$,*
(2) *all geodesics in Condition (1) have the same curvature k_p in $S^{n+1}(c)$.*

Then $k_p = k$ which is constant locally on M and M is locally congruent to either a totally umbilic hypersurface $S^n(c_1)$ with $k = \sqrt{c_1 - c}$, a Clifford hypersurface $M_{r,n-r}(2c, 2c)$ $(1 \leqq r \leqq n - 1)$ with $k = \sqrt{c}$, or a Cartan minimal hypersurface with $k = \sqrt{3c}$.

3. Homogeneous curves in a nonflat complex space form

The following lemma gives a necessary and sufficient condition for a Frenet (real) curve in $\widetilde{M}_n(c)$ to be a homogeneous curve in this space.

Lemma 3.1 ([35]). *For a Frenet curve $\gamma = \gamma(s)$ of proper order d in a nonflat complex space form $\widetilde{M}_n(c), n \geq 2$, it is a homogeneous curve in this space if and only if both of its all curvature functions $\kappa_1(s), \ldots, \kappa_{d-1}(s)$ and its all complex torsions $\tau_{ij}(s) := g(V_i(s), JV_j(s))$ $(1 \leq i < j \leq d)$ are constant functions. Here, J is the Kähler structure of $\widetilde{M}_n(c)$ and $\{V_1(s) := \dot{\gamma}(s), V_2(s), \ldots, V_d(s)\}$ is the Frenet frame along the curve γ.*

Helices of order 2 are usually called *circles*. Here, for a circle γ of positive curvature k in $\widetilde{M}_n(c)$, we set ordinary differential equations $\widetilde{\nabla}_{\dot{\gamma}}\dot{\gamma} = kV_2$, $\widetilde{\nabla}_{\dot{\gamma}}V_2 = -k\dot{\gamma}$. Then the complex torsion $\tau(s) := \tau_{12}(s) = g(\dot{\gamma}, JV_2(s))$ of the circle γ is constant along γ. That is, the complex torsion τ is a constant with $-1 \leqq \tau \leqq 1$. Hence it follows from Lemma 3.1 that every circle γ in $\widetilde{M}_n(c)$ is a homogeneous curve.

Since $\widetilde{M}_n(c)$ is a Riemannian symmetric space of rank one, all geodesics on this space are congruent to one another with respect to its isometry group $\mathrm{I}(\widetilde{M}_n(c))$. The following lemma is a congruence theorem on circles of positive curvature in $\widetilde{M}_n(c)$.

Lemma 3.2 ([35]). *For circles γ_i $(i = 1, 2)$ of curvature $k_i(> 0)$ and complex torsion τ_i in a nonflat complex space form $\widetilde{M}_n(c), n \geq 2$, they are congruent with respect to $\mathrm{I}(\widetilde{M}_n(c))$ if and only if the equation $k_1 = k_2, \tau_1 = \tau_2$ or $k_1 = k_2, \tau_1 = -\tau_2$ holds. These circles γ_1, γ_2 are congruent in the former case (resp. the latter case) by a holomorphic isometry (resp. an antiholomorphic isometry) of the ambient space $\widetilde{M}_n(c)$.*

So, in the following we take the complex torsion τ with $0 \leqq \tau \leqq 1$ for a circle in $\widetilde{M}_n(c)$. We call a circle a *totally real circle* (resp. *Kähler circle*) with complex torsion $\tau = 0$ (resp. $\tau = 1$). Note that the set of all congruence classes of circles in $\widetilde{M}_n(c)$ (with respect to $\mathrm{I}(\widetilde{M}_n(c))$) is a cone on $\mathbb{R}^2$.

We here study circles in $\mathbb{C}P^n(c)$. All of Geodesics (i.e., circles of null curvature), Kähler circles (i.e., circles lying on a totally geodesic complex curve $\mathbb{C}P^1(c)$) of curvature $k(> 0)$ and totally real circles (i.e., circles lying on a totally real totally geodesic $\mathbb{R}P^2(c/4)$) of curvature $k(> 0)$ are closed curves whose lengths are $2\pi/\sqrt{c}\,, 2\pi/\sqrt{k^2 + c}\,, 2\pi/\sqrt{k^2 + (c/4)}\,$, respectively.

We next study circles which are neither Kähler circles nor totally real circles in $\mathbb{C}P^n(c)$. As a matter of course, there are many open circles as well as many closed circles. A necessary and sufficient condition for a circle in $\mathbb{C}P^n(c)$ to be closed can be expressed in terms of its curvature k and complex torsion τ (for details, see [16]).

The following theorem gives information on the lengths of closed circles in $\mathbb{C}P^n(c)$.

Theorem 3.1 ([7]). *For all closed circles in* $\mathbb{C}P^n(c), n \geqq 2$ *we obtain the following properties.*

(1) *For all constants* $k(> 0)$ *and* τ $(0 < \tau < 1)$, *both the set of the lengths of closed circles with curvature* k *and the set of the lengths of closed circles with complex torsion* τ *are discrete unbounded sets on the real half line* $(0, \infty)$.

(2) *The set of the lengths of all closed circles in* $\mathbb{C}P^n(c)$ *coincides with the real half line* $(0, \infty)$. *This means that for each positive constant* λ *there exists a closed circle with length* λ *in* $\mathbb{C}P^n(c)$.

(3) *For every* $k(> 0)$, *the bottom of the set of the lengths of closed circles with curvature* k *is the length of a Kähler circle and the second lowest of this set is the length of a totally real circle. For every* $k > 0$, *if a closed circle* γ *of curvature* k *has the length* $2\pi/\sqrt{k^2 + c}$ *(resp.* $2\pi/\sqrt{k^2 + (c/4)}$ *), then it must be a Kähler circle (resp. totally real circle) of the same curvature* k.

(4) *For every* $\lambda > 0$, *denoting by* $m_c(\lambda)$ *the number of congruence classes (with respect to* $\mathrm{I}(\mathbb{C}P^n(c))$*) of closed circles with length* λ, *we find that* $m_c(\lambda)$ *is finite but not uniformly bounded. It satisfies* $\lim_{\lambda \to \infty} m_c(\lambda)/(\lambda^2 \log \lambda) = 9c/(8\pi^4)$.

(5) *There exists a unique closed circle with length* $\lambda(> 0)$ *with respect to* $\mathrm{I}(\mathbb{C}P^n(c))$ *if and only if* λ *satisfies* $2\pi/\sqrt{c} < \lambda \leqq 4\sqrt{5}\,\pi/(3\sqrt{c}\,)$.

(6) *For every* $\lambda > 0$, *denoting by* $m_c^k(\lambda)$ *the number of congruence classes*

(*with respect to* $\mathrm{I}(\mathbb{C}P^n(c))$) *of closed circles of curvature* k *whose length is* λ, *we see also that* $m_c^k(\lambda)$ *is finite but not uniformly unbounded. The growth of* $m_c^k(\lambda)$ *with respect to* λ *is not so rapid. It satisfies* $\lim_{\lambda\to\infty}\lambda^{-\delta}m_c^k(\lambda)=0$ *for each positive* δ. *That is, the growth of* $m_c^k(\lambda)$ *is smaller than every polynomial growth.*

Since every circle in $\mathbb{C}P^n(c)$ is a homogeneous curve in a totally geodesic $\mathbb{C}P^2(c)$, it is an orbit of a one-parameter subgroup of SU(3). Similarly, we find that all geodesics on a geodesic sphere $G(r)$ $(0 < r < \pi/\sqrt{c}\,)$ of $\mathbb{C}P^n(c)$ are homogeneous curves on a totally geodesic $\mathbb{C}P^2(c)$.

We next investigate geodesics on $G(r)$ of $\mathbb{C}P^n(c)$ in detail. We need a congruence theorem on those geodesics. For a geodesic $\gamma = \gamma(s)$ on $G(r)$, $\rho_\gamma := g(\dot{\gamma}(s), \xi_{\gamma(s)})$ is called a *structure torsion* of γ. By easy computation we see that ρ_γ is constant along γ.

Lemma 3.3 ([19]). *Geodesics* γ_1, γ_2 *on a geodesic sphere* $G(r)$ $(0 < r < \pi/\sqrt{c}\,)$ *in* $\mathbb{C}P^n(c)$ *are congruent with respect to the full isometry group* $\mathrm{I}(G(r))$ *if and only their complex torsions satisfy* $|\rho_{\gamma_1}| = |\rho_{\gamma_2}|$.

The following theorem gives information on the length spectrum of a geodesic sphere $G(r)$ $(0 < r < \pi/\sqrt{c}\,)$ in $\mathbb{C}P^n(c)$ in detail.

Theorem 3.2 ([19]). *For closed geodesics on a geodesic sphere* $G(r)$ $(0 < r < \pi/\sqrt{c}\,)$ *in* $\mathbb{C}P^n(c)$, *the following hold.*

(1) *There exist countably infinite congruence classes of closed geodesics on* $G(r)$ *with respect to* $\mathrm{I}(G(r))$. *The set of the lengths of all closed geodesics on* $G(r)$ *is a discrete unbounded subset in the real line* $\mathbb{R}$.

(2) *Denoting by* $K_{\max}$ *the maximal sectional curvature of* $G(r)$, *namely* $K_{\max} = c + (c/4)\cot^2(\sqrt{c}\,r/2)$, *we see that there exists a closed geodesic* γ *on* $G(r)$ *with length shorter than* $2\pi/\sqrt{K_{\max}}$ *if and only if* $\tan^2(\sqrt{c}\,r/2) > 2$ *holds. In this case, such a closed geodesic* γ *is an integral curve of the characteristic vector field* ξ. *Moreover, the length of every closed geodesic which is not congruent to an integral curve of* ξ *is longer than* $2\pi/\sqrt{K_{\max}}$ *for all* $r \in (0, \pi/\sqrt{c}\,)$.

(3) *If* $\tan^2(\sqrt{c}\,r/2)$ *is irrational, closed geodesics* γ_1, γ_2 *on* $G(r)$ *are congruent with respect to* $\mathrm{I}(G(r))$ *if and only if they have the common length.*

(4) *If* $\tan^2(\sqrt{c}\,r/2)$ *is rational, the number of congruence classes (with respect to* $\mathrm{I}(\mathbb{C}P^n(c))$) *of closed geodesics on* $G(r)$ *with common length* λ *is finite for each* $\lambda \in (0, \infty)$. *But it is not uniformly bounded. That*

is, denoting by $m(\lambda)$ the number of congruence classes (with respect to $I(\mathbb{C}P^n(c))$) of closed geodesics with length λ on $G(r)$, we find that $\limsup_{\lambda\to\infty} m(\lambda) = \infty$. Furthermore, for each positive number δ we have $\lim_{\lambda\to\infty} \lambda^{-\delta} m(\lambda) = 0$.

(5) *Denoting by $n(\lambda)$ the number of congruences (with respect to $I(\mathbb{C}P^n(c))$) of closed geodesics whose length is not longer than λ on $G(r)$, we find $\lim_{\lambda\to\infty} n(\lambda)/(\lambda^2) = 3c\sqrt{c}\, r/(8\pi^4 \sin(\sqrt{c}\, r))$.*

We here recall the following theorem related to Theorem 3.2(2).

Theorem 3.3 (Klingenberg). *Let M be a compact simply connected even dimensional Riemannian manifold. Suppose that there exists a positive constant K satisfying $0 < $ (every sectional curvature of M) $\leq K$ on M. Then M has no closed geodesics with length shorter than $2\pi/\sqrt{K}$.*

Remark 3.1. Theorem 3.3 does not hold in the case of odd dimensional Riemannian manifolds. In connection with this fact, we know that if we take the radius r of a geodesic sphere $G(r)$ ($0 < r < \pi/\sqrt{c}$) in $\mathbb{C}P^n(c)$ satisfying $\tan^2(\sqrt{c}\, r/2) > 2$, then every integral curve of ξ on $G(r)$ is a closed geodesic with length shorter than $2\pi/\sqrt{K}$, where $K = c + (c/4)\cot^2(\sqrt{c}\, r/2)$ (A. Weinstein, Distance spheres in complex projective spaces, Proc. Amer. Math. Soc. **39** (1973), 649-650). Note that $(c/4)\cot^2(\sqrt{c}\, r/2) \leq$ (every sectional curvature of $G(r)$) $\leq c + (c/4)\cot^2(\sqrt{c}\, r/2)$ for each $r \in (0, \pi/\sqrt{c})$. Theorem 3.2(2) gives detailed information concerning with the above fact.

4. Parallel immersions of real space forms into real space forms

We denote by $M^n(c)$ an n-dimensional *real space form* of constant sectional curvature c, which is an n-dimensional complete simply connected Riemannian manifold of constant sectional curvature c. Hence $M^n(c)$ is congruent to either a standard sphere $S^n(c)$, a Euclidean space $\mathbb{R}^n$ or a real hyperbolic space $H^n(c)$ according as c is positive, zero or negative.

An isometric immersion $f : M \to \widetilde{M}$ of an n-dimensional Riemannian manifold M into an $(n+p)$-dimensional Riemannian manifold $\widetilde{M}$ is said to be $(\lambda(x)\text{-})isotropic$ at $x \in M$ if $\|\sigma(X,X)\|/\|X\|^2 (= \lambda(x))$ does not depend on the choice of $X (\neq 0) \in T_x M$, where σ is the second fundamental form of the immersion f. If the immersion is $(\lambda(x)\text{-})$isotropic at every point x of M, then the immersion is said to be $(\lambda\text{-})$ isotropic (B. O'Neill, Isotropic and Kaehler immersions, Canadian J. Math. **17** (1965), 905-915). When $\lambda = 0$ on the submanifold M, this submanifold is totally geodesic in the ambient manifold $\widetilde{M}$. Note that a totally umbilic immersion is isotropic, but not *vice versa*.

The notion of isotropic immersions gives us valuable information in some cases. For example, we consider every parallel submanifold (M^n, f) of a standard sphere $S^{n+p}(\tilde{c})$. Then our manifold M^n is a compact Riemannian symmetric space. Moreover, by the classification theorem of parallel submanifolds in $S^{n+p}(\tilde{c})$ we can see that M^n is of rank one if and only if the isometric immersion $f : M^n \to S^{n+p}(\tilde{c})$ is isotropic (D. Ferus, Immersions with parallel second fundamental form, Math. Z. **140** (1974), 87-92 and M. Takeuchi, Parallel submanifolds of space forms, Manifolds and Lie groups (Notre Dame, Ind., 1980), in honor of Y. Matsushima, Birkhäuser, Boston, Mass. Progr. Math. **14** (1981), 429-447). That is, in this case the notion of isotropic immersions distinguishes symmetric spaces of rank one from symmetric spaces of higher rank.

Motivated by the above fact, we characterize all parallel immersions of real space forms into real space forms in terms of isotropic immersions.

Theorem 4.1 ([40]). *Let M^n be an $n(\geq 3)$-dimensional connected compact oriented isotropic submanifold whose mean curvature vector is parallel with respect to the normal connection in an $(n + p)$-dimensional real space form $\widetilde{M}^{n+p}(\tilde{c})$ of constant sectional curvature $\tilde{c}$ through an isometric immersion f. Suppose that every sectional curvature K of M^n satisfies $K \geqq (n/2(n + 1))(\tilde{c} + H^2)$, where H is the mean curvature of f. Then the immersion f has parallel second fundamental form and the submanifold M^n is congruent to one of the following:*

(1) *M^n is congruent to $S^n(K)$ of constant sectional curvature $K = \tilde{c} + H^2$ and f is a totally umbilic embedding;*

(2) *M^n is congruent to $S^n(K)$ of constant sectional curvature $K = (n/2(n+1))(\tilde{c} + H^2)$ and f is decomposed as*

$$f = f_2 \circ f_1 : \ S^n(K) \xrightarrow{f_1} S^{n+(n^2+n-2)/2}(2(n+1)K/n) \xrightarrow{f_2} \widetilde{M}^{n+p}(\tilde{c}),$$

where f_1 is a minimal immersion and f_2 is a totally umbilic embedding.

Remark 4.1. When $n = 2$, Theorem 4.1 also holds without the assumption that M^n is isotropic in $\widetilde{M}^{n+p}(\tilde{c})$ (see [40]).

References

1. T. Adachi, T. Bao and S. Maeda, *Congruence classes of minimal ruled real hypersurfaces in a nonflat complex space form*, a preprint.

2. T. Adachi, Y.H. Kim and S. Maeda, *A characterization of the homogeneous minimal ruled real hypersurface in a complex hyperbolic space*, J. Math. Soc. Japan **61** (2009), 315–325.

3. T. Adachi, Y.H. Kim and S. Maeda, *Characterizations of geodesic hyperspheres in a nonflat complex space form*, Glasgow Math. J. **55** (2013), 217–227.

4. T. Adachi, M. Kimura and S. Maeda, *A characterization of all homogeneous real hypersurfaces in a complex projective space by observing the extrinsic shape of geodesics*, Arch. Math. (Basel) **73** (1999), 303–310.

5. T. Adachi, M. Kimura and S. Maeda, *Real hypersurfaces some of whose geodesics are plane curves in nonflat complex space forms*, Tohoku Math. J. **57** (2005), 223–230.

6. T. Adachi and S. Maeda, *Holomorphic helices in a complex space form*, Proc. Amer. Math.. Soc. **125** (1997), 1197–1202.

7. T. Adachi and S. Maeda, *Length spectrum of circles in a complex projective space*, Osaka J. Math. **35** (1998), 553–565.

8. T. Adachi and S. Maeda, *Space forms from the viewpoint of their geodesic spheres*, Bull. Australian Math. Soc. **62** (2000), 205–210.

9. T. Adachi and S. Maeda, *Holomorphic helix of proper order 3 on a complex hyperbolic plane*, Topology and its Applications (Elsevier Science) **146/147** (2005), 201–207.

10. T. Adachi and S. Maeda, *Curves and submanifolds in rank one symmetric spaces*, Amer. Math. Soc., A Translation of Sugaku Expositions **19** (2006), 217–235.

11. T. Adachi and S. Maeda, *Integral curves of the characteristic vector field of real hypersurfaces in nonflat complex space forms*, Geometriae Dedicata **123** (2006), 65–72.

12. T. Adachi and S. Maeda, *Characterizations of type* (A₂) *hypersurfaces in a complex projective space*, Bull. Aust. Math. Soc. **77** (2008), 1–8.

13. T. Adachi and S. Maeda, *Extrinsic geodesics and hypersurfaces of type* (A) *in a complex projective space*, Tohoku Math. J. **60** (2008), 597–605.

14. T. Adachi and S. Maeda, *Sasakian curves on hypersurfaces in a nonflat complex space form*, Results in Math. **56** (2009), 489–499.

15. T. Adachi, S. Maeda and K. Ogiue, *Extrinsic shape of circles and standard imbeddings of projective spaces*, Manuscripta Math. **93** (1997), 267–272.

16. T. Adachi, S. Maeda and S. Udagawa, *Circles in a complex projective space*, Osaka J. Math. **32** (1995), 709–719.

17. T. Adachi, S. Maeda and S. Udagawa, *Ruled real hypersurfaces in a nonflat quaternionic space form*, Monatshefte Math. **145** (2005), 179–190.

18. T. Adachi, S. Maeda and S. Udagawa, *Schur's lemma for Kähler manifolds*, Arch. Math. (Basel) **90** (2008), 163–172.

19. T. Adachi, S. Maeda and M. Yamagishi, *Length spectrum of geodesic spheres in a non-flat complex space form*, J. Math. Soc. Japan **54** (2002), 373–408.

20. B.Y. Chen and S. Maeda, *Hopf hypersurfaces with constant principal curvatures in complex projective or complex hyperbolic spaces*, Tokyo J. Math. **24** (2001), 133–152.

21. B.Y. Chen and S. Maeda, *Real hypersurfaces in nonflat complex space forms are irreducible*, Osaka J. Math. **40** (2003), 121–138.

22. Y.H. Kim and S. Maeda, *Practical criterion for some submanifolds to be totally geodesic*, Monatshefte Math. **149** (2006), 233–242.

23. M. Kimura and S. Maeda, *On real hypersurfaces of a complex projective space*, Math. Z. **202** (1989), 299–311.

24. M. Kimura and S. Maeda, *Geometric meaning of isoparametric hypersurfaces in a real space form*, Canadian Math. Bull. **43** (2000), 74–78.

25. S. Maeda, *Imbedding of a complex projective space similar to Segre imbedding*, Arch. Math. (Basel) **37** (1981), 556–560.

26. S. Maeda, *Real hypersurfaces of complex projective spaces*, Math. Annalen **263** (1983), 473–478.

27. S. Maeda, *Isotropic immersions with parallel second fundamental form*, Canad. Math. Bull. **26** (1983), no.3, 291–296.

28. S. Maeda, *Isotropic immersions*, Canad. J. Math. **38** (1986), 416–430.

29. S. Maeda, *Real hypersurfaces of a complex projective space* II, Bull. Austra. Math. Soc. **30** (1984), 123–127.

30. S. Maeda, *Ricci tensors of real hypersurfaces in a complex projective space*, Proc. Amer. Math. Soc. **122** (1994), 1229–1235.

31. S. Maeda, *A characterization of constant isotropic immersions by circles*, Arch. Math. (Basel) **81** (2003), 90–95.

32. S. Maeda, *Geometry of the horosphere in a complex hyperbolic space*, Differential Geometry and its Applications **29** (2011), S246–S250.

33. S. Maeda and H. Naitoh, *Real hypersurfaces with ϕ-invariant shape operator in a complex projective space*, Glasgow Math. J. **53** (2011), 347–358.

34. S. Maeda and K. Ogiue, *Characterizations of geodesic hyperspheres in a complex projective space by observing the extrinsic shape of geodesics*, Math. Z. **225** (1997), 537–542.

35. S. Maeda and Y. Ohnita, *Helical geodesic immersions into complex space forms*, Geom. Dedicata **30** (1989), 93–114.

36. S. Maeda and K. Okumura, *Three real hypersurfaces some of whose geodesics are mapped to circles with the same curvature in a nonflat complex space form*, Geometriae Dedicata **156** (2012), 71–80.

37. S. Maeda and Y. Shimizu, *Imbedding of a complex projective space defined by monomials of same degree*, Math. Z. **179** (1982), 337–344.

38. S. Maeda and H. Tanabe, *Totally geodesic immersions of Kähler manifolds and Kähler Frenet curves*, Math. Z. **252** (2006), 787–795.

39. S. Maeda and K. Tsukada, *Isotropic immersions into a real space form*, Canad. Math. Bull. **37** (1994), 245–253.

40. S. Maeda and S. Udagawa, *Characterization of parallel isometric immersions of space forms into space forms in the class of isotropic immersions*, Canadian J. Math. **61** (2009), 641-655.

Received August 7, 2012.

Proceedings of the Workshop on
Differential Geometry of Submanifolds
and its Related Topics
Saga, August 4-6, 2012

INJECTIVITY PROPERTY OF REGULAR CURVES AND A SPHERE THEOREM

Osamu KOBAYASHI

Department of Mathematics, Osaka University,
Toyonaka, Osaka 560-0043, Japan
E-mail: kobayashi@math.sci.osaka-u.ac.jp

Dedicated to Professor Sadahiro Maeda on his 60th birthday

We have shown [3] that if the projective developing map of a regular curve in the sphere is injective then the curve has no self-intersection. In this paper we will show that among compact rank one symmetric spaces the only spheres have this injectivity property.

Keywords: scalar curvature, Schwarzian derivative, sphere theorem.

1. Introduction

The Schwarzian derivative of a regular curve $x\colon I \to (M, g)$ in a Riemannian n-manifold is defined [4] as

$$s_g x = (\nabla_{\dot x} \nabla_{\dot x} \dot x) \dot x^{-1} - \frac{3}{2}((\nabla_{\dot x} \dot x)\dot x^{-1})^2 - \frac{R_g}{2n(n-1)}\dot x^2,$$

where multiplications are the Clifford multiplications with respect to the metric g, and R_g denotes the scalar curvature. This is decomposed into the scalar part and the 2-form part:

$$s_q x(t) = s_q x^{(0)}(t) + s_g x^{(2)}(t) \in \mathbf{R} \oplus \Lambda^2 T_{x(t)}M.$$

For each point $t \in I$ there are a neighborhood $t \in U \subset I$ and a local parameter $u\colon U \to \mathbf{R}$ such that $s_g(x \circ u^{-1})^{(0)} = 0$. Such parameters define a projective structure [3] of the curve x, and we have a projective developing map

$$\mathrm{dev}_x\colon I \to \mathbf{RP}^1.$$

We say a compact Riemannian manifold M has *injectivity property* if $x: I \to M$ is injective provided dev_x is injective. The purpose of this paper is to show the following.

Theorem. *Suppose* (M, g) *is a compact rank one symmetric space. If* (M, g) *has the injectivity property, then it is the Euclidean sphere* S^n.

The sphere S^n has in fact the injectivity property [3].

2. Proof of theorem

Let (M, g) be a compact rank one symmetric space. Then we may assume g is a C_π-metric [1]. That is, all geodesics are closed with length π. Also the scalar curvature R_g is a positive constant. Let $x(s)$ be a geodesic with arclength parameter $s \in (-l/2, l/2)$. If we set $t(s) = \tan \sqrt{R_g/(4n(n-1))}\, s$, then with this parameter we have

$$s_g x^{(0)} = 2|\dot{x}|^2 \left(\frac{\frac{d^2}{ds^2}\sqrt{|\dot{x}|}}{\sqrt{|\dot{x}|}} + \frac{R_g}{4n(n-1)} \right) = 0.$$

This means that dev_x is injective for $l = \pi/\sqrt{R_g/(4n(n-1))}$ (*cf.* Proposition 2.5 of the reference [3]). Therefore, if $R_g < 4n(n-1)$ for a C_π-manifold, the injectivity property does not hold because $x(-\pi/2) = x(\pi/2)$.

For the C_π-metric of $\mathbf{RP}^n$, $R_g = n(n-1) < 4n(n-1)$ if $n > 1$. For the C_π-metric of $\mathbf{CP}^m$, $R_g = 4m(m+1) = n(n+2) < 4n(n-1)$ if $n = 2m > 2$. For the C_π-metric of $\mathbf{HP}^m$, $R_g = 16m(m+2) = n(n+8) < 4n(n-1)$ if $n = 4m > 4$. For the C_π-metric of $\mathbf{CaP}^2$, $\text{Ric}_g = 36g$ [5], and we have $R_g = 2^6 \cdot 3^2 < 4 \cdot 16(16-1)$.

These exhaust compact rank one symmetric spaces [1] except the spheres, and the proof is completed.

3. Concluding remarks

An argument similar to the above easily shows that *the product of Euclidean spheres* $S^p \times S^q$ *with possibly different radii has not the injectivity property unless* $pq = 0$.

If $R_g \leq 0$ and $n > 1$, then for every geodesic $x(s)$, $s \in \mathbf{R}$, we have $s_g x^{(0)} \leq 0$ and dev_x is injective [3]. On the other hand it is known [2] that every compact Riemannian manifold has a closed geodesic. Hence *a compact Riemannian manifold* (M, g) *with* $R_g \leq 0$ *and* $\dim M > 1$ *has not the injectivity property*.

These observations suggest that the class of compact Riemannian manifolds with the injectivity property is rather small. The author would like to raise the following question.

Question 3.1. If a compact connected (M, g) has the injectivity property, then is it the sphere?

References

1. A. Besse, *Manifolds all of whose geodesics are closed*, Springer, 1978.
2. W. Klingenberg, *Lectures on closed geodesics*, Springer, 1978.
3. O. Kobayashi, *Projective structures of a curve in a conformal space*, From Geometry to Quantum Mechanics, pp. 47–51, Progr. Math. 252, Birkhäuser, 2007.
4. O. Kobayashi and M. Wada, *Circular geometry and the Schwarzian*, Far East J. Math. Sci. Special Volume (2000), 335–363.
5. S. Maeda and K. Ogiue, *Geometry of submanifolds in terms of behavior of geodesics*, Tokyo J. Math. 17 (1994), 347–354.

Received October 5, 2012.

Proceedings of the Workshop on
Differential Geometry of Submanifolds
and its Related Topics
Saga, August 4-6, 2012

A FAMILY OF COMPLETE MINIMAL SURFACES
OF FINITE TOTAL CURVATURE WITH TWO ENDS

Shoichi FUJIMORI*

*Department of Mathematics, Faculty of Science, Okayama University,
Okayama, 700-8530 Japan
E-mail: fujimori@math.okayama-u.ac.jp*

Toshihiro SHODA†

*Faculty of Culture and Education, Saga University,
1 Honjo, Saga 840-8502 Japan
E-mail: tshoda@cc.saga-u.ac.jp*

Dedicated to Professor Sadahiro Maeda for his sixtieth birthday

In this paper, we construct a family of complete finite-total-curvature minimal
surfaces of positive genus with two ends.

Keywords: minimal surface, finite total curvature, two ends.

1. Introduction

One of recent developments of the theory of minimal surfaces has been
the discovery of the first new complete embedded minimal surface of finite topology [2]. The previous known examples were the plane, catenoid,
and helicoid. In 1984, Costa gave a complete finite-total-curvature minimal surface of genus one with three ends in Euclidean 3-space $\mathbb{R}^3$. After
that, many examples of complete finite-total-curvature minimal surfaces in
$\mathbb{R}^3$ have been found, even families of them. A main tool is the Weierstrass
representation formula. (See Theorem 1.1.) Nowadays, advanced analytic
methods like glueings to construct examples were established. In this paper,

---

*Partially supported by JSPS Grant-in-Aid for Young Scientists (B) 21740052.
†Partially supported by JSPS Grant-in-Aid for Young Scientists (B) 24740047.

we carry out the classical method, that is, using the Weierstrass representation formula. Now we refer to backgrounds and a motivation for our study.

Let $f : M \longrightarrow \mathbb{R}^3$ be a minimal immersion of a 2-manifold M into Euclidean 3-space $\mathbb{R}^3$. We usually call f a *minimal surface* in $\mathbb{R}^3$. The isothermal coordinates make M into a Riemann surface and f is called a *conformal minimal immersion*. For a conformal minimal immersion, it is well-known the following representation formula:

Theorem 1.1 (The Weierstrass representation, [11]). *Let $f : M \longrightarrow \mathbb{R}^3$ be a conformal minimal immersion. Then, up to translations, f can be represented by*

$$f(p) = \Re \int_{p_0}^{p} \begin{pmatrix} (1-g^2)\eta \\ i(1+g^2)\eta \\ 2g\eta \end{pmatrix} \tag{1}$$

where p_0 is a fixed point on M, g is a meromorphic function and η is a holomorphic differential on M satisfying the following: (i) η vanishes at the poles of g and the order of its zeros at such a point is exactly twice the order of the pole of g. (ii) If we set

$$\Phi := \begin{pmatrix} (1-g^2)\eta \\ i(1+g^2)\eta \\ 2g\eta \end{pmatrix} \tag{2}$$

then

$$\Re \oint_{\ell} \Phi = \mathbf{0} \quad \text{holds for any } \ell \in \pi_1(M). \tag{P}$$

Conversely, every path-integral (1) satisfying (i) and (ii) gives a conformal minimal immersion $f : M \longrightarrow \mathbb{R}^3$.

Remark 1.1. (i) (P) is called *the period condition* of the minimal surface and guarantees the well-definedness of the path-integral (1). (ii) $g : M \longrightarrow \mathbb{C} \cup \{\infty\}$ coincides with the composition of the Gauss map $G : M \longrightarrow S^2$ of the minimal surface and the stereographic projection $\sigma : S^2 \longrightarrow \mathbb{C} \cup \{\infty\}$, that is, $g = \sigma \circ G$. So we call g the Gauss map of the minimal surface.

Next, we assume that a minimal surface is complete and of finite total curvature. These two conditions give rise to restrictions on the topological and conformal type of minimal surfaces.

Theorem 1.2 ([6, 11]). *Let $f : M \to \mathbb{R}^3$ be a conformal minimal immersion. Suppose that f is complete and of finite total curvature. Then, we have the following:*

(1) M is conformally equivalent to a compact Riemann surface $\overline{M}_\gamma$ of genus γ punctured at a finite number of points $p_1, \ldots, p_n$.

(2) The Gauss map g extends to a holomorphic mapping $\hat{g} : \overline{M}_\gamma \to \mathbb{C} \cup \{\infty\}$.

Removed points $p_1, \ldots, p_n$ correspond to the ends of the minimal surface.

Now we consider a family of complete minimal surfaces of finite total curvature and focus on a few ends case in terms of their genus. For $n = 1$, there is a family of minimal surfaces with genus γ. In fact, a family of minimal surfaces derived from Enneper's surface is given by Chen and Gackstatter [1], Sato [12], Thayer [14], Weber and Wolf [15]. (See Figure 1.) The family is obtained by adding handles to Enneper's surface.

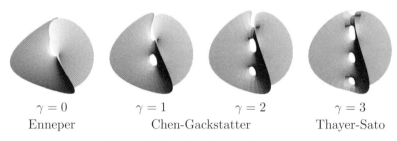

$\gamma = 0$ $\gamma = 1$ $\gamma = 2$ $\gamma = 3$

Enneper   Chen-Gackstatter   Thayer-Sato

Fig. 1.   Minimal surfaces with $n = 1$.

On the other hand, few families of minimal surfaces of positive genus for $n = 2$ are well-investigated. One of difficulties is that we cannot apply the same technique as above to solve the period problem (P) in Theorem 1.1. Catenoid is well-known example of minimal surfaces with two ends, and so, one may want to try adding handles to catenoid firstly. But, the following holds:

Theorem 1.3 ([13]). *Let $f : M \longrightarrow \mathbb{R}^3$ be a complete conformal minimal surface of finite total curvature. If f has two embedding ends, then f must be a catenoid.*

Since catenoid has two embedding ends, the surface given by adding handles to catenoid must have two embedding ends. Unfortunately, by Theorem 1.3, we cannot construct positive genus examples with two embedding ends. These are our motivation for constructing a family of minimal surfaces with an arbitrary genus for $n = 2$. Our main result is the following:

Theorem 1.4. *There exists a family of complete conformal minimal surfaces of finite total curvature with two ends for an arbitrary genus. (See Figure 2.)*

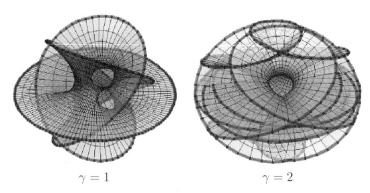

$$\gamma = 1 \qquad\qquad \gamma = 2$$

Fig. 2.   Minimal surfaces in Theorem 1.4.

The paper is organized as follows: Section 2 contains a construction of concrete examples to prove Theorem 1.4. In Section 3, we refer to our previous work related to complete minimal surfaces with the lowest total absolute curvature. Finally, we mention the remaining problems in Section 4.

Acknowledgments

This proceedings is published in honor of Professor Sadahiro O'Maeda's sixtieth birthday. He is one of authorities on the theory of real hypersurfaces. On the other hand, he is a good supervisor. In fact, his students got permanent positions at University and National colledge in Japan. The authors would like to thank him for his friendship and giving us an opportunity to submit the present paper to this memorial book.

Finally, we remark using O'Maeda in stead of Maeda as his surname. It is his nickname and the namer was Professor Koichi Ogiue who was his supervisor when he was a doctoral student at Tokyo Metropolitan University. He likes this nickname and we always call him O'Maeda. So we write his nickname in this paper as well.

2. Construction

In this section we will construct the family as in Theorem 1.4. We will use Theorem 1.1 for which we need a Riemann surface M, a meromorphic function g, and a holomorphic differential η.

For an arbitrary positive integer γ, let $\overline{M}_\gamma$ be the Riemann surface defined by

$$\overline{M}_\gamma = \left\{ (z, w) \in (\mathbb{C} \cup \{\infty\})^2 \,\middle|\, w^{\gamma+1} = z^\gamma (z^2 - 1) \right\}.$$

This $\overline{M}_\gamma$ was treated by Hoffman and Meeks [4] and hence we can apply the same arguments. But we have to improve these arguments to our situation. The surface we will consider is

$$M = \overline{M}_\gamma \setminus \{(0,0), (\infty, \infty)\},$$

a Riemann surface of genus γ from which two points have been removed. We want to define a complete conformal minimal immersion of finite total curvature from M into $\mathbb{R}^3$. To do this, set

$$g = cz^{2\gamma} w^\gamma, \qquad \eta = e^{-\frac{\pi i}{\gamma+1}} \frac{dz}{z^{2\gamma} w^\gamma},$$

where $c \in \mathbb{C}$ is a constant to be determined. Let Φ be the $\mathbb{C}^3$-valued differential as in (2). We shall prove that (1) gives a conformal minimal immersion of M.

It is easy to verify (i) in Theorem 1.1 and the completeness of f. (See the table below.)

(z, w)	$(0, 0)$	$(1, 0)$	$(-1, 0)$	(∞, ∞)
g	$0^{\gamma(3\gamma+2)}$	0^γ	0^γ	$\infty^{\gamma(3\gamma+4)}$
η	$\infty^{\gamma(3\gamma+1)}$			$0^{3\gamma^2+3\gamma-2}$

Hence, it suffices to show (P) in Theorem 1.1. Firstly,

$$\Re \oint_\ell g\eta = 0 \quad (\ell \in \pi_1(M))$$

follows from the exactness of $g\eta$, and so, we will check (P) for other components. To do so, we use the following symmetry of f.

Lemma 2.1 (Symmetries of the surface). *Consider the following conformal mapping of M:*

$$\kappa(z, w) = \left(-z, -e^{-\frac{\pi i}{\gamma+1}} w\right).$$

Then,

$$\kappa^* \Phi = \begin{pmatrix} (-1)^\gamma \cos \frac{\pi}{\gamma+1} & -(-1)^\gamma \sin \frac{\pi}{\gamma+1} & 0 \\ (-1)^\gamma \sin \frac{\pi}{\gamma+1} & (-1)^\gamma \cos \frac{\pi}{\gamma+1} & 0 \\ 0 & 0 & -1 \end{pmatrix} \Phi.$$

We next check the residues of η, $g^2\eta$ at the ends $(0,0)$, (∞,∞). Let $\tilde{\alpha}(t) = e^{it}$ $(0 \le t \le 2\pi)$ and α be the lift of $\tilde{\alpha}$ to M. The curve α winds once around the end $(0,0)$. Note that κ^2 is a deck transformation for a branched covering $(z, w) \longmapsto z$. Since $\tilde{\alpha}([0, 2\pi]) = -\tilde{\alpha}([0, 2\pi])$, there exists a number m satisfying $\kappa(\alpha) = (\kappa^2)^m(\alpha)$. Lemma 2.1 gives

$$\Re \oint_\alpha \Phi = \begin{pmatrix} (-1)^\gamma \cos \frac{\pi}{\gamma+1} & -(-1)^\gamma \sin \frac{\pi}{\gamma+1} & 0 \\ (-1)^\gamma \sin \frac{\pi}{\gamma+1} & (-1)^\gamma \cos \frac{\pi}{\gamma+1} & 0 \\ 0 & 0 & -1 \end{pmatrix}^{2m-1} \Re \oint_\alpha \Phi,$$

which implies that

$$\Re \oint_\alpha \Phi = \mathbf{0}.$$

Combining this and the residue theorem yields that they have no residues at two ends.

Finally, we consider path-integrals along topological 1-cycles on $\overline{M}_\gamma$. We will give the following convenient path on $\overline{M}_\gamma$:

$$\ell = \left\{ (z, w) = \left(t, e^{\frac{\pi i}{\gamma+1}} \sqrt[\gamma+1]{t^\gamma(1-t^2)} \right) \,\middle|\, 0 \le t \le 1 \right\}.$$

For an arbitrary number k $(1 \le k \le 2\gamma + 1)$, $\ell \cup \{-\kappa^k(\ell)\}$ defines 1-cycle on $\overline{M}_\gamma$. In this way, by the actions of κ and suitable linear combinations, we can obtain all of 1-cycles on M from ℓ.

We now calculate path-integrals of η and $g^2\eta$ along $\ell \cup \{-\kappa^k(\ell)\}$. Note that η has a pole at $(0,0)$. To avoid divergent integrals, here we add an exact 1-form which contains principal parts of η. After adding the exact 1-form, we can calculate path-integrals along $\ell \cup \{-\kappa^k(\ell)\}$ directly. Moreover, if (P) holds for ℓ, then

$$\Re \int_{\kappa\circ\ell} \Phi = \Re \int_\ell \kappa^* \Phi = K \Re \int_\ell \Phi = \mathbf{0}$$

for some orthogonal matrix K by Lemma 2.1. Hence all that remains to be done is to show that (P) holds for ℓ.

It is straightforward to check

$$\frac{dz}{z^{2\gamma}w^\gamma} - \frac{\gamma+1}{3\gamma^2+\gamma-1} d\left(\frac{w}{z^{3\gamma-1}}\right) = \frac{3\gamma^2+\gamma-3}{3\gamma^2+\gamma-1} \frac{dz}{z^{2(\gamma-1)}w^\gamma}.$$

As a consequence, up to exact 1-forms,

$$\frac{dz}{z^{2\gamma}w^{\gamma}} \equiv \prod_{k=1}^{2\gamma} \frac{3k^2 + k - 3}{3k^2 + k - 1} \frac{z^{2\gamma}}{w^{\gamma}} dz.$$

Setting

$$\delta_{\gamma} := \prod_{k=1}^{2\gamma} \frac{3k^2 + k - 3}{3k^2 + k - 1} \tag{3}$$

for a simplicity, we have

$$\int_{\ell} \eta = e^{-\frac{\pi i}{\gamma+1}} \int_{\ell} \frac{dz}{z^{2\gamma}w^{\gamma}} = \delta_{\gamma} e^{-\frac{\pi i}{\gamma+1}} \int_{\ell} \frac{z^{2\gamma}}{w^{\gamma}} dz = -\delta_{\gamma} \int_{0}^{1} \frac{t^{2\gamma}}{\{t^{\gamma}(1-t^2)\}^{\frac{\gamma}{\gamma+1}}} dt,$$

$$\int_{\ell} g^2 \eta = c^2 e^{-\frac{\pi i}{\gamma+1}} \int_{\ell} z^{2\gamma} w^{\gamma} dz = -c^2 e^{-\frac{2\pi}{\gamma+1}i} \int_{0}^{1} t^{2\gamma} \{t^{\gamma}(1-t^2)\}^{\frac{\gamma}{\gamma+1}} dt.$$

By setting

$$A_{\gamma} = \delta_{\gamma} \int_{0}^{1} \frac{t^{2\gamma}}{\{t^{\gamma}(1-t^2)\}^{\frac{\gamma}{\gamma+1}}} dt, \quad B_{\gamma} = \int_{0}^{1} t^{2\gamma} \{t^{\gamma}(1-t^2)\}^{\frac{\gamma}{\gamma+1}} dt,$$

(P) is reduced to

$$\Re\{-A_{\gamma} + c^2 e^{-\frac{2\pi}{\gamma+1}i} B_{\gamma}\} = 0, \quad \Re\{i(-A_{\gamma} - c^2 e^{-\frac{2\pi}{\gamma+1}i} B_{\gamma})\} = 0.$$

Let us set

$$c = e^{\frac{\pi i}{\gamma+1}} \sqrt{\frac{A_{\gamma}}{B_{\gamma}}}.$$

This choice of c gives (P) and this completes the proof.

3. Further progress

In this section, we consider complete minimal surfaces of finite total curvature with two ends from other point of view and introduce our previous work [3].

For a complete minimal surface of finite total curvature in $\mathbb{R}^3$, a stronger inequality of the classical inequality of Cohn-Vossen holds. Recall that if M is a complete Riemannian 2-manifold of finite total curvature $\int_M K \, dA$ and finite Euler characteristic $\chi(M)$, then $\int_M K \, dA \leq 2\pi\chi(M)$ (Cohn-Vossen inequality). For minimal surfaces, a stronger statement can be made:

Theorem 3.1 ([7,11,13]). *Let $f : M \to \mathbb{R}^3$ be a minimal surface as in Theorem 1.2. Then, the following inequality holds:*

$$\frac{1}{2\pi} \int_M K dA \leq \chi(M) - n = \chi(\overline{M}_\gamma) - 2n = 2(1 - \gamma - n). \tag{4}$$

Furthermore, equality in (4) holds if and only if each end is properly embedded.

By this fact, the total absolute curvature $\int_M |K| dA = -\int_M K dA$ has the lower bound. In this point of view, it is natural to ask whether there is a minimal surface which attains the minimum value of the total absolute curvature or not.

If $n = 1$, then a minimal surface satisfying $\int_M K dA = -4\pi\gamma$ must be a plane [4]. Thus on a non-planar minimal surface,

$$-\int_M K dA \geq 4\pi(\gamma + 1) \tag{5}$$

holds. We have examples satisfying equality in (5). In fact, examples in Figure 1 are typical ones.

Thus we now focus on the case of $n = 2$. It follows from Theorem 1.3 that on a non-catenoidal minimal surface with two ends,

$$-\int_M K dA \geq 4\pi(\gamma + 2) \tag{6}$$

holds. As a consequence, it is reasonable to ask the following problem:

Problem 3.1. *For an arbitrary genus γ, does there exist a complete conformal minimal surface of finite total curvature with two ends which satisfies equality in (6)?*

In the case $\gamma = 0$ such minimal surfaces exist, and moreover, these minimal surfaces have been classified by López [8]. (See Figure 3.) However, for the case $\gamma > 0$ no answer to Problem 3.1 is known. For this, we obtain the following partial answer:

Theorem 3.2 ([3]). *If γ is equal to 1 or even number, there exists a complete conformal minimal surface of finite total curvature with two ends which satisfies equality in (6).*

We will skip the details of Theorem 3.2 but we introduce their Weierstrass representation formulae, respectively.

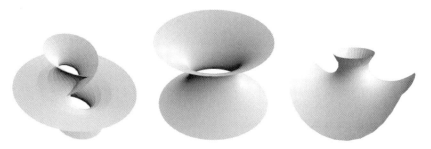

Fig. 3.   Examples for $\gamma = 0$. The surface in the middle is a double cover of a catenoid.

For $\gamma = 1$ in Theorem 3.2, $\overline{M}_1$ is defined by

$$\overline{M}_\gamma = \left\{ (z, w) \in (\mathbb{C} \cup \{\infty\})^2 \mid w^{\gamma+1} = z(z^2 - 1)^\gamma \right\},$$

and the surface we will consider is

$$M = \overline{M}_\gamma \setminus \{(0, 0), (\infty, \infty)\}.$$

Its Weierstrass representation is given by

$$g = cw, \qquad \eta = i \frac{dz}{z^2 w},$$

where $c \in \mathbb{R}$ is a suitable constant. (See Figure 4.)

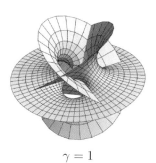

$\gamma = 1$

Fig. 4.   A minimal surface of genus 1 with two ends which satisfies equality in (6).

Next we refer to examples for even γ in Theorem 3.2. (See Figure 5.) For an even number $\gamma \geq 2$, let $\overline{M}_\gamma$ be the Riemann surface defined by

$$\overline{M}_\gamma = \left\{ (z, w) \in (\mathbb{C} \cup \{\infty\})^2 \, \middle| \, w^{\gamma+1} = z^2 \left(\frac{z-1}{z-a} \right)^\gamma \right\},$$

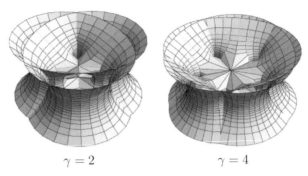

$$\gamma = 2 \qquad\qquad \gamma = 4$$

Fig. 5. Minimal surfaces of genus k with two ends which satisfy equality in (6).

where $a \in (1, \infty)$ is a suitable constant, and

$$M = \overline{M}_\gamma \setminus \{(0,0), (\infty, \infty)\}\,.$$

Its Weierstrass representation is given by

$$g = cw \quad \left(c = a^{(k-2)/(2k+2)} \in \mathbb{R}_{>0}\right), \quad \eta = \frac{dz}{zw}\,.$$

4. Remaining problems

In this section we introduce remaining problems related to the work in Section 3.

4.1. *The case γ odd and greater than 1*

For the case that the genus γ is odd and greater than 1, a complete minimal surface of finite total curvature $f : M = \overline{M}_\gamma \setminus \{p_1, p_2\} \to \mathbb{R}^3$ which satisfies equality in (6) is yet to be found. However, M. Weber has constructed the following examples numerically. (See Figure 6.)

Example 4.1 (Weber). *Let γ be a positive integer. Define*

$$F_1(z; a_1, a_3, \ldots, a_{2\gamma-1}) - \prod_{i=1}^{\gamma}(z - a_{2i-1}), \quad F_2(z; a_2, a_4, \ldots, a_{2\gamma}) = \prod_{i=1}^{\gamma}(z - a_{2i}),$$

where $1 = a_1 < a_2 < \cdots < a_{2\gamma}$ are constants to be determined. Define a compact Riemann surface $\overline{M}_\gamma$ of genus γ by

$$\overline{M}_\gamma = \left\{ (z, w) \in (\mathbb{C} \cup \{\infty\})^2 \,\middle|\, w^2 = z\,\frac{F_1(z; a_1, a_3, \ldots, a_{2\gamma-1})}{F_2(z; a_2, a_4, \ldots, a_{2\gamma})} \right\}.$$

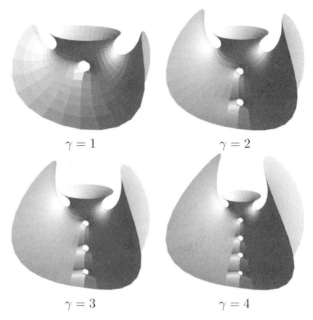

$\gamma = 1$

$\gamma = 2$

$\gamma = 3$

$\gamma = 4$

Fig. 6.   Minimal surfaces of genus γ with two ends which satisfy $\deg(g) = \gamma + 2$.

We set

$$M = \overline{M}_\gamma \setminus \{(0,0), (\infty,\infty)\}, \quad g = c\frac{w}{z+1} \ (c > 0), \quad \eta = \frac{(z+1)^2}{zw} dz.$$

Then there exist constants $c, a_2, a_3, \ldots, a_{2\gamma}$ such that (P) *holds.*

For $\gamma = 1$, we can prove the existence of the surface rigorously. However, for other cases, since the surface does not have enough symmetry, the rigorous proof of the existence still remains an open problem.

4.2. *Existence of non-orientable minimal surfaces*

Recall that the total absolute curvature of a minimal surface in $\mathbb{R}^3$ is just the area under the Gauss map $g : M \to \mathbb{C} \cup \{\infty\} \cong S^2$, that is,

$$-\int_M K dA = (\text{the area of } S^2) \deg(g) = 4\pi \deg(g) \in 4\pi \, \mathbb{Z}.$$

(See for example (3.11) in Hoffman and Osserman [5] for details.)

From a point of view of the Gauss map, our work is devoted to minimal surfaces satisfying $\deg(g) = \gamma + 2$. On the other hand, it is important to consider the existence of minimal surfaces with $\deg(g) = \gamma + 3$ in terms of non-orientable minimal surfaces. Now, we review non-orientable minimal surfaces in $\mathbb{R}^3$.

Let $f' : M' \to \mathbb{R}^3$ be a minimal immersion of a non-orientable surface into $\mathbb{R}^3$. Then the oriented two sheeted covering space M of M' inherits naturally a Riemann surface structure and we have a canonical projection $\pi : M \to M'$. We can also define a map $I : M \to M$ such that $\pi \circ I = \pi$, and which is an antiholomorphic involution on M without fixed points. Here M' can be identified with $M/\langle I \rangle$. In this way, if $f : M \to \mathbb{R}^3$ is a conformal minimal surface and there is an antiholomorphic involution $I : M \to M$ without fixed points so that $f \circ I = f$, then we can define a non-orientable minimal surface $f' : M' = M/\langle I \rangle \to \mathbb{R}^3$. Conversely, every non-orientable minimal surface is obtained in this procedure.

Suppose that $f' : M' = M/\langle I \rangle \to \mathbb{R}^3$ is complete and of finite total curvature. Then, we can apply Theorems 1.2 and 3.1 to the conformal minimal immersion $f : M \to \mathbb{R}^3$. Furthermore, we have a stronger restriction on the topology of M' or M. In fact, Meeks [10] showed that the Euler characteristic $\chi(\overline{M}_\gamma)$ and $2\deg(g)$ are congruent modulo 4, where g is the Gauss map of f. By these facts, we can observe that for every complete non-orientable minimal surface of finite total curvature, $\deg(g) \geq \gamma + 3$ holds.

For $\gamma = 0$ and $\gamma = 1$, Meeks' Möbius strip [10] and López' Klein bottle [9] satisfy $\deg(g) = \gamma + 3$, respectively. But, for $\gamma \geq 2$, no examples with $\deg(g) = \gamma + 3$ are known. So, it is interesting to give a minimal surface satisfying $\deg(g) = \gamma + 3$ with an antiholomorphic involution without fixed points.

References

1. C. C. Chen and F. Gackstatter, *Elliptische und hyperelliptische Funktionen und vollstandige Minimalflachen vom Enneperschen Typ*, Math. Ann. **259** (1982), 359-369.
2. A. Costa, *Examples of a Complete Minimal Immersion in of Genus One and Three Embedded Ends*, Bil. Soc. Bras. Mat. **15** (1984), 47-54.
3. S. Fujimori and T. Shoda, *Minimal surfaces with two ends which have the least total absolute curvature*, preprint.
4. D. Hoffman and W. H. Meeks III, *Embedded minimal surfaces of finite topology*, Ann. of Math. (2) **131** (1990), 1–34.
5. D. Hoffman and R. Osserman, *The geometry of the generalized Gauss map*, Mem. Amer. Math. Soc. **28** (1980), iii+105 pp.
6. A. Huber, *On subharmonic functions and differential geometry in the large*, Comment. Math. Helv. **32** (1957), 13–72.

7. L. Jorge and W. H. Meeks III, *The topology of complete minimal surfaces of finite total Gaussian curvature*, Topology, (2) **22** (1983), 203–221.
8. F. J. López, *The classification of complete minimal surfaces with total curvature greater than* −12π, Trans. Amer. Math. Soc. **334** (1992), 49–74.
9. F. J. López, *A complete minimal Klein bottle in* R^3, Duke Math. J. **71** (1993), 23–30.
10. W. H. Meeks III, *The classification of complete minimal surfaces in* R^3 *with total curvature greater than* −8π, Duke Math. J. **48** (1981), 523–535.
11. R. Osserman, *A survey of minimal surfaces*, Dover Publications, Inc.
12. K. Sato, *Construction of higher genus minimal surfaces with one end and finite total curvature*, Tohoku Math. J. **48** (1996), 229–246.
13. R. Schoen, *Uniqueness, symmetry and embeddedness of minimal surfaces*, J. Diff. Geom. **18** (1983), 791–809.
14. E. C. Thayer, *Higher-Genus Chen-Gackstatter Surfaces and The Weierstrass Representation for Surfaces of Infinite Genus*, Exper. Math. **4** (1995), 19-39.
15. M. Weber and M. Wolf, *Minimal surfaces of least total curvature and moduli spaces of plane polygonal arcs*, Geom. Funct. Anal. **8** (1998), 1129–1170.

Received November 26, 2012.

Proceedings of the Workshop on
Differential Geometry of Submanifolds
and its Related Topics
Saga, August 4-6, 2012

MINIMAL SURFACES IN THE ANTI-DE SITTER
SPACETIME

Toshiyuki ICHIYAMA

Faculty of Economics, Asia University,
Sakai, Musashino, 180-8629, Japan
E-mail: ichiyama@asia-u.ac.jp

Seiichi UDAGAWA

School of Medicine, Nihon University,
Itabashi, Tokyo 173-0032, Japan
E-mail: udagawa.seiichi@nihon-u.ac.jp

Dedicated to Professor Sadahiro Maeda for his sixtieth birthday

We give a modern differential geometric treatment of oriented spacelike minimal surface in H_1^3, which is recently studied by the theoretical physicists in [1,
2, 4, 5]. We also give a constructive method of an example of oriented spacelike
minimal surface in H_1^3 from the viewpoints of the extended framing coming
from the twisted loop group.

Keywords: anti-de Sitter spacetime, spacelike minimal surface, sinh-Gordon
equation, twisted loop group.

1. Introduction

In the literature of theoretical physics, minimal surfaces in anti-de Sitter spacetime AdS_5 are studied from the points of views with respect to
the AdS/CFT correspondence, which asserts an equivalence between 4-
dimensional supersymmetric gauge theory and superstring theory on anti-
de Sitter spacetime. The relation between gluon scattering amplitudes and
minimal surfaces in anti-de Sitter spacetime is a new example of this correspondence (see [1]). In this paper, we study an oriented spacelike minimal
surface in the anti-de Sitter spacetime $H_1^3(= AdS_3)$. The integrability condition for the surface is given by the sinh-Gordon equation. We write down

the Jacobi elliptic function solutions. These constitute the 1-parameter family and contain the trivial solution. This trivial solution gives an interesting example of oriented spacelike minimal surface in H_1^3, which is already known in the literature of theoretical physics ([4, 5]). However, our method depends on the twisted loop group formalism. We discuss the fundamental materials of the twisted loop group and loop algebra and also discuss the relation between the spectral curves and the Jacobi elliptic function solution for the sinh-Gordon equation.

2. anti-de Sitter spacetime

Let $\mathbf{R}^4$ be a 4-dimensional real linear vector space. We endow $\mathbf{R}^4$ an indefinite metric $<,>$ defined by

$$< \mathbf{x}, \mathbf{y} >= x_1 y_1 + x_2 y_2 - x_3 y_3 - x_4 y_4,$$

where $\mathbf{x} = (x_1, \cdots, x_4), \mathbf{y} = (y_1, \cdots, y_4) \in \mathbf{R}^4$. We then denote by $\mathbf{R}^{2,2}$ the metric space $(\mathbf{R}^4, <, >)$. A quadratic hypersurface $\{\mathbf{x} \in \mathbf{R}^{2,2} | < \mathbf{x}, \mathbf{x} >= -1\}$ is called *anti-de Sitter spacetime* and denoted by H_1^3. The tangent space is given by

$$T_\mathbf{x} H_1^3 = \left\{ X \in \mathbf{R}^{2,2} | < X, \mathbf{x} >= 0 \right\}.$$

We have an orthogonal sum decomposition $\mathbf{R}^{2,2} = T_\mathbf{x} H_1^3 \oplus \mathbf{R}\{\mathbf{x}\}$ with respect to $<,>$. Since the signature of $<,>|_{T_\mathbf{x} H_1^3}$ is $(2,1)$, the induced metric on H_1^3 is a Lorentzian metric, which is denoted by g. Define a vector field ξ along H_1^3 by $\xi_\mathbf{x} = -\mathbf{x}$. We then have $< \xi, \xi >= -1$. Let D be the covariant differentiation of $\mathbf{R}^{2,2}$ and $\widetilde{\nabla}$ be the induced covariant differentiation of H_1^3. We have

$$D_X Y = \widetilde{\nabla}_X Y - g(X, Y)\xi \tag{1}$$

We see that the curvature tensor $\widetilde{R}$ with respect to $\widetilde{\nabla}$ satisfies the equation $\widetilde{R}(X, Y)Z = -(X \wedge Y)Z$. Therefore, we see that H_1^3 has a constant curvature -1.

3. Oriented spacelike surfaces

Let M be an oriented spacelike surface in H_1^3. Using a conformal coordinate system (x, y) of M, we may express the induced metric g_M as $g_M = 4e^{2u}(dx^2 + dy^2)$, where u is a smooth function on M. We denote by N the unit normal vector to M. Since M is a spacelike, we have

$<N, N> = g(N, N) = -1$. We denote by ∇ the induced Levi-Civita connection on M. We use the notation that $u_x = \dfrac{\partial u}{\partial x}, u_y = \dfrac{\partial u}{\partial y}$. By a simple calculation we obtain the following.

Lemma 3.1.

$$
\begin{cases}
\nabla_{\frac{\partial}{\partial x}} \dfrac{\partial}{\partial x} = u_x \dfrac{\partial}{\partial x} - u_y \dfrac{\partial}{\partial y}, \\
\nabla_{\frac{\partial}{\partial x}} \dfrac{\partial}{\partial y} = u_y \dfrac{\partial}{\partial x} + u_x \dfrac{\partial}{\partial y} = \nabla_{\frac{\partial}{\partial y}} \dfrac{\partial}{\partial x}, \\
\nabla_{\frac{\partial}{\partial y}} \dfrac{\partial}{\partial y} = -u_x \dfrac{\partial}{\partial x} + u_y \dfrac{\partial}{\partial y}.
\end{cases}
\tag{2}
$$

Using Lemma (3.1), we calculate the Hessian of the immersion $f : M \longrightarrow H_1^3$.

$$
\begin{aligned}
D_{\frac{\partial}{\partial x}} f_* \left(\dfrac{\partial}{\partial x} \right) &= \tilde{\nabla}_{\frac{\partial}{\partial x}} f_* \left(\dfrac{\partial}{\partial x} \right) + g \left(\dfrac{\partial}{\partial x}, \dfrac{\partial}{\partial x} \right) f \\
&= u_x f_x - u_y f_y + h_{xx} N + 4e^{2u} f,
\end{aligned}
\tag{3}
$$

$$
\begin{aligned}
D_{\frac{\partial}{\partial x}} f_* \left(\dfrac{\partial}{\partial y} \right) &= \tilde{\nabla}_{\frac{\partial}{\partial x}} f_* \left(\dfrac{\partial}{\partial y} \right) + g \left(\dfrac{\partial}{\partial x}, \dfrac{\partial}{\partial y} \right) f \\
&= u_y f_x + u_x f_y + h_{xy} N,
\end{aligned}
\tag{4}
$$

$$
\begin{aligned}
D_{\frac{\partial}{\partial y}} f_* \left(\dfrac{\partial}{\partial y} \right) &= \tilde{\nabla}_{\frac{\partial}{\partial y}} f_* \left(\dfrac{\partial}{\partial y} \right) + g \left(\dfrac{\partial}{\partial y}, \dfrac{\partial}{\partial y} \right) f \\
&= -u_x f_x + u_y f_y + h_{yy} N + 4e^{2u} f,
\end{aligned}
\tag{5}
$$

where h is the second fundamental form of the immersion f and

$$
h_{xx} = h \left(\dfrac{\partial}{\partial x}, \dfrac{\partial}{\partial x} \right), h_{xy} = h \left(\dfrac{\partial}{\partial x}, \dfrac{\partial}{\partial y} \right), h_{yy} = h \left(\dfrac{\partial}{\partial y}, \dfrac{\partial}{\partial y} \right).
$$

Consider a framing $F = (f \ f_x \ f_y \ N)$. Before calculating the differential of F, we here remember the role of the shape operator A_N of f :

$$
D_{\frac{\partial}{\partial x}} N = \tilde{\nabla}_{\frac{\partial}{\partial x}} N = -A_N \left(\dfrac{\partial}{\partial x} \right), \quad
D_{\frac{\partial}{\partial y}} N = \tilde{\nabla}_{\frac{\partial}{\partial y}} N = -A_N \left(\dfrac{\partial}{\partial y} \right),
$$

where we remark that $h(X, Y) = -g(A_N X, Y)$ holds. Hence, we have

$$
\begin{cases}
A_N \left(\dfrac{\partial}{\partial x} \right) = -\dfrac{1}{4} e^{-2u} h_{xx} \dfrac{\partial}{\partial x} - \dfrac{1}{4} e^{-2u} h_{xy} \dfrac{\partial}{\partial y}, \\
A_N \left(\dfrac{\partial}{\partial y} \right) = -\dfrac{1}{4} e^{-2u} h_{xy} \dfrac{\partial}{\partial x} - \dfrac{1}{4} e^{-2u} h_{yy} \dfrac{\partial}{\partial y}.
\end{cases}
\tag{6}
$$

It follows from (6) that

$$D_{\frac{\partial}{\partial x}} N = -A_N (f_x) = \frac{1}{4} e^{-2u} h_{xx} f_x + \frac{1}{4} e^{-2u} h_{xy} f_y, \tag{7}$$

$$D_{\frac{\partial}{\partial y}} N = -A_N (f_y) = \frac{1}{4} e^{-2u} h_{xy} f_x + \frac{1}{4} e^{-2u} h_{yy} f_y. \tag{8}$$

We find $\partial_x F = FA, \partial_y F = FB$, where

$$A = \begin{pmatrix} 0 & 4e^{2u} & 0 & 0 \\ 1 & u_x & u_y & \frac{e^{-2u}}{4} h_{xx} \\ 0 & -u_y & u_x & \frac{e^{-2u}}{4} h_{xy} \\ 0 & h_{xx} & h_{xy} & 0 \end{pmatrix}, \tag{9}$$

$$B = \begin{pmatrix} 0 & 0 & 4e^{2u} & 0 \\ 0 & u_y & -u_x & \frac{e^{-2u}}{4} h_{xy} \\ 1 & u_x & u_y & \frac{e^{-2u}}{4} h_{yy} \\ 0 & h_{xy} & h_{yy} & 0 \end{pmatrix}. \tag{10}$$

We here use the notation $h_{ab,x} = \frac{\partial}{\partial x} h_{ab}, h_{ab,y} = \frac{\partial}{\partial y} h_{ab}$, where $a, b = x$ or y. On the other hand, we denote by $h_{ab;x}$ the covariant differentiation $(\nabla_{\frac{\partial}{\partial x}} h) \left(\frac{\partial}{\partial a}, \frac{\partial}{\partial b} \right)$ of h_{ab} and similarly for $h_{ab;y}$. We may verify that the integrability condition for the framing F is equivalent to $\partial_y A - \partial_x B = [A, B]$. The latter condition is also equivalent to the following.

Lemma 3.2.

$$\begin{cases} u_{xx} + u_{yy} & = 4e^{2u} + \frac{e^{-2u}}{4} \left(h_{xx} h_{yy} - (h_{xy})^2 \right), \\ h_{xx,y} - h_{xy,x} = u_y (h_{xx} + h_{yy}), \\ h_{xy,y} - h_{yy,x} = -u_x (h_{xx} + h_{yy}). \end{cases} \tag{11}$$

Proof. This is a direct calculation and easily verified. $\qquad\square$

Remark 3.1. The 1st equation in (11) is nothing but the condition that the Gaussian curvature of H_1^3 is identically equal to -1. The 2nd and 3rd equations in (11) are respectively equivalent to the Codazzi equations $h_{xx;y} = h_{xy;x}$ and $h_{xy;y} = h_{yy;x}$.

We now assume that M is an oriented spacelike *minimal* surface in H_1^3. M is minimal if and only if

$$h_{xx} + h_{yy} = 0, \tag{12}$$

which gives

$$h_{xx}h_{yy} - (h_{xy})^2 = -\left((h_{xx})^2 + (h_{xy})^2\right).$$

Next, we introduce a local complex coordinate system $z = x + iy$ for M. Set

$$\frac{\partial}{\partial z} = \frac{1}{2}\left(\frac{\partial}{\partial x} - i\frac{\partial}{\partial y}\right), \quad \frac{\partial}{\partial \bar{z}} = \frac{1}{2}\left(\frac{\partial}{\partial x} + i\frac{\partial}{\partial y}\right).$$

Set

$$u_{z\bar{z}} = \frac{\partial}{\partial \bar{z}}\frac{\partial}{\partial z}u = \frac{1}{4}(u_{xx} + u_{yy}), \tag{13}$$

$$h_{zz} = h\left(\frac{\partial}{\partial z}, \frac{\partial}{\partial z}\right) = \frac{1}{4}(h_{xx} - h_{yy} - 2ih_{xy}) = \frac{1}{2}(h_{xx} - ih_{xy}), \tag{14}$$

where we used (12) in the last equality. Denote by $h_{zz;\bar{z}}$ the covariant differential $(\nabla_{\frac{\partial}{\partial \bar{z}}}h)\left(\frac{\partial}{\partial z}, \frac{\partial}{\partial z}\right)$ of h_{zz}. It follows from (11) and (12) that

$$h_{zz;\bar{z}} = \nabla_{\frac{\partial}{\partial \bar{z}}}\left(h\left(\frac{\partial}{\partial z}, \frac{\partial}{\partial z}\right)\right) - 2h\left(\nabla_{\frac{\partial}{\partial \bar{z}}}\frac{\partial}{\partial z}, \frac{\partial}{\partial z}\right)$$

$$= \frac{1}{4}(h_{xx;x} + ih_{xx;y} - ih_{xy;x} + h_{xy;y})$$

$$= \frac{1}{4}(h_{xx;x} + h_{yy;x} + i(h_{xx;y} - h_{xx;y})) = 0.$$

We then have the following.

Lemma 3.3. *If M is minimal then $h_{zz}(dz)^2$ is a holomorphic differential.*

Set $P(z) = \frac{1}{2}h_{zz}$. We then have

$$|P(z)|^2 = \frac{1}{4}|h_{zz}|^2 = \frac{1}{16}\left((h_{xx})^2 + (h_{xy})^2\right)$$

$$= -\frac{1}{16}\left(h_{xx}h_{yy} - (h_{xy})^2\right). \tag{15}$$

Substituting (13) and (15) into the 1st equation in Lemma 3.2 we obtain

$$u_{z\bar{z}} - e^{2u} + |P(z)|^2 e^{-2u} = 0. \tag{16}$$

We here may assume that $|P(z)| \equiv 1$. In fact, since $P(z)dz^2 = \frac{1}{2}h_{zz}dz^2$ is a holomorphic differential by Lemma 3.3, it is invariant under a holomorphic transformation. Let $z = \varphi(w)$ be a holomorphic transformation of the local coordinate system. We have $P(w)dw^2 = P(z)dz^2$. We choose a holomorphic function φ so that φ satisfies $(\varphi'(w))^2 = \frac{1}{2}h_{ww}$. If $\varphi'(w) \neq 0$, which is the case for non totally geodesic minimal surface M, we see that

$$P(z) = \frac{1}{2}h_{zz} = \frac{1}{2}\frac{1}{(\varphi'(w))^2}h_{ww} = 1.$$

Consequently, we may assume that $P(z) \equiv 1$ for non totally geodesic M by choosing a local complex coordinate system z. Now, the equation (16) becomes to

$$u_{z\bar{z}} = e^{2u} - e^{-2u} = 2\sinh(2u). \tag{17}$$

This is a sinh-Gordon equation.

4. A special solution of sinh-Gordon equation

In this section, we give a special solution of the sinh-Gordon equation (17). To begin with, we rewrite (17) in terms of the real variables x and y.

$$u_{xx} + u_{yy} = 4e^{2u} - 4e^{-2u}. \tag{18}$$

We now assume that $u = u(x)$ depends only the variable x. Then, we have $u_{xx} = 4e^{2u} - 4e^{-2u}$. Multiplying this by u_x and integrating it with respect to x we obtain the following.

$$(u_x)^2 = 4e^{2u} + 4e^{-2u} + C,$$

where C is a constant of the integration. Under the initial condition $e^{u(0)} = \frac{\alpha}{2}, u_x(0) = 0$, where we suppose $\alpha \geq 2$. Setting $Y = e^{2u}$, we have

$$\left(\frac{dY}{dx}\right)^2 = 16Y^3 + 4CY^2 + 16Y, \tag{19}$$

$$C = -\left(\alpha^2 + \frac{16}{\alpha^2}\right). \tag{20}$$

Define $\hat{Y}$ by $\hat{Y} = 4Y + \frac{C}{3}$. We then have

$$\left(\frac{d\hat{Y}}{dx}\right)^2 = 4\left(\hat{Y} - \mathbf{e}_1\right)\left(\hat{Y} - \mathbf{e}_2\right)\left(\hat{Y} - \mathbf{e}_3\right), \tag{21}$$

where

$$\mathbf{e}_1 = -\frac{\alpha^4 + 16}{3\alpha^2}, \mathbf{e}_2 = \frac{-\alpha^4 + 32}{3\alpha^2}, \mathbf{e}_3 = \frac{2\alpha^4 - 16}{3\alpha^2}. \tag{22}$$

The solution of (21) may be described in terms of the Jacobi elliptic function as follows.

$$\hat{Y}(x) = \mathbf{e}_1 + (\mathbf{e}_2 - \mathbf{e}_1)\mathrm{sn}^2\left(\sqrt{\mathbf{e}_3 - \mathbf{e}_1}x, k\right),\qquad(23)$$

$$k = \sqrt{\frac{\mathbf{e}_2 - \mathbf{e}_1}{\mathbf{e}_3 - \mathbf{e}_1}}.\qquad(24)$$

5. Extended framing and spectral curves

Let $SL(2, \mathbf{R})$ be a real special linear group of order 2. Consider a map $\Phi : SL(2, \mathbf{R}) \times SL(2, \mathbf{R}) \ni (f, g) \longrightarrow f \cdot g^{-1} \in SL(2, \mathbf{R})$. Since $\mathrm{Ker}\Phi = \{(f, f)|f \in SL(2, \mathbf{R})\}$, we have an isomorphism $SL(2, \mathbf{R}) \cong SL(2, \mathbf{R}) \times SL(2, \mathbf{R})/\mathrm{Ker}\Phi$. On the other hand, since it is known that $SL(2, \mathbf{R}) \times SL(2, \mathbf{R}) \cong SO(2, 2)$, we construct a framing which has values in $SO(2, 2)$. First of all, consider $F = (f\ f_x\ f_y\ N)$. Since $< f, f > = -1, < N, N > = -1, < f_x, f_x > = 4e^{2u}$ and $< f_y, f_y > = 4e^{2u}$, define a new framing G by

$$G = F \begin{pmatrix} 1 & 0 & 0 & 0 \\ 0 & \frac{1}{2}e^{-u} & 0 & 0 \\ 0 & 0 & \frac{1}{2}e^{-u} & 0 \\ 0 & 0 & 0 & 1 \end{pmatrix}.\qquad(25)$$

We then find G has values in $SO(2, 2)$. We see that $G^{-1}\partial_z G = \frac{1}{2}\left(G^{-1}\partial_x G - iG^{-1}\partial_y G\right), G^{-1}\partial_{\bar{z}} G = \frac{1}{2}\left(G^{-1}\partial_x G + iG^{-1}\partial_y G\right)$ and

$$\begin{cases} G^{-1}\partial_z G = \begin{pmatrix} 0 & e^u & -ie^u & 0 \\ e^u & 0 & \frac{i}{2}u_z & \frac{1}{2}e^{-u}h_{zz} \\ -ie^u & -\frac{i}{2}u_z & 0 & \frac{i}{2}e^{-u}h_{zz} \\ 0 & \frac{1}{2}e^{-u}h_{zz} & \frac{i}{2}e^{-u}h_{zz} & 0 \end{pmatrix}, \\[3em] G^{-1}\partial_{\bar{z}} G = \begin{pmatrix} 0 & e^u & ie^u & 0 \\ e^u & 0 & -\frac{i}{2}u_{\bar{z}} & \frac{1}{2}e^{-u}h_{\bar{z}\bar{z}} \\ ie^u & \frac{i}{2}u_{\bar{z}} & 0 & -\frac{i}{2}e^{-u}h_{\bar{z}\bar{z}} \\ 0 & \frac{1}{2}e^{-u}h_{\bar{z}\bar{z}} & -\frac{i}{2}e^{-u}h_{\bar{z}\bar{z}} & 0 \end{pmatrix}. \end{cases}\qquad(26)$$

For a diagonal matrix $J = \mathrm{diag}(-1, 1, 1, -1)$, $SO(2, 2)$ may be represented as $SO(2, 2) = SL(4, \mathbf{R}) \cap \{A \in GL(4, \mathbf{R})|^t AJA = J\}$. Denote by $SO^0(2, 2)$ the connected component of $SO(2, 2)$ containing the identity. Its Lie algebra is given by $so(2, 2) = \{X \in \mathfrak{gl}(4, \mathbf{R})|^t X = -JXJ\}$ and it is seen that $G^{-1}\partial_x G$ and $G^{-1}\partial_y G$ has values in $so(2, 2)$. We embed $SO(2)$ into

$SO^0(2,2)$ by

$$\begin{pmatrix} 1 & & \\ & SO(2) & \\ & & 1 \end{pmatrix} \subset SO^0(2,2).$$

Moreover, for $\widetilde{J} = \mathrm{diag}(1,1,-1)$, $SO(2,1)$ may be represented as $SO(2,1) = SL(3,\mathbf{R}) \cap \{A \in GL(3,\mathbf{R})|^tA\widetilde{J}A = \widetilde{J}\}$. Its connected component $SO^0(2,1)$ containing the identity may be also embedded into $SO^0(2,2)$ by

$$\begin{pmatrix} 1 & \\ & SO^0(2,1) \end{pmatrix} \subset SO^0(2,2).$$

In the same way, $SO(2)$ may be embedded into $SO^0(2,1)$ naturally. We may then express H_1^3 as a homogeneous space $H_1^3 \cong SO^0(2,2)/SO^0(2,1)$.

Proposition 5.1. *A homogeneous pace $SO^0(2,2)/SO(2)$ may be endowed with a regular 4-symmetric space structure. For an natural homogeneous projection $\pi : SO^0(2,2)/SO(2) \longrightarrow SO^0(2,2)/SO^0(2,1)$, any oriented spacelike minimal surface $f : M \longrightarrow H_1^3 \cong SO^0(2,2)/SO^0(2,1)$ may be locally lifted to a primitive map $g : M \longrightarrow SO^0(2,2)/SO(2)$ such that $f = \pi \circ g$.*

Proof. Set $\tau = \mathrm{Ad}Q$, where

$$Q = \begin{pmatrix} -1 & 0 & 0 & 0 \\ 0 & 0 & 1 & 0 \\ 0 & -1 & 0 & 0 \\ 0 & 0 & 0 & 1 \end{pmatrix}.$$

It satisfies $\tau^4 = 1$. Moreover, we find that $+1$-eigenspace of τ is

$$\left\{ \begin{pmatrix} 0 & 0 & 0 & 0 \\ 0 & 0 & -a & 0 \\ 0 & a & 0 & 0 \\ 0 & 0 & 0 & 0 \end{pmatrix} \middle| a \in \mathbf{R} \right\} \subset so(2,2),$$

which is a Lie algebra $so(2)$ of $SO(2)$ embedded into $SO^0(2,2)$. Therefore, we see that τ gives $SO^0(2,2)/SO(2)$ a regular 4-symmetric space structure. Set $\mathfrak{g} = so(2,2)$ and let $\mathfrak{g}^{\mathbf{C}}$ be the complexification of $\mathfrak{g}$. Let $\mathfrak{g}_j$ be the i^j-eigenspace of τ for $j = 0,1,2,3$. It is enough to verify the property

$G^{-1}\partial_z G \in \mathfrak{g}_0 \oplus \mathfrak{g}_1$. In fact, since we see that

$$\mathfrak{g}_1 = \mathrm{span}_{\mathbf{C}} \left\{ \begin{pmatrix} 0 & 1 & -i & 0 \\ 1 & 0 & 0 & 0 \\ -i & 0 & 0 & 0 \\ 0 & 0 & 0 & 0 \end{pmatrix}, \begin{pmatrix} 0 & 0 & 0 & 0 \\ 0 & 0 & 0 & 1 \\ 0 & 0 & 0 & i \\ 0 & 1 & i & 0 \end{pmatrix} \right\},$$

it follows from (26) that g is a primitive map.   □

[Spectral curves] Set

$$A_1 = G^{-1}\partial_z G \cap \mathfrak{g}_1, \ A_{-1} = G^{-1}\partial_{\bar{z}}G \cap \mathfrak{g}_3, \ A_0 = \left(G^{-1}\partial_z G + G^{-1}\partial_{\bar{z}}G\right) \cap \mathfrak{g}_0.$$

For $(\mu, \lambda) \in \mathbf{C}^2$, the spectral curve is defined by

$$\det\left(\lambda A_1 + \lambda^{-1}A_{-1} + A_0 - \mu I\right) = 0, \tag{27}$$

where I is the identity matrix of order 4. For example, we shall calculate the spectral curve corresponding to the solution in terms of Jacobi elliptic function. Since the spectral curve is independent of the variables x and y, we let $x = y = 0$. We then have $e^{u(0)} = \frac{\alpha}{2}, u_x(0) = 0, u_y(0) = 0$. Since $h_{xx}^2 + h_{xy}^2 \equiv 16$ from (15) and $P(z) = 1$, we have $|h_{zz}|^2 = 4$. We then may put $h_{zz} = 2e^{i\theta}$ for some real constant θ. Setting $\hat{\mu} = \dfrac{\mu}{2}$, we see that the spectral curve is given by the equation

$$4\hat{\mu}^4 - \left(\alpha^2 + \frac{16}{\alpha^2}\right)\hat{\mu}^2 + 4 = \left(\lambda^2 e^{i\theta} + \lambda^{-2}e^{-i\theta}\right)^2.$$

If we put $\hat{Y} = \hat{\mu}, \hat{B} = \lambda^2 e^{i\theta} + \lambda^{-2}e^{-i\theta}$, the equation of the spectral curve becomes

$$\hat{B}^2 = 4\hat{Y}^4 - \left(\alpha^2 + \frac{16}{\alpha^2}\right)\hat{Y}^2 + 4, \tag{28}$$

which is equivalent to (19). The equivalence is given by $Y = \hat{Y}^2$ and $\hat{B} = \dfrac{d\hat{Y}}{dx}$.

6. Twisted loop group and twisted loop algebra

Set $G = SO^0(2,2)$ and let $\mathfrak{g} = so(2,2)$ be the Lie algebra of G. Recall that $\tau : G \longrightarrow G$ is the automorphism of order 4 as in the former section. Denote by S^1 the set of all complex numbers c's with $|c| = 1$. Set

$$\Lambda \mathfrak{g}^{\mathbf{C}} = \left\{ \xi : S^1 \longrightarrow \mathfrak{g}^{\mathbf{C}} \ \mid \ ||\xi|| < \infty \right\}, \tag{29}$$

where ξ is a continuous map. With respect to the Fourier expansion $\xi = \sum_{n \in \mathbf{Z}} \lambda^n \xi_n$, we have

$$\begin{cases} ||\xi|| = \sum_{n \in \mathbf{Z}} ||\xi_n|| < \infty, \\ ||\xi_n|| = \text{Max}_{j \in \{1,2,3,4\}} \sum_{i=1}^{4} \left| (\xi_n)_{ij} \right|. \end{cases} \tag{30}$$

Then, $\Lambda \mathfrak{g}^{\mathbf{C}}$ is a Banach Lie algebra with respect to the Wiener topology. Let $\Lambda G^{\mathbf{C}}$ be the set of all elements $g : S^1 \longrightarrow G^{\mathbf{C}}$ with g being absolutely convergent with respect to the Wiener topology. Set

$$\Lambda G = \{ g \in \Lambda G^{\mathbf{C}} | g : S^1 \longrightarrow G \},$$
$$\Lambda \mathfrak{g} = \{ \xi \in \Lambda \mathfrak{g}^{\mathbf{C}} | \xi : S^1 \longrightarrow \mathfrak{g} \}.$$

We now introduce the twisted condition on the loop group and loop algebra using τ. Define a twisted loop algebra $\Lambda \mathfrak{g}^{\mathbf{C}}_{\tau}$ by

$$\Lambda \mathfrak{g}^{\mathbf{C}}_{\tau} = \left\{ \xi \in \Lambda \mathfrak{g}^{\mathbf{C}} \ \middle| \ \xi(i\lambda) = \tau \xi(\lambda) \text{ for any } \lambda \in S^1 \right\}. \tag{31}$$

Define its real form as follows.

$$\Lambda \mathfrak{g}_{\tau} = \left\{ \xi \in \Lambda \mathfrak{g}^{\mathbf{C}}_{\tau} | \xi : S^1 \longrightarrow \mathfrak{g} \right\}. \tag{32}$$

The twisted loop group $\Lambda G^{\mathbf{C}}_{\tau}$ corresponding to $\Lambda \mathfrak{g}^{\mathbf{C}}_{\tau}$ is defined by

$$\Lambda G^{\mathbf{C}}_{\tau} = \left\{ g \in \Lambda G^{\mathbf{C}} \ \middle| \ g(i\lambda) = \tau(g(\lambda)) \text{ for any } \lambda \in S^1 \right\}. \tag{33}$$

Define its real form as follows.

$$\Lambda G_{\tau} = \left\{ g \in \Lambda G^{\mathbf{C}}_{\tau} | g : S^1 \longrightarrow G \right\}. \tag{34}$$

We need the Iwasawa decomposition of $\Lambda G^{\mathbf{C}}_{\tau}$ into ΛG_{τ} and the complementary space $\Lambda^{-} G^{\mathbf{C}}_{\tau}$ defined by

$$\Lambda^{-} G^{\mathbf{C}}_{\tau} = \left\{ g \in \Lambda G^{\mathbf{C}}_{\tau} | g \text{ extends analytically when } \lambda \to \infty, g(\infty) = I \right\},$$

where D is the unit disk with $\partial D = S^1$. Since G is non-compact, it is not so easy to obtain such a decomposition. However, in some open dense subset, called a *big cell*, it is often possible (see [3, 6]). We will discuss these materials elsewhere.

7. An example of oriented spacelike minimal surface in H_1^3

We will produce an immersion f corresponding to the trivial solution $u = 0$ of the sinh-Gordon equation. Define a matrix A by

$$A = \begin{pmatrix} 0 & 1 & -i & 0 \\ 1 & 0 & 0 & 1 \\ -i & 0 & 0 & i \\ 0 & 1 & i & 0 \end{pmatrix} \in \mathfrak{g}_1.$$

We then see that $\overline{A} \in \mathfrak{g}_3 = \mathfrak{g}_{-1}$. We have $[A, \overline{A}] = 0$. Now, the extended framing $F(\lambda)$ is given by $F(\lambda) = \exp\left(\lambda z A + \lambda^{-1}\overline{z}\overline{A}\right) \in \Lambda G_\tau$. The reason for that relating to the Iwasawa decomposition of the twisted loop group stated in the former section is that $\exp(\lambda z A) \in \Lambda G_\tau^{\mathbb{C}}$ and

$$\exp(\lambda z A) = \exp\left(\lambda z A + \lambda^{-1}\overline{z}\overline{A}\right) \cdot \exp\left(-\lambda^{-1}\overline{z}\overline{A}\right) \in \Lambda G_\tau \cdot \Lambda^- G_\tau^{\mathbb{C}}.$$

Letting $\lambda = 1$, we see that $F = \exp\left(z A + \overline{z}\overline{A}\right)$ is a $SO^0(2,2)$-framing for f and the immersion f is given by the first raw of F. We may set $\exp(z A) = \varphi_0 I + \varphi_1 A + \varphi_2 A^2 + \varphi_3 A^3$, where

$$\begin{cases} \varphi_0 = \dfrac{1}{4}\left(e^{\sqrt{2}z} + e^{-\sqrt{2}z} + e^{i\sqrt{2}z} + e^{-i\sqrt{2}z}\right), \\ \varphi_1 = \dfrac{1}{4\sqrt{2}}\left(e^{\sqrt{2}z} - e^{-\sqrt{2}z} - ie^{i\sqrt{2}z} + ie^{-i\sqrt{2}z}\right), \\ \varphi_2 = \dfrac{1}{8}\left(e^{\sqrt{2}z} + e^{-\sqrt{2}z} - e^{i\sqrt{2}z} - e^{-i\sqrt{2}z}\right), \\ \varphi_3 = \dfrac{1}{8\sqrt{2}}\left(e^{\sqrt{2}z} - e^{-\sqrt{2}z} + ie^{i\sqrt{2}z} - ie^{-i\sqrt{2}z}\right). \end{cases} \tag{35}$$

It follows from these observations that we obtain the following.

Theorem 7.1. *The following f gives an example of an oriented spacelike minimal surface in H_1^3 :*

$$f = \begin{pmatrix} \dfrac{1}{2}\left(\cosh(2\sqrt{2}x) + \cosh\left(2\sqrt{2}y\right)\right) \\ \dfrac{\sqrt{2}}{2}\sinh\left(2\sqrt{2}x\right) \\ \dfrac{\sqrt{2}}{2}\sinh\left(2\sqrt{2}y\right) \\ \dfrac{1}{2}\left(\cosh\left(2\sqrt{2}x\right) - \cosh\left(2\sqrt{2}y\right)\right) \end{pmatrix}. \tag{36}$$

Moreover, the timelike normal vector N is given by

$$N = \begin{pmatrix} \frac{1}{2}\left(\cosh(2\sqrt{2}x) - \cosh\left(2\sqrt{2}y\right)\right) \\ \frac{\sqrt{2}}{2}\sinh\left(2\sqrt{2}x\right) \\ -\frac{\sqrt{2}}{2}\sinh\left(2\sqrt{2}y\right) \\ \frac{1}{2}\left(\cosh\left(2\sqrt{2}x\right) + \cosh\left(2\sqrt{2}y\right)\right) \end{pmatrix}. \tag{37}$$

N *also defines an oriented spacelike minimal surface in H_1^3, which is a suitable $SO^0(2,2)$-rotation of f.*

8. Problems

Problem 8.1. *Give the immersion $f : M \longrightarrow H_1^3$ corresponding to the Jacobi elliptic function solution for sinh-Gordon equation.*

Problem 8.2. *Give the DPW(Dorfmeister-Pedit-Wu)-method for obtaining the immersion f.*

References

1. L. F. ALDAY AND J. MALDACENA, *Minimal surfaces in AdS and the eight-gluon scattering amplitude at strong coupling*, arXive:0903.4707 [hep-th]. hep-th.
2. L. F. ALDAY AND J. MALDACENA, *Null polygonal Wilson loops and minimal surfaces in Anti-de-Sitter space*, arXiv:0904.0663 [hep-th].
3. V. BALAN AND J. DORFMEISTER, *Birkhoff decompositions and Iwasawa decompositions for loop groups*, Tohoku Math. J. **53**(2000), 593-615.
4. H. DORN, *Some comments on spacelike minimal surfaces with null polygonal boundaries in AdS$_m$*, JHEP 1002(2010) 013 arXiv:0910.0934 [hep-th].
5. H. DORN, G. JORJADZE AND S. WUTTKE, *On spacelike and timelike minimal surfaces in AdS$_n$*, JHEP 0905 (2009) 064 arXiv:0903.0977 [hep-th].
6. P. KELLERSCH, *The Iwasawa decomposition for the untwisted group of loops in semisimple Lie groups*, Ph. D. Thesis, Technische Universität München, 1999.

Received March 29, 2013.

44

Proceedings of the Workshop on
Differential Geometry of Submanifolds
and its Related Topics
Saga, August 4-6, 2012

EXTRINSIC CIRCULAR TRAJECTORIES
ON GEODESIC SPHERES
IN A COMPLEX PROJECTIVE SPACE

Toshiaki ADACHI*

*Department of Mathematics, Nagoya Institute of Technology,
Nagoya 466-8555, Japan
E-mail: adachi@nitech.ac.jp*

Dedicated to Professor Sadahiro MAEDA on the occasion of his 60th birthday

In this paper we explain circles of positive geodesic curvature on a complex projective space as extrinsic shapes of trajectories for Sasakian magnetic fields on geodesic spheres in this space. On the "moduli space" of extrinsic circular trajectories on geodesic spheres, we give a foliation which corresponds to the lamination on the moduli space of circles of positive geodesic curvature on a complex projective space.

Keywords: Sasakian magnetic fields, trajectories, structure torsions, extrinsic shapes, circles, foliations and laminations, geodesic spheres, complex projective spaces.

1. Introduction

A smooth curve γ parameterized by its arclength is said to be a *circle* if it satisfies the system of ordinary differential equations $\nabla_{\dot{\gamma}}\dot{\gamma} = kY$, $\nabla_{\dot{\gamma}}Y = -k\dot{\gamma}$ with a nonnegative constant k and a unit vector field Y along γ. We call k and $\{\dot{\gamma}, Y\}$ the *geodesic curvature* and the *Frenet frame* of γ, respectively. When $k = 0$ it is a geodesic. Therefore there is no doubt about the saying that circles are simplest curves next to geodesics from the viewpoint of Frenet formula (cf. [14]).

---
*The author is partially supported by Grant-in-Aid for Scientific Research (C) (No. 24540075) Japan Society of Promotion Science.

We call two smooth curves γ_1, γ_2 on a Riemannian manifold which are parameterized by their arclength *congruent* to each other if there exist an isometry φ of the base manifold and a constant t_0 satisfying $\gamma_2(t) = \varphi \circ \gamma_1(t+t_0)$ for all t. On a real space form, which is one of a standard sphere, a Euclidean space and a real hyperbolic space, two circles are congruent to each other if and only if they have the same geodesic curvature. The moduli space, the set of all congruence classes, of circles of positive geodesic curvature on a real space form is hence identified with a half line $(0, \infty)$. For circles on a nonflat complex space form of complex dimension greater than 1, which is either a complex projective space or a complex hyperbolic space, we need another invariant. Given a circle γ of positive geodesic curvature with Frenet frame $\{\dot\gamma, Y\}$, we define its *complex torsion* τ by $\tau = \langle \dot\gamma, JY \rangle$. Clearly it is constant along γ and satisfies $|\tau| \le 1$. On a nonflat complex space form, two circles are congruent to each other if and only if they have the same geodesic curvature and have the same absolute value of complex torsion. The moduli space of circles of positive geodesic curvature is hence set theoretically identified with the band $[0, 1] \times (0, \infty)$. If we consider congruency of curves on a nonflat complex space form by the action of holomorphic isometries, as isometries on this space are either holomorphic isometries or anti-holomorphic isometries, the moduli space of circles of positive geodesic curvature is identified with the band $[-1, 1] \times (0, \infty)$.

We are interested in structures of this moduli space. A smooth curve γ parameterized by its arclength is said to be *closed* if there is a positive t_c with $\gamma(t+t_c) = \gamma(t)$ for all t. The minimum positive t_c with this property is called the *length* of γ and is denoted by length(γ). For an open curve γ, a curve which is not closed, we set length$(\gamma) = \infty$. It is clear that we can define a length function on the moduli space of curves parameterized by their arclength. On the moduli space $\mathcal{M}(\mathbb{C}P^n) \cong [0, 1] \times (0, \infty)$ of circles of positive geodesic curvature on a complex projective space $\mathbb{C}P^n$, we have a lamination structure $\mathcal{F}$ which is closely concerned with this length function (see Figure 1): On each of its leaf, the length function is smooth with respect to the induced differential structure as a subset of $\mathbb{R}^2$, and each leaf is maximal with respect to this property (see [2, 8] for more detail). If we write down this lamination structure $\mathcal{F} - \{\mathcal{F}_\nu\}_{0 \le \nu \le 1}$ explicitly for a complex projective space $\mathbb{C}P^n(4)$ of constant holomorphic sectional curvature 4, it is as follows;

$$
\mathcal{F}_\nu = \begin{cases} \{[\gamma_{k,0}] \mid k > 0\}, & \text{if } \nu = 0, \\ \{[\gamma_{k,\tau}] \mid 3\sqrt{3}k\tau(k^2 + 1)^{-3/2} = 2\nu,\ 0 < \tau < 1\}, & \text{if } 0 < \nu < 1, \\ \{[\gamma_{k,1}] \mid k > 0\}, & \text{if } \nu = 1, \end{cases}
$$

where $[\gamma_{k,\tau}]$ denotes the congruence class containing a circle of geodesic curvature k and of complex torsion τ.

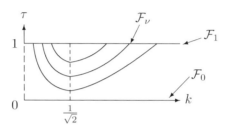

Figure 1.   Lamination on the moduli space of circles on $\mathbb{C}P^n(4)$

This lamination has singularities only on the leaf of circles of complex torsions ± 1. Such a property does not change even if we consider the moduli space by the action of holomorphic isometries, because the corresponding lamination is symmetric with respect to the half line $\{0\} \times (0, \infty)$. This suggests us there is a difference between circles of complex torsions ± 1 and other circles. In this paper we attempt to explain this difference by use of magnetic fields. We shall consider circles on a complex projective space as trajectories for some magnetic fields on geodesic spheres. From this point of view, we construct a foliation on a corresponding moduli space which is concerned with the length function.

2. Sasakian magnetic fields on geodesic spheres

A closed 2-form on a Riemannian manifold is said to be a *magnetic field*. For a magnetic field $\mathbb{B}$ on M we take a skew-symmetric operator $\Omega_{\mathbb{B}} : TM \to TM$ defined by $\mathbb{B}(v, w) = \langle v, \Omega_{\mathbb{B}}(w) \rangle$ for all $v, w \in T_pM$ at an arbitrary point $p \in M$. A smooth curve parameterized by its arclength is said to be a *trajectory* for $\mathbb{B}$ if it satisfies $\nabla_{\dot{\gamma}}\dot{\gamma} = \Omega_{\mathbb{B}}(\dot{\gamma})$. On a Kähler manifold with complex structure J as we have canonical 2-form $\mathbb{B}_J$ which is called the Kähler form, we have natural magnetic fields which are constant multiples of the Kähler form. We call them *Kähler magnetic fields* (cf. [1, 6]). A smooth curve γ parameterized by its arclength is then a trajectory for a Kähler magnetic field $\mathbb{B}_\kappa = \kappa \mathbb{B}_J$ if it satisfies $\nabla_{\dot\gamma}\dot\gamma = \kappa J\dot\gamma$. Since complex structure J on a Kähler manifold is parallel, we find a trajectory for $\mathbb{B}_\kappa$ is a circle of geodesic curvature $|\kappa|$ and of complex torsion $-\mathrm{sgn}(\kappa)$ and vice versa. Here, $\mathrm{sgn}(\kappa)$ denotes the signature of κ.

In order to explain circles on $\mathbb{C}P^n$ other than circles of complex torsion ± 1, we prepare magnetic fields of another kind. We take a geodesic sphere $G(r)$ of radius r in $\mathbb{C}P^n$ with inward unit normal vector field $\mathcal{N}$. We then have canonical almost contact metric structure $(\phi, \xi, \eta, \langle\ ,\ \rangle)$, which is a quadruple of a skew-symmetric operator, a vector field, a 1-form and the induced metric defined by $\xi = -J\mathcal{N}$, $\eta(v) = \langle v, \xi\rangle$ and $\phi v = Jv - \eta(v)\mathcal{N}$ for arbitrary $v \in TG(r)$. If we define a 2-form $\mathbb{F}_\phi$ on $G(r)$ by $\mathbb{F}_\phi(v, w) = \langle v, \phi w\rangle$, we find it is closed (see [9]). We call constant multiples of this closed 2-form *Sasakian magnetic fields*. For a Sasakian magnetic field $\mathbb{F}_\kappa = \kappa \mathbb{F}_\phi$ of *strength* κ, a smooth curve γ parameterized by its arclength is then a trajectory if it satisfies $\nabla_{\dot\gamma}\dot\gamma = \kappa\phi\dot\gamma$. Though Sasakian magnetic fields are quite resemble to Kähler magnetic fields, Sasakian magnetic fields are not uniform. That is, the Lorentz force of $\mathbb{F}_\kappa$ depends on velocity vectors: For a trajectory γ we have $\|\nabla_{\dot\gamma}\dot\gamma\| = |\kappa|\sqrt{1 - \eta(\dot\gamma)^2}$, and find that it depends upon $\dot\gamma$. We put $\rho_\gamma = \eta(\dot\gamma)\ (= \langle\dot\gamma, \xi\rangle)$ and call it the *structure torsion* of γ. We can show the constancy of structure torsion ρ_γ along a trajectory γ on $G(r)$ in the following manner. The shape operator A of $G(r)$ in $\mathbb{C}P^n$ satisfies $A\phi = \phi A$. Since we have $\nabla_X\xi = \phi AX$ for an arbitrary vector field X on $G(r)$, and A is symmetric and ϕ is skew-symmetric, we see

$$\rho'_\gamma = \nabla_{\dot\gamma}\langle\dot\gamma, \xi\rangle = \langle\kappa\phi\dot\gamma, \xi\rangle + \langle\dot\gamma, \phi A\dot\gamma\rangle$$
$$= \langle\dot\gamma, \phi A\dot\gamma\rangle = -\langle A\phi\dot\gamma, \dot\gamma\rangle = -\langle\phi A\dot\gamma, \dot\gamma\rangle,$$

hence $\rho' = 0$.

We here consider a trajectory γ for $\mathbb{F}_\kappa$ with initial vector $\pm\xi_p$ at an arbitrary point $p \in G(r)$. We then find $\rho_\gamma = \pm 1$ and $\nabla_{\dot\gamma}\dot\gamma = 0$. Thus, it does not depend on κ and is a geodesic on $G(r)$.

3. Extrinsic circular trajectories on geodesic spheres

We are interested in how trajectories for Sasakian magnetic fields on geodesic spheres can be seen in a complex projective space. We take an isometric immersion $\iota : G(r) \to \mathbb{C}P^n(4)$ of a geodesic sphere $G(r)$ of radius r into a complex projective space of constant holomorphic sectional curvature 4. For a curve γ on $G(r)$, we call the curve $\iota \circ \gamma$ on $\mathbb{C}P^n(4)$ its *extrinsic shape*. When the extrinsic shape of a curve is a circle, we call it *extrinsic circular.* The shape operator A of $G(r)$ in $\mathbb{C}P^n(4)$ has two principal curvatures: It satisfies $Av = \cot r\, v$, $A\xi = 2\cot 2r\, \xi$ for each tangent vector v orthogonal to ξ. We denote by ∇ and $\widetilde{\nabla}$ the Riemannian connections of a geodesic sphere $G(r)$ and $\mathbb{C}P^n$, respectively. We study extrinsic shapes of trajectories by use of the formulas of Gauss and Weingarten which are given as

$$\widetilde{\nabla}_X Y = \nabla_X Y + \langle AX, Y\rangle \mathcal{N}, \qquad \widetilde{\nabla}_X \mathcal{N} = -AX$$

for vector fields X, Y tangent to M.

Given a trajectory γ for a Sasakian magnetic field $\mathbb{F}_\kappa$ on $G(r)$ in $\mathbb{C}P^n(4)$ we have

$$\widetilde{\nabla}_{\dot\gamma}\dot\gamma = \nabla_{\dot\gamma}\dot\gamma + \langle A\dot\gamma, \dot\gamma\rangle \mathcal{N} = \kappa\phi\dot\gamma + \left(\cot r - \rho_\gamma^2 \tan r\right)\mathcal{N},$$

and

$$\begin{aligned}
\widetilde{\nabla}_{\dot\gamma}&\left\{\kappa\phi\dot\gamma + \left(\cot r - \rho_\gamma^2 \tan r\right)\mathcal{N}\right\}\\
&= \widetilde{\nabla}_{\dot\gamma}\left\{\kappa J\dot\gamma + \left(\cot r - \kappa\rho_\gamma - \rho_\gamma^2 \tan r\right)\mathcal{N}\right\}\\
&= -\left(\kappa^2 + \cot^2 r - \kappa\rho_\gamma \cot r - \rho_\gamma^2\right)\dot\gamma\\
&\quad - \left(\kappa\cot r - \kappa^2\rho_\gamma - \rho_\gamma + \rho_\gamma^3 \tan^2 r\right)\xi\\
&= -\left\{\kappa^2(1-\rho_\gamma^2) + (\cot r - \rho_\gamma^2 \tan r)^2\right\}\dot\gamma\\
&\quad - (\kappa\rho_\gamma - \cot r + \rho_\gamma^2 \tan r)(\kappa - \rho_\gamma \tan r)(\rho_\gamma\dot\gamma - \xi).
\end{aligned}$$

Therefore we find that a trajectory γ for $\mathbb{F}_\kappa$ is extrinsic geodesic if and only if $r = \pi/4$ and $\rho_\gamma = \pm 1$ and that γ is extrinsic circular if and only if it satisfies either $\rho_\gamma = \pm 1$ or $\rho_\gamma \neq \pm 1$ and $(\kappa\rho_\gamma - \cot r + \rho_\gamma^2 \tan r)(\kappa - \rho_\gamma \tan r) = 0$. In the former case, the geodesic curvature and complex torsion of the extrinsic shape are $k_1 = |2\cot 2r|$ and $\tau_{12} = \mp\mathrm{sgn}(\cot 2r)$. Its Frenet frame is $\{\dot\gamma,\ \mathrm{sgn}(\cot 2r)\mathcal{N}\}$ when $r \neq \pi/4$. In the latter case, they are

$$k_1 = \sqrt{\kappa^2(1-\rho_\gamma^2) + (\cot r - \rho_\gamma^2 \tan r)^2}\ \ (>0),$$

$$\tau_{12} = -\frac{1}{k_1}\left\{\kappa + \rho_\gamma(\cot r - \kappa\rho_\gamma - \rho_\gamma^2 \tan r)\right\},$$

and its Frenet frame is

$$\left\{\dot\gamma,\ \frac{1}{k_1}\left\{\kappa J\dot\gamma + \left(\cot r - \kappa\rho_\gamma - \rho_\gamma^2 \tan r\right)\mathcal{N}\right\}\right\}.$$

We hence obtain the following.

Proposition 3.1. *Let γ be a trajectory for a Sasakian magnetic field $\mathbb{F}_\kappa$ with $\rho_\gamma = \pm 1$ on a geodesic sphere $G(r)$ in $\mathbb{C}P^n(4)$.*

(1) When $r = \pi/4$, its extrinsic shape is a geodesic.
(2) Otherwise, its extrinsic shape is a circle of geodesic curvature $k_1 = |2\cot 2r|$ and of complex torsion $\tau_{12} = \mp\mathrm{sgn}(\cot 2r)$.

Proposition 3.2 ([3]). *Let γ be a trajectory for a Sasakian magnetic field $\mathbb{F}_\kappa$ with $\rho_\gamma \neq \pm 1$ on a geodesic sphere $G(r)$ in $\mathbb{C}P^n(4)$.*

(1) *It is extrinsic circular if and only if $\kappa = \rho_\gamma^{-1}\cot r - \rho_\gamma \tan r$ or $\kappa = \rho_\gamma \tan r$.*
(2) *When $\kappa = \rho_\gamma^{-1}\cot r - \rho_\gamma \tan r$, its extrinsic shape is a trajectory for the Kähler magnetic field $\mathbb{B}_\kappa$. That is, it is a circle of geodesic curvature $|\kappa|$ and of complex torsion $-\mathrm{sgn}(\kappa)$.*
(3) *When $\kappa = \rho_\gamma \tan r$, its extrinsic shape is a circle of geodesic curvature k_1 and of complex torsion τ_{12} which are given as*

$$k_1 = \sqrt{\cot^2 r + \rho_\gamma^2 \tan^2 r - 2\rho_\gamma^2} = \frac{1}{|\kappa|}\sqrt{\rho_\gamma^2 + \kappa^4 - 2\rho_\gamma^2 \kappa^2},$$

$$\tau_{12} = \frac{\rho_\gamma\{(2\rho_\gamma^2 - 1)\tan r - \cot r\}}{\sqrt{\cot^2 r + \rho_\gamma^2 \tan^2 r - 2\rho_\gamma^2}} = \mathrm{sgn}(\kappa)\frac{2\rho_\gamma^2 \kappa^2 - \kappa^2 - \rho_\gamma^2}{\sqrt{\rho_\gamma^2 + \kappa^4 - 2\rho_\gamma^2 \kappa^2}}.$$

Remark 3.1. We may consider that the case $\rho_\gamma = \pm 1$ is included in the case (3) of Proposition 3.2. This is because trajectories of structure torsion ± 1 do not depend on strengths of Sasakian magnetic fields.

We here study the converse of the above propositions, particularly focus our mind on the converse of the third assertion of Proposition 3.2. On a geodesic sphere $G\big(\tan^{-1}\sqrt{2}\big)$ in $\mathbb{C}P^n(4)$, a trajectory γ for $\mathbb{F}_\kappa$ is extrinsic circular if and only if $\kappa = -\rho_\gamma + 1/(\sqrt{2}\,\rho_\gamma)$ or $\kappa = \sqrt{2}\,\rho_\gamma$. In the latter case the geodesic curvature and complex torsion are $k_1 = 1/\sqrt{2}$ and $\tau_{12} = 4\rho_\gamma^3 - 3\rho_\gamma$. We find this circular trajectory corresponds to an image of a geodesic through the parallel embedding given by Naitoh [13]. As ρ_γ runs over the interval $[-1, 1]$, we can guess that all circles of geodesic curvature $1/\sqrt{2}$ are extrinsic shapes of trajectories. We shall show this for circles of arbitrary geodesic curvature.

Theorem 3.1. *Every circle γ of positive geodesic curvature on a complex projective space $\mathbb{C}P^n(4)$ is the extrinsic shape of a trajectory for some Sasakian magnetic field on some geodesic sphere. For each circle of complex torsion $|\tau_{12}| < 1$, exactly three trajectories correspond to it.*

Proof. In the first place we show the existence. We take a circle of geodesic curvature k and of complex torsion τ. Following Remark 3.1, we would like to find a solution of the system of equations on $\lambda\ (= \cot r)$ and ρ

$$\begin{cases} \lambda^4 - 2\rho^2\lambda^2 + \rho^2 = k^2\lambda^2, \\ \rho(2\rho^2 - 1 - \lambda^2) = k\tau\lambda, \end{cases} \tag{3.1}$$

under the assumption that $\lambda > 0$ and $|\rho| \leq 1$. Here, r shows the radius of a geodesic sphere.

We first consider the range of λ by the first equality of (3.1). To do this we first suppose $\lambda \leq 1/\sqrt{2}$. The condition $|\rho| \leq 1$ shows that $\lambda \leq k \leq \lambda^{-1} - \lambda$. Hence we have $0 < \lambda \leq k$ and $\lambda \leq (\sqrt{4+k^2} - k)/2$. Thus we see

$$\begin{cases} 0 < \lambda \leq k, & \text{when } k < 1/\sqrt{2}, \\ 0 < \lambda \leq (\sqrt{4+k^2} - k)/2, & \text{when } k \geq 1/\sqrt{2}. \end{cases}$$

Next we suppose $1/\sqrt{2} < \lambda \leq 1$. The condition $|\rho| \leq 1$ shows that $\lambda^{-1} - \lambda \leq k \leq \lambda$. Hence we have

$$\begin{cases} (\sqrt{4+k^2} - k)/2 \leq \lambda \leq 1, & \text{when } k \leq 1/\sqrt{2}, \\ k \leq \lambda \leq 1, & \text{when } 1/\sqrt{2} < k \leq 1, \\ \text{no range}, & \text{when } k > 1. \end{cases}$$

Finally we suppose $\lambda > 1$. The condition $|\rho| \leq 1$ shows that $\lambda - \lambda^{-1} \leq k \leq \lambda$. Hence we have

$$\begin{cases} 1 < \lambda \leq (\sqrt{4+k^2} + k)/2, & \text{when } k \leq 1, \\ k \leq \lambda \leq (\sqrt{4+k^2} + k)/2, & \text{when } k > 1. \end{cases}$$

Thus we find that the first equation of (3.1) shows that the range of λ is the following union of intervals:

$$\begin{cases} (0, k] \cup \left[(\sqrt{4+k^2} - k)/2, (\sqrt{4+k^2} + k)/2\right], & \text{when } k < 1/\sqrt{2}, \\ (0, \sqrt{2}\,], & \text{when } k = 1/\sqrt{2}, \\ \left(0, (\sqrt{4+k^2} - k)/2\right] \cup \left[k, (\sqrt{4+k^2} + k)/2\right], & \text{when } k > 1/\sqrt{2}. \end{cases}$$

Next we study solutions of the system (3.1) of equations. By the two equalities we have

$$\tau^2 = \frac{(k^2 - \lambda^2)(2k^2\lambda^2 + \lambda^2 - 1)^2}{k^2(1 - 2\lambda^2)^3} \tag{3.2}$$

when $\lambda \neq 1/\sqrt{2}$. We consider the right hand side of (3.2) and denote it by $f_k(\lambda)$ for $k \neq 1/\sqrt{2}$. When $k < 1/\sqrt{2}$, we find that

1) $f_k(\lambda)$ is monotone decreasing on the union of intervals
   $(0, k] \cup \left[(\sqrt{4+k^2} - k)/2, \, 1/\sqrt{2k^2+1}\right]$,
2) $f_k(\lambda)$ is monotone increasing on the interval
   $\left[1/\sqrt{2k^2+1}, \, (\sqrt{4+k^2} + k)/2\right]$.

When $k > 1/\sqrt{2}$, we find that

1) $f_k(\lambda)$ is monotone decreasing on the interval $\left(0, 1/\sqrt{2k^2+1}\right]$,
2) $f_k(\lambda)$ is monotone increasing on the union of intervals
   $\left[1/\sqrt{2k^2+1}, \, (\sqrt{4+k^2} - k)/2\right] \cup \left[k, \, (\sqrt{4+k^2} + k)/2\right]$.

Since we have

$$f_k(0) = f_k\left((\sqrt{4+k^2} - k)/2\right) = f_k\left((\sqrt{4+k^2} + k)/2\right) = 1,$$

$$f_k(k) = f_k\left(1/\sqrt{2k^2+1}\right) = 0,$$

for τ with $|\tau| < 1$, there exists $\lambda = \lambda_1, \lambda_2, \lambda_3$ $\left(\lambda_1 \leq \lambda_2 \leq \lambda_3, \, \lambda_i \neq 1/\sqrt{2}\right)$ satisfying $f_k(\lambda) = \tau^2$, which is the equation (3.2), in case $k \neq 1/\sqrt{2}$. Here, either $\lambda_1 = \lambda_2 \left(= 1/\sqrt{2k^2+1}\right)$ or $\lambda_2 = \lambda_3 \left(= 1/\sqrt{2k^2+1}\right)$ holds if and only if $\tau = 0$. For such λ we set $\rho = \pm\lambda\sqrt{(k^2 - \lambda^2)/(1 - 2\lambda^2)}$. In the case $\tau \neq 0$, the signature of ρ is the opposite to that of τ when $\lambda = \lambda_1$ or $\lambda = \lambda_3$, and is the same as that of τ when $\lambda = \lambda_2$. In the case $\tau = 0$, we have $\rho = 0$ when $\lambda = k$, and we consider two $\rho = \pm\sqrt{(k^2+1)/(2k^2+1)}$ when $\lambda = 1/\sqrt{2k^2+1}$. We then obtain that the pair (λ, ρ) satisfies the system of equations (3.1). We here note the following: We find that only $\lambda_\pm = (\sqrt{4 + k^2}\pm k)/2$ satisfies $f_k(\lambda_\pm) = 1$, hence find that $\rho^2 = 1$ if and only if $\tau^2 = 1$ because $\lambda = \lambda_\pm$ if and only if $\rho^2 = 1$ by the previous argument on the range of λ.

When $k = 1/\sqrt{2}$, we find that the system (3.1) of equations turns to

$$\begin{cases} \lambda^4 - 2\left(\rho^2 + \dfrac{1}{4}\right)\lambda^2 + \rho^2 = 0, \\ \sqrt{2}\rho\,(2\rho^2 - 1 - \lambda^2) = \lambda\tau. \end{cases} \tag{3.3}$$

The first equality of (3.3) shows that $\lambda = \sqrt{2}\,|\rho|$ or $\lambda = 1/\sqrt{2}$. In the case $\lambda = \sqrt{2}\,|\rho|$, the second equality of (3.3) shows that $\tau = -\mathrm{sgn}(\rho)$. In the case $\lambda = 1/\sqrt{2}$, the second equality of (3.3) turns to $4\rho^3 - 3\rho = \tau$. We shall study the latter case. If we consider $\tau(\rho) = 4\rho^3 - 3\rho$ as a function of ρ, it is monotone increasing on the union of intervals $[-1, -1/2] \cup [1/2, 1]$ and is monotone decreasing on the interval $[-1/2, 1/2]$. Since $\tau(-1) = \tau(1/2) = -1$ and $\tau(-1/2) = \tau(1) = 1$, we get solutions of the system (3.3) of equations for a given τ.

We shall now find out a geodesic sphere and a trajectory for some Sasakian magnetic field on this geodesic sphere whose extrinsic shape coincides with γ. When $k \neq 1/\sqrt{2}$, we take a solution (λ, ρ) of the system (3.1) of equations and set $\kappa = \rho/\lambda$. With the Frenet frame $\{\dot{\gamma}, Y\}$ of γ, we set $v = \{k\lambda Y(0) - \rho J\dot{\gamma}(0)\}/(\lambda^2 - 2\rho^2)$. We should note that the assumption $k \neq 1/\sqrt{2}$ and the first equation in (3.1) guarantee $\lambda^2 \neq 2\rho^2$. We take a geodesic σ of initial vector v and take a geodesic sphere G of radius $\cot^{-1} \lambda$ centered at $\sigma(\cot^{-1} \lambda)$. Then the inward unit normal $\mathcal{N}$ of G satisfies $\mathcal{N}_{\gamma(0)} = v$ at $\gamma(0)$. We therefore have $\dot{\gamma}(0)$ is tangent to G and have

$$\langle \dot{\gamma}(0), -J\mathcal{N}_{\gamma(0)} \rangle = \frac{-k\lambda\tau - \rho}{\lambda^2 - 2\rho^2} = \rho$$

by use of the second equation of (3.1). If we take a trajectory for $\mathbb{F}_\kappa = \mathbb{F}_{\rho/\lambda}$ on G with initial vector $\dot{\gamma}(0)$, its structure torsion is ρ and its extrinsic shape $\hat{\gamma}$ is a circle of geodesic curvature $\sqrt{\lambda^2 - 2\rho^2 + (\rho^2/\lambda^2)} = k$ and of complex torsion $\rho(2\rho^2 - 1 - \lambda^2)/(k\lambda) = \tau$ by Proposition 3.2 and (3.1). Since its Frenet frame at $\hat{\gamma}(0)$ consists of $\dot{\gamma}(0)$ and $\frac{1}{k\lambda}\{\rho J\dot{\gamma}(0) + (\lambda^2 - 2\rho^2)v\} = Y(0)$, we find $\hat{\gamma}$ coincides with γ.

When $k = 1/\sqrt{2}$, we take ρ satisfying $4\rho^3 - 3\rho = \tau$. We set $\lambda = 1/\sqrt{2}$ and $\kappa = \sqrt{2}\rho$. If $\tau \neq \pm 1$, as we have $\rho \neq \pm 1/2$, we put $v = \{Y(0) - 2\rho J\dot{\gamma}(0)\}/(1 - 4\rho^2)$ and take the geodesic σ of initial vector v. If we consider the geodesic sphere G of radius $\tan^{-1}\sqrt{2}$ centered at $\sigma(\tan^{-1}\sqrt{2})$, its inward unit normal $\mathcal{N}$ satisfies $\mathcal{N}_{\gamma(0)} = v$, hence we have $\dot{\gamma}(0)$ is tangent to G and have

$$\langle \dot{\gamma}(0), -J\mathcal{N}_{\gamma(0)} \rangle = \frac{-\tau - 2\rho}{1 - 4\rho^2} = \rho.$$

Take the trajectory for $\mathbb{F}_\kappa = \mathbb{F}_{\sqrt{2}\rho}$ on G with initial vector $\dot{\gamma}(0)$. Then its structure torsion is ρ, and its extrinsic shape $\hat{\gamma}$ is a circle of geodesic curvature $k = 1/\sqrt{2}$ and of complex torsion $4\rho^2 - 3\rho = \tau$ by Proposition 3.2. Since its Frenet frame at $\hat{\gamma}(0)$ consists of $\dot{\gamma}(0)$ and $2\rho J\dot{\gamma}(0) + (1 - 4\rho^2)v = Y(0)$, we find $\hat{\gamma}$ coincides with γ.

When $k = 1/\sqrt{2}$ and $\tau = \pm 1$, we take an arbitrary unit tangent vector $w \in T_{\gamma(0)}\mathbb{C}P^n$ which is orthogonal to both $\dot{\gamma}(0)$ and $J\dot{\gamma}(0)$, and set $v = (\mp J\dot{\gamma}(0) + \sqrt{3}w)/2$. Here, the double sign takes the opposite signature of τ. We then have $\langle \dot{\gamma}(0), -Jv \rangle = \mp 1/2$, which is a solution of the equation $4\rho^3 - 3\rho = \tau$. By taking a geodesic sphere G of radius $\tan^{-1}\sqrt{2}$ by just the same way as above, we consider the trajectory for $\mathbb{F}_{\mp 1/\sqrt{2}}$ on G with initial vector $\dot{\gamma}(0)$. Its structure torsion is $\pm 1/2$, and its extrinsic shape $\hat{\gamma}$ is a circle of geodesic curvature $1/\sqrt{2}$ and of complex torsion ± 1. Since its

Frenet frame at $\hat{\gamma}(0)$ consists of $\dot{\gamma}(0)$ and $\mp J\dot{\gamma}(0)$, we find $\hat{\gamma}$ coincides with γ. This completes the proof of showing the existence.

We now study the number of trajectories whose extrinsic shape is a circle γ of geodesic curvature k and of complex torsion τ ($\neq \pm 1$). We here consider the number of solutions of (3.1). First, we study the case $k \neq 1/\sqrt{2}$. The above argument guarantees that we have 3 solutions for the equation (3.2) on λ when $\tau \neq 0$ and have 2 solutions when $\tau = 0$. When $\tau \neq 0$, for each solution of (3.2) the solution ρ is uniquely determined. When $\tau = 0$, for the solution $\lambda = k$ we have $\rho = 0$, but for the solution $\lambda = 1/\sqrt{2k^2 + 1}$ we have two solutions of ρ satisfying (3.1). Thus we find the system (3.1) of equations has exactly 3 solutions in the case $k \neq 1/\sqrt{2}$. Second, we study the case $k = 1/\sqrt{2}$. Since $\tau \neq \pm 1$, the behavior of the function $\tau(\rho)$ shows that we have 3 solutions of type $(1/\sqrt{2}, \rho)$.

We now suppose the extrinsic shape of a trajectory ς for a Sasakian magnetic field $\mathbb{F}_\kappa$ on a geodesic sphere M of radius R coincides with a given circle γ in $\mathbb{C}P^n$ whose geodesic curvature is k, complex torsion τ ($\neq \pm 1$) and Frenet frame $\{\dot{\gamma}, Y\}$. We then have $\kappa = \rho_\varsigma \tan R$ and $kY = \kappa J\dot{\varsigma} + (\cot R - \kappa\rho_\varsigma + \rho_\varsigma^2 \tan R)\mathcal{N}_M$, where $\mathcal{N}_M$ is the inward unit normal of M. Also, we find that $(\cot R, \rho_\varsigma)$ satisfies (3.1) when $k \neq 1/\sqrt{2}$ and that $\cot R = 1/\sqrt{2}$ and ρ_ς satisfies $4\rho_\varsigma^3 - 3\rho_\varsigma = \tau$ when $k = 1/\sqrt{2}$. Thus we find M and ς are those we investigate to show the existence. This completes the proof. $\qquad\square$

The proof of Theorem 3.1 shows the following.

Proposition 3.3. *Every circle γ of positive geodesic curvature $1/\sqrt{2}$ on $\mathbb{C}P^n(4)$ is the extrinsic shape of a trajectory for some Sasakian magnetic field on some geodesic sphere of radius $\tan^{-1}\sqrt{2}$. For each circle of complex torsion $|\tau_{12}| < 1$ exactly three trajectories correspond to it.*

Proposition 3.4. *For each circle of positive geodesic curvature and of complex torsion ± 1 on $\mathbb{C}P^n$, we have infinitely many trajectories for some Sasakian magnetic fields on some geodesic spheres whose extrinsic shapes coincide with this circle.*

We should note that every circle on a standard sphere is the extrinsic shape of a geodesic on some geodesic hypersphere which is totally umbilic in this standard sphere. We also note that there are no totally umbilic real hypersurfaces in a complex projective space. We may say that Theorem 3.1 corresponds to this fact on a standard sphere.

4. Moduli space of extrinsic circular trajectories

Since we find all circles of positive geodesic curvature on a complex projective space are obtained as extrinsic shapes of trajectories for Sasakian
magnetic fields on geodesic spheres, we here consider a lamination which
is concerned with the length function on the moduli space of extrinsic circular trajectories. It is well known that isometries on a geodesic sphere in
$\mathbb{C}P^n$ are equivariant in $\mathbb{C}P^n$. That is, if we denote by $\iota : G(r) \to \mathbb{C}P^n$ an
isometric immersion of a geodesic sphere $G(r)$, then for every isometry φ
of $G(r)$ there is an isometry $\tilde{\varphi}$ of $\mathbb{C}P^n$ satisfying $\tilde{\varphi} \circ \iota = \iota \circ \varphi$. Thus we
see that if two curves on $G(r)$ are congruent to each other as curves on
$G(r)$ then their extrinsic shapes are congruent to each other as curves in
$\mathbb{C}P^n$. For congruence of trajectories for Sasakian magnetic fields, we have
the following.

Lemma 4.1 ([3]). *Let γ_1 and γ_2 are trajectories for Sasakian magnetic
fields $\mathbb{F}_{\kappa_1}$ and $\mathbb{F}_{\kappa_2}$ on a geodesic sphere $G(r)$ in $\mathbb{C}P^n$, respectively. They
are congruent to each other if and only if one of the following conditions
holds:*

i) $|\rho_{\gamma_1}| = |\rho_{\gamma_2}| = 1$,
ii) $\rho_{\gamma_1} = \rho_{\gamma_2} = 0$ *and* $|\kappa_1| = |\kappa_2|$,
iii) $0 < |\rho_{\gamma_1}| = |\rho_{\gamma_2}| < 1$ *and* $\kappa_1 \rho_{\gamma_1} = \kappa_2 \rho_{\gamma_2}$.

We now investigate the moduli space $\mathcal{E}_2(\mathbb{C}P^n(4))$ of all extrinsic circular trajectories for Sasakian magnetic fields on geodesic spheres in $\mathbb{C}P^n(4)$
of the type (3) in Proposition 3.2. Here, we consider that two circular trajectories γ_1, γ_2 for Sasakian magnetic fields on geodesic spheres $G(r_1)$ and
$G(r_2)$ are congruent to each other if and only if the following conditions
hold:

i) $r_1 = r_2$, that is, there is an isometry $\tilde{\varphi}$ of $\mathbb{C}P^n$ with $\tilde{\varphi}(G(r_1)) = G(r_2)$;
ii) $(\tilde{\varphi}|_{G(r_1)}) \circ \gamma_1$ and γ_2 are congruent to each other as curves on $G(r_2)$.

In order to represent this moduli space in a Euclidean plane, we study this
set by use of the parameters $(\kappa, \rho) \in [0, \infty) \times (-1, 1)$ on structure torsions
of trajectories and on strengths of Sasakian magnetic fields. Though these
parameters are natural, we need to take care in treating the case $\kappa = 0$.
When $\kappa = 0$ (or when $\rho = 0$) the relationship $\kappa = \rho \tan r$, which is a
condition on extrinsic circular trajectories given in Proposition 3.2, leads
us to $\rho = 0$ (or $\kappa = 0$) and gives no restrictions on radius r of a geodesic
sphere. Since the geodesic curvature of extrinsic shape of extrinsic circular

trajectory for $\mathbb{F}_0$ on $G(r)$ is $\cot r$, hence depends on r, we find that infinitely many "congruence classes" correspond to the point $(\kappa, \rho) = (0,0)$. We hence omit this point for now to bring forward our study. Also, as extrinsic circular trajectories corresponding to (κ, ρ) and $(\kappa, -\rho)$ with $\rho \neq 0$ have the same geodesic curvature and same absolute value of complex torsion, we shall identify them along the same context that we studied the moduli space $\mathcal{M}(\mathbb{C}P^n)$ of circles under the action of all isometries. Therefore we consider a lamination structure on the open band $\mathcal{E}_{\kappa,\rho} = (0, \infty) \times (0,1)$. In view of the construction of the lamination $\mathcal{F}$ on $\mathcal{M}(\mathbb{C}P^n)$ explained in §1, with the aid of Proposition 3.2, we have a lamination $\mathcal{G} = \{\mathcal{G}_\nu\}_{0 \leq \nu \leq 1}$ on this band which is given as

$$
\mathcal{G}_\nu = \begin{cases} \left\{(\kappa, \rho) \mid 2\rho^2\kappa^2 - \kappa^2 - \rho^2 = 0, \ 0 < \rho < 1\right\}, & \nu = 0, \\[2mm] \left\{(\kappa, \rho) \mid \dfrac{3\sqrt{3}\kappa^2(2\rho^2\kappa^2 - \kappa^2 - \rho^2)}{2(\rho^2 + \kappa^4 - 2\rho^2\kappa^2 + \kappa^2)^{3/2}} = \nu, \ 0 < \rho < 1\right\}, & 0 < \nu < 1, \\[2mm] \left\{(1/\sqrt{2}, \rho) \mid 0 < \rho < 1\right\}, & \nu = 1. \end{cases}
$$

As we do not consider trajectories of structure torsion ± 1 whose extrinsic shapes are circles of complex torsion ± 1, this structure $\mathcal{G}$ on $\mathcal{E}_{\kappa,\rho}$ is like Figure 2 and does not have singularities. Thus $\mathcal{G}$ is a foliation on $\mathcal{E}_{\kappa,\rho}$. Since our construction is based on the construction of the lamination $\mathcal{F}$, this foliation has the following properties:

i) On each of its leaf, the length function is smooth with respect to the induced differential structure as a subset of $\mathbb{R}^2$;

ii) Each leaf is maximal with respect to this property.

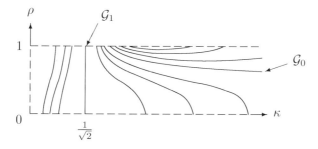

Figure 2. Foliation $\mathcal{G}$ on $\mathcal{E}_{\kappa,\rho}$

In order to complete our study on a foliation concerned with the length function we need to consider the omitted "congruence classes" of extrinsic circular geodesics (i.e. trajectories for $\mathbb{F}_0$) of null structure torsion. Their extrinsic shapes are circles of null complex torsion whose geodesic curvature runs over the interval $(0, \infty)$. If we consider circles corresponding to congruence classes on each leaf $\mathcal{G}_\nu$, we see the behavior of their geodesic curvature is

$$\lim_{\rho\uparrow 1} k_1 = |2\cot 2r| = \left| \frac{1}{\kappa} - \kappa \right|$$

by Proposition 3.2. In view of Figure 2, it is natural to consider that these classes form the set $\{0\} \times (0, 1)$. Adding leaves corresponding to this set, we obtain a foliation on the moduli space $\mathcal{E}_2(\mathbb{C}P^n) \cong [0, \infty) \times \{(-1, 0) \cup (0, 1)\}$ of all extrinsic circular trajectories of the type (3) in Proposition 3.2. Summarizing up, we get the following.

Theorem 4.1. *On the moduli space $\mathcal{E}_2(\mathbb{C}P^n)$ of extrinsic circular trajectories for Sasakian magnetic fields on geodesic spheres in $\mathbb{C}P^n$, we have a foliation satisfying the following properties*:

i) *On each of its leaf, the length function is smooth with respect to the induced differential structure as a subset of $\mathbb{R}^2$*;

ii) *Each leaf is maximal with respect to this property.*

To represent $\mathcal{E}_2(\mathbb{C}P^n)$ in a Euclidean plane, we have another pair (ρ, λ) of parameters on structure torsion and on radius $\lambda = \cot r$ of geodesic spheres. With these parameters we do not have exceptional leaves. But unfortunately, it is not easy to draw graphs of leaves for the lamination.

Our results Theorems 3.1 and 4.1 suggest us that from the viewpoint of the length function it is much more natural to treat circles on a complex projective space as extrinsic shapes of trajectories for Sasakian magnetic fields on geodesic spheres.

5. Other trajectories on geodesic spheres

To reinforce our idea on treating circles as extrinsic shapes of trajectories, we come to the position to study extrinsic shapes of other trajectories on geodesic spheres. Here, we only make mention of a result corresponding to Proposition 3.2 and give some problems.

A smooth curve γ on a Riemannian manifold M is said to be a *helix* of proper order d if it satisfies the system of differential equations $\nabla_{\dot\gamma} Y_j = -k_{j-1} Y_{j-1} + k_j Y_{j+1}$ $(j = 1, \ldots, d)$ with positive constants $k_1, \ldots, k_{d-1}$

and an orthonormal frame field $\{Y_1 = \dot{\gamma}, Y_2, \ldots, Y_d\}$ along γ. Here, we set Y_0, Y_{d+1} to be null vector fields along γ and $k_0 = k_d = 0$. These positive constants and the frame field are called the geodesic curvatures and the Frenet frame of γ, respectively. We say γ *Killing* if it is generated by some Killing vector field on M. On $\mathbb{C}P^n$, if we define complex torsions τ_{ij} ($1 \leq i < j \leq d$) of a helix γ of proper order d with Frenet frame $\{Y_i\}$ by $\tau_{ij} = \langle Y_i, JY_j \rangle$, then it is Killing if and only if all its complex torsions are constant along γ (see [12] and also [5]). All circles on $\mathbb{C}P^n$ are Killing, but are not necessarily Killing for helices of proper order greater than 2. We call a helix of proper order $2m - 1$ or $2m$ on $\mathbb{C}P^n$ *essential* if it lies on a totally geodesic $\mathbb{C}P^m$ $(\subset \mathbb{C}P^n)$. For circles, only circles of complex torsion ± 1 are essential.

It is known that the moduli space of essential Killing helices of proper order 3 and 4 on $\mathbb{C}P^n$ is identified with the set $(0, \infty) \times (0, \infty) \times \mathbb{R}$ of geodesic curvatures and a relation of complex torsions: For a point $(k_1, k_2, k_3) \in (0, \infty) \times (0, \infty) \times (0, \infty)$, we correspond it to the congruence classes of essential Killing helices of proper order 4 satisfying $\tau_{34} = \tau_{12}$ and have geodesic curvatures k_1, k_2, k_3; we correspond $(k_1, k_2, -k_3)$ to the congruence classes of essential Killing helices of proper order 4 satisfying $\tau_{34} = -\tau_{12}$ and have geodesic curvatures k_1, k_2, k_3; and we correspond to $(k_1, k_2, 0)$ the congruence classes of essential Killing helices of proper order 3 of geodesic curvatures k_1, k_2 (see [5]).

We here consider the length function on the moduli space of helices. Even if we restrict ourselves to essential Killing helices on $\mathbb{C}P^n$, from the viewpoint of the length function, the moduli space of circles of complex torion ± 1, which is identified with a half line $(0, \infty)$, and that of essential Killing helices of proper order 3 and 4 do not form a "building structure". At this point, the moduli space of essential Killing helices on $\mathbb{C}P^n$ is quite different from the moduli space of helices on a real space form (c.f. [4]).

Thus we are interested in extrinsic shapes of other trajectories for Sasakian magnetic fields on geodesic spheres in $\mathbb{C}P^n$. By continuing the calculation in §3 we can get the following.

Proposition 5.1 ([3]). *Let γ be a trajectory for a Sasakian magnetic field $\mathbb{F}_\kappa$ on a geodesic sphere $G(r)$ in $\mathbb{C}P^n(4)$ satisfying*

$$\rho_\gamma \neq \pm 1, \ \left(\sqrt{\kappa^2 \tan^2 r + 4} - \kappa \tan r\right)/2.$$

Its extrinsic shape is as follows:

(1) *When $\kappa\rho_\gamma(\cot r - \tan r) = \cot^2 r - \rho_\gamma^2$, it is an essential Killing helix of proper order 3;*

(2) *Otherwise, it is an essential Killing helix of proper order* 4.

In these cases, their geodesic curvatures are

$$k_1 = \sqrt{\kappa^2(1-\rho_\gamma^2) + \cot^2 r - 2\rho_\gamma^2 + \rho^4 \tan^2 r},$$

$$k_2 = \left|(\kappa\rho_\gamma - \cot r + \rho_\gamma^2 \tan r)(\kappa - \rho_\gamma \tan r)\right|\sqrt{1-\rho_\gamma^2}/k_1,$$

$$k_3 = \left|\kappa\rho_\gamma(\cot r - \tan r) + \rho_\gamma^2 - \cot^2 r\right|/k_1,$$

and their complex torsions are

$$\tau_{12} = \left\{\rho_\gamma^3 \tan r - \rho_\gamma \cot r - \kappa(1-\rho_\gamma^2)\right\}/k_1,$$

$$\tau_{23} = -\operatorname{sgn}(\kappa - \rho_\gamma \tan r)\left|\kappa\rho_\gamma - \cot r + \rho_\gamma^2 \tan r\right|\sqrt{1-\rho_\gamma^2}/k_1,$$

$$\tau_{14} = \operatorname{sgn}(\alpha)\tau_{23}, \quad \tau_{34} = \operatorname{sgn}(\alpha)\tau_{12}, \quad \tau_{13} = \tau_{24} = 0$$

with $\alpha = \kappa\rho_\gamma(\cot r - \tan r) + \rho_\gamma^2 - \cot^2 r$. *Here, in the case* (1) *we do not consider* τ_{14}, τ_{24} *and* τ_{34}.

The author considers that the converse of the above proposition holds (cf. [7]). Our problems are as follows.

1) For a pair of positive constants (k_1, k_2), find a trajectory for some Sasakian magnetic field on some geodesic sphere whose extrinsic shape is an essential Killing helix of proper order 3 of geodesic curvatures k_1, k_2.

2) For a triplet of positive constants (k_1, k_2, k_3), find two trajectories for some Sasakian magnetic fields on some geodesic spheres having the following properties: Their extrinsic shapes are essential Killing helices of proper order 4; their geodesic curvatures are k_1, k_2, k_3; one of them has complex torsions with $\tau_{12} = \tau_{34}$ and the other has complex torsions with $\tau_{12} = -\tau_{34}$.

Once we can solve these problems positively, we consider the moduli space of trajectories on geodesic spheres in $\mathbb{C}P^n$ which is parameterized by strength κ of Sasakian magnetic field, structure torsion ρ and radius $\lambda = \cot r$ of a geodesic sphere. The author considers that we have a foliation which is concerned with the length function in our sense.

References

1. T. Adachi, *Kähler magnetic flows on a manifold of constant holomorphic sectional curvature*, Tokyo J. Math. 18(1995), 473-483.

2. _____, *Lamination of the moduli space of circles and their length spectrum for a non-flat complex space form*, Osaka J. Math. 40(2003), 895–916.
3. _____, *Trajectories on geodesic spheres in a non-flat complex space form*, J. Geom. 90(2008), 1–29.
4. _____, *Foliation on the moduli space of helices on a real space form*, Int. Math. Forum 4(2009), 1699–1707.
5. _____, *Essential Killing helices of order less than five on a non-flat complex space form*, J. Math. Soc. Japan 64(2012), 969–984.
6. _____, *Kähler magnetic fields on Kähler manifolds of negative curvature*, Diff. Geom. its Appl. 29(2011), S2–S8.
7. _____, *Dynamical systematic aspect of horocyclic circles in a complex hyperbolic space*, in *Prospects of Differential Geometry and its Related Fields*, T. Adachi, H. Hashimoto & M. Hristov eds., World Scientific, 99–114, 2013.
8. T. Adachi, S. Maeda & S. Udagawa, *Circles in a complex projective space*, Osaka J. Math. 21(1997), 29–42.
9. T. Bao & T. Adachi, *Circular trajectories on real hypersurfaces in a nonflat complex space form*, J. Geom. 96(2009), 41–55.
10. B.-y. Chen & S. Maeda, *Extrinsic characterizations of circles in a complex projective space imbedded in a Euclidean space*, Tokyo J. Math. 19(1986), 169–185.
11. S. Maeda & T. Adachi, *Sasakian curves on hypersurfaces of type (A) in a nonflat complex space form*, Results Math. 56(2009), 489–499.
12. S. Maeda & Y. Ohnita, *Helical geodesic immersion into complex space form*, Geom. Dedicata, 30(1989), 93–114.
13. H. Naitoh, *Isotropic submanifolds with parallel second fundamental form in $P^m(c)$*, Osaka J. Math. 18(1981), 427–464.
14. K. Nomizu & K. Yano, *On circles and spheres in Riemannian geometry*, Math. Ann. 210(1974), 163–170.

Received April 1, 2013.
Revised April 15, 2013.

Proceedings of the Workshop on
Differential Geometry of Submanifolds
and its Related Topics
Saga, August 4-6, 2012

GEOMETRY OF CERTAIN LAGRANGIAN
SUBMANIFOLDS
IN HERMITIAN SYMMETRIC SPACES

Yoshihiro OHNITA

Department of Mathematics, Osaka City University,
& Osaka City University Advanced Mathematical Institute,
Sugimoto, Sumiyoshi-ku, Osaka, 558-8585, Japan
E-mail: ohnita@sci.osaka-cu.ac.jp

Dedicated to Professor Sadahiro Maeda for his sixtieth birthday

In this paper we discuss a method to construct a family of Lagrangian sub-manifolds in an $(n+1)$-dimensional complex projective space $\mathbb{C}P^{n+1}$ from an arbitrary given Lagrangian submanifold in an n-dimensional complex hyper-quadric $Q_n(\mathbb{C}) \subset \mathbb{C}P^{n+1}$. We use the moment map and symplectic quotient techniques involving with the isoparametric function on $\mathbb{C}P^{n+1}$ associated to a rank 2 Riemannian symmetric pair $(SO(n+4), SO(2) \times SO(n+2))$. Moreover by using this method we also show similar results for Lagrangian submanifolds in $Q_{n+1}(\mathbb{C})$ and $Q_n(\mathbb{C})$, and Lagrangian submanifolds in $\mathbb{C}P^{n+1}$ and $\mathbb{C}P^n$. This paper is an improved version and a continuation of our previous article [18].

Keywords: submanifold geometry, Lagrangian submanifolds, Hermitian symmetric spaces, isoparametric hypersurface.

Introduction

Submanifold geometry is one of major subjects in classical and modern differential geometry. In submanifold geometry it is of especial interest to study various types of submanifolds in symmetric spaces and related geometric variational problems such as *minimal submanifold theory* and *harmonic map theory*. Interest in geometry of Lagrangian submanifolds in Kähler manifolds increases in recent years. The notion of *Hamiltonian minimality* initiated by Y. G. Oh ([14]) plays a fundamental role in the study of Lagrangian submanifolds in Kähler manifolds as well as minimal submanifolds in Riemannain manifolds (see Section 1). The Lie theoretic technique in symplectic geometry such as Hamiltonian group actions, the moment maps and symplectic reductions also provides a powerful tool in the study of Lagrangian submanifolds in Kähler manifolds involving with the symmetry.

The purpose of this paper is to study a geometric construction of Lagrangian submanifolds in a Hermitian symmetric space of compact type from given Lagrangian submanifolds in a lower dimensional Hermitian symmetric space, by the method of moment maps and symplectic reductions involved with *isoparametric hypersurface* geometry. It is fundamental and interesting to study the geometric construction and the classification problem of Lagrangian submanifolds in individual Hermitian symmetric spaces. However we should also give an attention to a relationship among Lagrangian submanifolds in different Hermitian symmetric spaces.

In Sections 2 and 3 we shall also review briefly known results on compact homogeneous Lagrangian submanifolds in complex projective spaces and the Gauss map construction of Lagrangian submanifolds in complex hyperquadrics $Q_n(\mathbb{C})$ from isoparametric hypersurfaces in $S^{n+1}(1)$. By applying our result to nice Lagrangian submanifolds in complex hyperquadrics, we can obtain many interesting examples of Lagrangian submanifolds in complex projective space.

In Section 4 first we provide a method to construct a family of Lagrangian submanifolds in an $(n+1)$-dimensional complex projective space $\mathbb{C}P^{n+1}$ from an arbitrary given Lagrangian submanifold in an n-dimensional complex hyperquadric $Q_n(\mathbb{C})$. We use a family of homogeneous isoparametric hypersurfaces in the unit standard sphere $S^{2n+3}(1) \subset \mathbb{C}^{n+2}$ associated to a rank 2 Riemannian symmetric pair $(SO(n+4), SO(2) \times SO(n+2))$. It induces a family of homogeneous isoparametric real hypersurfaces in a complex projective space $\mathbb{C}P^{n+1}$ whose focal manifolds are $Q_n(\mathbb{C})$ and $\mathbb{R}P^{n+1}$. Such a homogeneous real hypsersurface of a complex projective space is a *tube* over $Q_n(\mathbb{C})$ and $\mathbb{R}P^{n+1}$ and is called *of type B* by R. Takagi ([23,24]). For each $t \in (-1,1)$, we have the positive focal map $\nu_+ : M_t \to Q_n(\mathbb{C})$ and the negative focal map $\nu_- : M_t \to \mathbb{R}P^{n+1}$. By using the positive focal map $\nu_+ : M_t \to Q_n(\mathbb{C})$, we shall show that *for any Lagrangian submanifold L in $Q_n(\mathbb{C})$ we have a family of Lagrangian submanifolds $\hat{L}^{n+1} = \nu_+^{-1}(L) (\subset M_t)$ in $\mathbb{C}P^{n+1}$ for $t \in (-1,1)$.* We shall discuss their structure and properties in detail. In the case of $L = S^n \subset Q_n(\mathbb{C})$, the author also will correct an error in the statement of his previous article ([18, Theorem 5.6 (1)]).

In Sections 5 and 6, by using this method we shall also study the constructions of a family of Lagrangian submanifolds in $Q_{n+1}(\mathbb{C})$ from an arbitrary given Lagrangian submanifold in $Q_n(\mathbb{C})$, and a family of Lagrangian submanifolds in $\mathbb{C}P^{n+1}$ from an arbitrary given Lagrangian submanifold in $\mathbb{C}P^n$. We shall mention about some common or different properties in these cases.

The method of this paper is expected to be generalized to study Lagrangian submanifolds. in more general compact Hermitian symmetric spaces. This paper is an improved version and a continuation of our previous article [18]. We assume that any manifold in this paper is smooth and connected.

1. Lagrangian submanifolds in Kähler manifolds

A smooth immersion (resp. embedding) $\varphi : L \to P$ of a smooth manifold L into a $2n$-dimensional symplectic manifold (P, ω) is called a *Lagrangian immersion (resp. embedding)* if $\varphi^*\omega = 0$ and $\dim L = n$. A submanifold $\varphi : L \to S$ of a symplectic manifold (P, ω) is called *isotropic* if S satisfies only the condition $\varphi^*\omega = 0$, which implies $\dim S \leq n$. For each $v \in (\varphi^{-1}TP)_x = T_{\varphi(x)}P$, we define $\alpha_v \in T_x^*L$ by $(\alpha_v)_x(X) := \omega_{\varphi(x)}(v, (d\varphi)_x(X))$ for each $x \in L$ and each $X \in T_xL$. If φ is a Lagrangian immersion, then we have the canonical isomorphisms $\varphi^{-1}TP/\varphi_*TL \cong T^*L$ and $C^\infty(\varphi^{-1}TP/\varphi_*TL) \cong \Omega^1(L)$. A smooth family of smooth immersions $\varphi_t : L \to P$ with $\varphi_0 = \varphi$ is called a *Lagrangian deformation* if $\varphi_t : L \to P$ is a Lagrangian immersion for each t. A Lagrangian deformation $\varphi_t : L \to P$ is characterized by the condition that α_{V_t} is closed for each t, where $V_t = \frac{\partial \varphi_t}{\partial t} \in C^\infty(\varphi_t^{-1}TP)$ is the variational vector field of $\{\varphi_t\}$. If α_{V_t} is exact for each t, then we call $\{\varphi_t\}$ a *Hamiltonian deformation* of φ.

Suppose that (P, ω, g, J) is a Kähler manifold with a Kähler metric g and the complex structure J. Let $\varphi : L \to P$ be a Lagrangian immersion. Denote by σ the second fundamental form of φ. We define a symmetric tensor field S on L by

$$S(X, Y, Z) := \omega(\sigma(X, Y), Z) = g(J\sigma(X, Y), Z)$$

for each $X, Y, Z \in TL$. The mean curvature vector field H of φ is defined by

$$H := \sum_{i=1}^n B(e_i, e_i),$$

where $\{e_i\}$ is an orthonormal basis of $T_x L$. The 1-form α_H on L corresponding to the mean curvature vector field H is called the *mean curvature form* of φ, which satisfies the identity ([5])

$$d\alpha_H = \varphi^* \rho_P, \tag{1}$$

where ρ_P denotes the Ricci form of a Käher manifold P.

The notion of Hamiltonian minimality for Lagrangian submanifolds in a Käher manifold P was introduced and investigated first by Y. G. Oh ([14]). A Lagrangian immersion $\varphi : L \to P$ into a Käher manifold P is called *Hamiltonian minimal* if for each compactly supported Hamiltonian deformation $\{\varphi_t\}$ of $\varphi_0 = \varphi$ the first variation of the volume vanishes:

$$\frac{d}{dt} \text{Vol}(L, \varphi^* g)|_{t=0} = 0 \,.$$

The Hamiltonian minimal Lagrangian submanifold equation is

$$\delta\alpha_H = 0. \tag{2}$$

The usual minimal submanifold equation $H = 0$ is equivalent to $\delta\alpha_H = 0$ and in this case we call φ a *minimal Lagrangian immersion*. A Hamiltonian Lagrangian immersion $\varphi : L \to P$ into a Käher manifold P is called *Hamiltonian stable* if for each compactly supported Hamiltonian deformation $\{\varphi_t\}$ of $\varphi_0 = \varphi$ the second variation of the volume is nonnegative:

$$\frac{d^2}{dt^2} \text{Vol}(L, \varphi^* g)|_{t=0} \geq 0 \,.$$

The Lagrangian version of the second variational formula was given in [14]. Moreover, the notion of *strict Hamiltonian stability* is also defined (cf. [8]).

Let G be a connected Lie group which has the Hamiltonian group action on a symplectic manifold (P, ω) with the moment map $\mu : P \to \mathfrak{g}^*$. It is well-known that

Lemma 1.1. *If an isotropic submanifold S of (P, ω) is invariant under the group action of G, then $S \subset \mu^{-1}(\zeta)$ for some $\zeta \in \mathfrak{z}(\mathfrak{g})$, where $\mathfrak{z}(\mathfrak{g}) := \{\eta \in \mathfrak{g}^* \mid \text{Ad}^*(a)(\eta) = \eta \ (\forall a \in G)\}$.*

A Lagrangian submanifold L embedded in a symplectic manifold (P, ω) obtained as an orbit of G is called a *Lagrangian orbit* of G.

Assume that (P, ω, J, g) is a Kähler manifold and the action of G on P preserves the Kähler structure (ω, g, J), or G is a connected Lie subgroup of the automorphism group $\text{Aut}(P, \omega, J, g)$ of the Kähler structure (ω, g, J). If a Lagrangian submanifold L embedded in P is obtained as an orbit of

G, then L is called a *homogeneous Lagrangian submanifold* in a Kähler manifold P. Moreover if L is compact without boundary, then L is said to be a *compact homogeneous Lagrangian submanifold* in a Kähler manifold P. In this case, as $\delta\alpha_H$ is a constant function on L, by the divergence theorem we have $\delta\alpha_H = 0$. Hence we know the following (cf. [8]):

Proposition 1.1. *Any compact homogeneous Lagrangian submanifold in a Kähler manifold is always Hamiltonian minimal.*

It is also interesting to ask when a noncompact homogeneous Lagrangian submanifold in a Käher manifold is Hamiltonian minimal or not.

Suppose that (P, ω, J, g) is a compact Kähler manifold with $\dim_{\mathbb{C}} H^{1,1}(P, \mathbb{C}) = 1$ and G is a compact connected Lie subgroup of the automorphism group $\mathrm{Aut}(P, \omega, J, g)$. Let $G^{\mathbb{C}} \subset \mathrm{Aut}(P, J)$ be the complexified Lie group of G. Then the following is a fundamental result on the existence of compact homogeneous Lagrangian submanifolds in Kähler manifolds, which was shown by Bedulli and Gori ([4]):

Theorem 1.1. *If $G \cdot x$ is a Lagrangian orbit of G, then the complexified orbit $G^{\mathbb{C}} \cdot x$ is Stein and Zariski open in P. Conversely if the complexified orbit $G^{\mathbb{C}} \cdot x$ is Stein and Zariski open in P, then $G^{\mathbb{C}} \cdot x$ contains a Lagrangian orbit of G.*

2. Homogeneous Lagrangian submanifolds in complex projective spaces

Let $\mathbb{C}^{n+1}$ be the complex Euclidean space and $S^{2n+1}(1)$ be the unit standard hypersphere of $\mathbb{C}^{n+1} \cong \mathbb{R}^{2n+2}$. Let $\pi : S^{2n+1}(1) \to \mathbb{C}P^n$ be the standard Hopf fibration over the n-dimensional complex projective space $\mathbb{C}P^n$ equipped with the standard Fubini-Study metric of constant holomorphic sectional curvature 4. Note that the standard S^1-action on $S^{2n+1}(1)$ preserves the induced metrics. and $\pi : S^{2n+1}(1) \to \mathbb{C}P^n$ is a Riemannian submersion. The energy function $f : \mathbb{C}^{n+1} \to \mathbb{R}$ defined by $f(\mathbf{x}) := \|\mathbf{x}\|^2$ ($\mathbf{x} \in \mathbb{C}^{n+1}$) can be considered as a Hamiltonian function or a moment map with respect to the standard S^1-action on $\mathbb{C}^{n+1}$. Since $f^{-1}(1) = S^{2n+1}(1)$ and $f^{-1}(1)/S^1 = \mathbb{C}P^n$, the complex projective space $\mathbb{C}P^n$ is a symplectic Kähler quotient of $\mathbb{C}^{n+1}$ by the moment map f. Then we know the following properties (see [6] for more general results):

Proposition 2.1. *Let L be a submanifold (immersed) in $\mathbb{C}P^n$. Then*

(1) *L is a Lagrangian submanifold (immersed) in $\mathbb{C}P^n$ if and only if the inverse image $\tilde{L} = \pi^{-1}(L)(\subset S^{2n+1}(1))$ by the Hopf fibration π is a Lagrangian submanifold (immersed) in $\mathbb{C}^{n+1}$.*

(2) *L is a Hamiltonian minimal in $\mathbb{C}P^n$ if and only if $\tilde{L} = \pi^{-1}(L)(\subset S^{2n+1}(1))$ is Hamiltonian minimal in $\mathbb{C}^{n+1}$.*

$$
\begin{array}{ccccc}
\tilde{L}^{n+1} = \pi^{-1}(L) & \subset & S^{2n+1}(1) = f^{-1}(1) & \subset & \mathbb{C}^{n+1} \\
\pi \downarrow S^1 & & \pi \downarrow S^1 & & \\
L^n & \subset & \mathbb{C}P^n = f^{-1}(1)/S^1 & &
\end{array}
$$

We mention about known results on the construction and classification of compact homogeneous Lagrangian submanifolds in complex projective spaces. In the case when L has parallel second fundamental form, that is, $\nabla S = 0$, $\pi^{-1}(L)$ also has parallel second fundamental form in both $S^{2n+1}(1)$ and $\mathbb{C}^{n+1}$. Such Lagrangian submanifolds with $\nabla S = 0$ in complex space forms were completely classified by Hiroo Naitoh and Mararu Takeuchi in 1980's by theory of symmetric R-spaces (of type $U(r)$) and in particular they showed that any Lagrangian submanifold with $\nabla S = 0$ in a complex space form is homogeneous. Compact Lagrangian submanifolds with $\nabla S = 0$ embedded in complex projective spaces and complex Euclidean spaces are strictly Hamiltonian stable (Amarzaya-Ohnita [1], [2]).

There are many examples of compact homogeneous Lagrangian submanifolds with $\nabla S \neq 0$ embedded in complex projective spaces and complex Euclidean spaces. The simplest non-trivial example can be given by a 3-dimensional minimal Lagrangian orbit in $\mathbb{C}P^2$ under an irreducible unitary representation of $SU(2)$ of degree 3 (cf. [3], [15]).

Bedulli-Gori ([4]) classified compact homogeneous Lagrangian submanifolds in complex projective spaces obtained as orbits of simple compact Lie groups of $PU(n+1)$. It is a crucial to use Theorem 1.1 and theory of prehomogeneous vector spaces due to Mikio Sato and Tatsuo Kimura. It is still an open problem to classify compact homogeneous Lagrangian submanifolds in $\mathbb{C}P^n$ obtained as orbits of *non-simple* compact Lie groups of $PU(n+1)$ (see also [22]).

3. Homogeneous Lagrangian submanifolds in complex hyperquadrics and hypersurface geometry in spheres

Let $Q_n(\mathbb{C})$ be a complex hyperquadrics of $\mathbb{C}P^{n+1}$ defined by the homogeneous quadratic equation $z_0^2 + z_1^2 + \cdots + z_{n+1}^2 = 0$. Let $\widetilde{Gr}_2(\mathbb{R}^{n+2})$ be the real Grassmann manifold of all oriented 2-dimensional vector subspaces of $\mathbb{R}^{n+2}$ and $Gr_2(\mathbb{R}^{n+2})$ the real Grassmann manifold of all 2-dimensional vector subspaces of $\mathbb{R}^{n+2}$. We denote by $[W]$ a 2-dimensional vector subspace W of $\mathbb{R}^{n+2}$ equipped with an orientation and by $-[W]$ the same vector subspace W of $\mathbb{R}^{n+2}$ equipped with an orientation opposite to $[W]$. The map $\widetilde{Gr}_2(\mathbb{R}^{n+2}) \ni [W] \longmapsto W \in Gr_2(\mathbb{R}^{n+2})$ is the universal covering with the deck transformation group $\mathbb{Z}_2$ defined by $\widetilde{Gr}_2(\mathbb{R}^{n+2}) \ni [W] \longmapsto -[W] \in \widetilde{Gr}_2(\mathbb{R}^{n+2})$. Then we have the identification

$$Q_n(\mathbb{C}) \ni [\mathbf{a} + \sqrt{-1}\mathbf{b}] \longleftrightarrow [W] = \mathbf{a} \wedge \mathbf{b} \in \widetilde{Gr}_2(\mathbb{R}^{n+2}),$$

where $\{\mathbf{a}, \mathbf{b}\}$ denotes an orthonormal basis of W compatible with the orientation of $[W]$.

Let N^n be an oriented hypersurface immersed in the unit standard sphere $S^{n+1}(1)$. Let $\mathbf{x}(p)$ denote the position vector of a point $p \in N^n$ and $\mathbf{n}(p)$ denote the unit normal vector at $p \in N^n$ in $S^{n+1}(1)$. The *Gauss map* $\mathcal{G} : N^n \to Q_n(\mathbb{C})$ of the oriented hypersurface $N^n \subset S^{n+1}(1)$ is defined by

$$\mathcal{G} : N^n \ni p \longmapsto [\mathbf{x}(p) + \sqrt{-1}\mathbf{n}(p)] = \mathbf{x}(p) \wedge \mathbf{n}(p) \in Q_n(\mathbb{C}) = \widetilde{Gr}_2(\mathbb{R}^{n+2}).$$

It is well-known that the Gauss map $\mathcal{G} : N^n \to Q_n(\mathbb{C})$ is always a Lagrangian immersion.

We shall assume that N^n is an oriented *isoparametric* hypersurface immersed in the unit standard sphere $S^{n+1}(1)$, that is, a hypersurface with g distinct constant principal curvatures.

In general, an *isoparametric hypersurface* on a Riemannian manifold M is defined as a regular level hypersurface of an *isoparametric function*. A smooth function f is called an *isoparametric function* on a Riemannian manifold M if f satisfies the system of partial differential equations:

$$\Delta^M f = a(f), \quad \|\mathrm{grad} f\|^2 = b(f), \tag{3}$$

where a is a continuous function and b is a function of class C^2. It is well-known that each regular level hypersurface of an isoparametric function has constant mean curvature. Elie Cartan showed that a hypersurface immersed in real space forms is an isoparametric hypersurface if and only if it has constant principal curvatures. Let N^n be an isoparametric hypersurface in the standard hypersphere $S^{n+1}(1) \subset \mathbb{R}^{n+2}$ with g distinct principal curvatures $k_1 > k_2 > \cdots > k_g$ and the corresponding multiplicities m_α ($\alpha = 1, \cdots, g$). We express the principal curvatures $k_1 > k_2 > \cdots > k_g$ as $k_\alpha = \cot\theta_\alpha$ ($\alpha = 1, \cdots, g$) with $\theta_1 < \cdots < \theta_g$. Münzner ([11]) showed that $\theta_\alpha = \theta_1 + (\alpha - 1)\frac{\pi}{g}$ and $m_\alpha = m_{\alpha+2}$ indexed modulo g. Moreover he proved that N^n extends to a regular level hypersurface $f^{-1}(t)$ for some $t \in (-1, 1)$ defined by an isoparametric function $f : S^{n+1}(1) \to [-1, 1]$ which is a restriction of a homogeneous polynomial F of degree g satisfying

$$\Delta^{\mathbb{R}^{n+2}} F = c\, r^{g-2}, \quad \|\mathrm{grad}^{\mathbb{R}^{n+2}} F\|^2 = g^2\, r^{2g-2}, \tag{4}$$

where $r = \sum_{i=1}^{n+2}(x_i)^2$ and $c = g^2(m_2 - m_1)/2$. Here the isoparametric function f is defined in a neighborhood of N^n as

$$t = f(q) = \cos(g\theta(q)) = \cos(g\theta_1), \tag{5}$$

where $\theta(q) := \theta_1$ for the maximal principal curvature $k_1 = \cot\theta_1$ of $f^{-1}(q)$. Such a homogeneous polynomial is called the *Cartan-Münzner polynomial*.

Remark 3.1. Recently the relationship of the Cartan-Münzner polynomial of $g = 4$ with the moment maps was studied by R. Miyaoka ([10]).

Let $k_1 > \cdots > k_g$ denote its distinct constant principal curvatures and $m_1, \cdots, m_g$ the corresponding multiplicities. Münzner ([11]) showed that $\theta_\alpha = \theta_1 + (\alpha - 1)\frac{\pi}{g}$ and $m_\alpha = m_{\alpha+2}$ indexed modulo g. Hence we have

$$\frac{2n}{g} = \begin{cases} m_1 + m_2 & \text{if } g \geq 2, \\ 2m_1 & \text{if } g = 1. \end{cases} \tag{6}$$

The famous result of Münzner is that g must be $1, 2, 3, 4$ or 6 ([11], [12]).

The Lagrangian immersions $\mathcal{G}$ and the Gauss image $\mathcal{G}(N^n)$ of isoparametric hypersurfaces have the following properties. It follows from the mean curvature form formula of Palmer ([21]) that

Proposition 3.1. *The Gauss map $\mathcal{G} : N^n \to Q_n(\mathbb{C})$ is a minimal Lagrangian immersion.*

It follows from [7], [16] that

Proposition 3.2.

(1) *The Gauss image $\mathcal{G}(N^n)$ is a compact smooth minimal Lagrangian submanifold embedded in $Q_n(\mathbb{C})$.*
(2) *The Gauss map $\mathcal{G} : N^n \to \mathcal{G}(N^n)$ is a covering map with the deck transformation group $\mathbb{Z}_g$.*

(3) $\mathcal{G}(N^n)$ is invariant under the deck transformation group $\mathbb{Z}_2$ of the universal covering $Q_n(\mathbb{C}) = \widetilde{Gr}_2(\mathbb{R}^{n+2}) \to Gr_2(\mathbb{R}^{n+2})$.

(4) $2n/g$ is even (resp. odd) if and only if $\mathcal{G}(N^n)$ is orientable (resp. non-orientable).

(5) $\mathcal{G}(N^n)$ is a monotone and cyclic Lagrangian submanifold in $Q_n(\mathbb{C})$ with minimal Maslov number $2n/g$.

In [7] we classified explicitly all compact homogeneous Lagrangian submanifolds in $Q_n(\mathbb{C})$. In particular we showed that all compact homogeneous *minimal* Lagrangian submanifolds in $Q_n(\mathbb{C})$ are obtained as the Guass images of homogeneous isoparametric hypersurfaces in $S^{n+1}(1)$. Palmer ([21]) showed that $\mathcal{G} : N^n \to Q_n(\mathbb{C})$ is Hamiltonian stable if and only if $g = 1$, that is, N^n is a great or small sphere of $S^{n+1}(1)$. In [7], [9] we completely determined the strict Hamiltonian stability of the Gauss images $\mathcal{G}(N^n)$ of all homogeneous isoparametric hypersurfaces N^n in $S^{n+1}(1)$. The main result in these papers was that the Gauss image $\mathcal{G}(N^n)$ of homogeneous isoparametric hypersurfaces N^n is Hamiltonian stable if and only if $|m_1 - m_2| \leq 2$ or N^n is a principal orbit of the isotropy representation of the Riemannian symmetric pair of type EIII (in this case $(m_1, m_2) = (6, 9)$).

4. Construction of Lagrangian submanifolds in $\mathbb{C}P^{n+1}$ from a Lagrangian submanifold in $Q_n(\mathbb{C})$

We shall recall the argument of [18]. Let us consider *tubes* M_t ($t \in (-1, 1)$) over a complex hyperquadric $Q_n(\mathbb{C})\,(= M_1 = M_+)$ in $\mathbb{C}P^{n+1}$, which are also tubes over a real projective subspace $\mathbb{R}P^n\,(= M_{-1} = M_-)$ in $\mathbb{C}P^{n+1}$ at the same time. The tube M_t is a real hypersurface embedded in $\mathbb{C}P^{n+1}$, which is a homogeneous isoparametric real hypersurface of $\mathbb{C}P^{n+1}$ and its focal manifolds are $Q_n(\mathbb{C})$ and $\mathbb{R}P^n$. Let ν denote the unit normal vector field to M_t in $\mathbb{C}P^{n+1}$. The focal maps from M_t to focal manifolds are defined by using geodesics of $\mathbb{C}P^{n+1}$ normal to M_t. Let $\nu_+ : M_t \to Q_n(\mathbb{C})$ denote the positive focal map from M_t onto $Q_n(\mathbb{C})$ and $\nu_- : M_t \to \mathbb{R}P^n$ the negative focal map from M_t onto $\mathbb{R}P^n$:

$$
\begin{array}{ccccc}
\mathbb{C}P^{n+1} & = & \mathbb{C}P^{n+1} & = & \mathbb{C}P^{n+1} \\
\cup & & \cup & & \cup \\
M_{-1} = \mathbb{R}P^{n+1} & \xleftarrow{\;\nu_-\;} & M_t & \xrightarrow{\;\nu_+\;} & Q_n(\mathbb{C}) = M_1
\end{array}
$$

The homogeneous isoparametric real hypersurfaces M_t in $\mathbb{C}P^{n+1}$ are obtained as the projection of homogeneous isoparametric real hypersurfaces $\tilde{M}_t$ in $S^{2n+3}(1)$ by the Hopf fibration $\pi : S^{2n+3}(1) \to \mathbb{C}P^{n+1}$.

Let $(\tilde{G}, \tilde{U}) = (SO(n+4), SO(2) \times SO(n+2))$ be a compact rank 2 Hermitian symmetric pair of type BDII_2. Let $\tilde{\mathfrak{g}} = \tilde{\mathfrak{u}} + \tilde{\mathfrak{p}}$ be the corresponding canonical decomposition of its symmetric Lie algebra, where $\tilde{\mathfrak{g}} = \mathfrak{o}(n+4)$, $\tilde{\mathfrak{u}} = \mathfrak{o}(2) \oplus \mathfrak{o}(n+2)$ and the vector subspace $\tilde{\mathfrak{p}}$ is

$$\tilde{\mathfrak{p}} = \left\{ Z = \begin{bmatrix} 0 & \xi \\ -\xi^t & 0 \end{bmatrix} \mid \xi \in M(2, n+2; \mathbb{R}) \right\}.$$

Here $M(2, n+2; \mathbb{R})$ denotes the vector space of all real $2 \times (n+2)$-matrices. The standard complex structure $\tilde{J}$ of $\tilde{\mathfrak{p}}$ invariant under the isotropy representation of $\tilde{U}$ is given by

$$J\left(\begin{bmatrix} 0 & \xi \\ -\xi^t & 0 \end{bmatrix} \right) = \mathrm{Ad}_{\tilde{\mathfrak{p}}} \begin{bmatrix} 0 & -1 & 0 \\ 1 & 0 & 0 \\ 0 & 0 & 0 \end{bmatrix} \left(\begin{bmatrix} 0 & \xi \\ -\xi^t & 0 \end{bmatrix} \right) = \begin{bmatrix} 0 & \begin{bmatrix} 0 & -1 \\ 1 & 0 \end{bmatrix}\xi \\ -\xi^t \begin{bmatrix} 0 & -1 \\ 1 & 0 \end{bmatrix}^t & 0 \end{bmatrix}.$$

Relative to the standard complex structure, the vector subspace $\tilde{\mathfrak{p}}$ is identified with $M(2, n+2; \mathbb{R}) \cong \mathbb{R}^{2n+4} \cong \mathbb{C}^{n+2}$. In particular, the standard S^1-action on $\mathbb{C}^{n+2}$ coincides with the isotropy action of $SO(2) \times \{I_{n+2}\} \subset SO(2) \times SO(n+2)$ on $\tilde{\mathfrak{p}}$.

Let $S^{2n+3}(1) \subset \tilde{\mathfrak{p}}$ denote the unit standard hypersphere of $\tilde{\mathfrak{p}}$. Each principal orbit of $\tilde{U} = SO(2) \times SO(n+2)$ on $S^{2n+3}(1)$ is of codimension 1. Thus we have a family $\{\tilde{M}_t\}$ of homogeneous isoparametric hypersurfaces of $S^{2n+3}(1)$ with $g = 4$ distinct constant principal curvatures and multiplicities $(m_1, m_2) = (1, n)$. This isoparametric hypersurfaces with four distinct constant principal curvatures was found first by K. Nomizu ([13]), which affirmatively solved one of Elie Cartan's problems in the isoparametric hypersurface theory.

Let $\tilde{f} : S^{2n+3}(1) \to [-1, 1] \subset \mathbb{R}$ be such an isoparametric function defining $\{\tilde{M}_t\}$ and $F : \mathbb{C}^{n+2} \to \mathbb{R}$ denote the corresponding Cartan-Münzner polynomial of degree 4 so that $F|_{S^{2n+3}(1)} = \tilde{f}$. Explicitly the Cartan-Münzner polynomial F is given as follows ([20]):

$$F(Z) = \frac{3}{4} (\mathrm{Tr}(Z^2))^2 - 2\,\mathrm{Tr}(Z^4)$$

for each $Z \in \tilde{\mathfrak{p}}$, in other words $F = r^4 - 2\,F_0$, where the functions r and F are defined as

$$r(\xi) := \left(\sum_{i=1}^{n+2} |\xi_i|^2 \right)^{\frac{1}{2}}, \quad F_0(\xi) := \left| \sum_{i=1}^{n+2} \xi_i^2 \right|^2 = \left(\sum_{i=1}^{n+2} \xi_i^2 \right) \left(\overline{\sum_{i=1}^{n+2} \xi_i^2} \right)$$

for each $\xi = \begin{bmatrix} \xi_1 \\ \vdots \\ \xi_{n+2} \end{bmatrix} \in \mathbb{C}^{n+2}$.

Set $\tilde{M}_t := \tilde{f}^{-1}(t)$ for each $t \in [-1,1]$ and $\tilde{M}_\pm := \tilde{M}_{\pm 1}$. Then for each $t \in (-1,1)$ the level regular hypersurface $\tilde{M}_t$ is a compact homogeneous isoparametric hypersurface embedded in $S^{2n+3}(1)$ and $\tilde{M}_t \cong \dfrac{SO(2) \times SO(n+2)}{\mathbb{Z}_2 \times SO(n)}$. For $t = \pm 1$, the focal manifolds $\tilde{M}_\pm$ are compact minimal submanifolds embedded in $S^{2n+3}(1)$ and $\tilde{M}_+ \cong \dfrac{SO(2) \times SO(n+2)}{SO(2) \times SO(n)}$, $\tilde{M}_- \cong \dfrac{SO(2) \times SO(n+2)}{O(n+1)} \cong S^1 \cdot S^{n+1} \cong Q_{2,n+2}(\mathbb{R})$ (a real hyperquadric).

Let $t \in (-1,1)$. Let $k_1 > k_2 > k_3 > k_4$ be four distinct principal curvatures of $\tilde{M}_t$ in $S^{2n+3}(1)$. Here $k_1 = \cot \theta_1$ and $t = \cos(4\theta_1)$. Denote by $\mathbf{x}$ the position vector of points on $\tilde{M}_t$ and by $\mathbf{n} := \dfrac{\mathrm{grad}\tilde{f}}{\|\mathrm{grad}\tilde{f}\|}$ the unit normal vector to $\tilde{M}$ in $S^{2n+3}(1)$. Let $A_{\mathbf{n}}$ denote the shape operator of $\tilde{M}_t$ in $S^{2n+3}(1)$ in the direction of $\mathbf{n}$. The principal curvatures are nothing but eigenvalues of $A_{\mathbf{n}}$. For each $p \in \tilde{M}_t$, we express the eigenspace decomposition of the tangent vector space $T_p\tilde{M}_t$ corresponding to the four constant principal curvatures as follows:

$$T_p\tilde{M} = \tilde{V}_1(p) \oplus \tilde{V}_2(p) \oplus \tilde{V}_3(p) \oplus \tilde{V}_4(p)$$
$$= \mathbb{R}\,J\mathbf{x}(p) \oplus \mathbb{R}\,J\mathbf{n}(p) \oplus \tilde{V}_2(p) \oplus J(\tilde{V}_2(p)),$$

where $\tilde{V}_1(p) = \mathbb{R}(-k_3 J\mathbf{x}(p) + J\mathbf{n}(p))$, $\tilde{V}_3 = \mathbb{R}(-k_1 J\mathbf{x}(p) + J\mathbf{n}(p))$, $\tilde{V}_4(p) = J(\tilde{V}_2(p))$.

The positive focal map $\tilde{\nu}_+ : \tilde{M}_t \to \tilde{M}_+$ is defined by

$$\tilde{\nu}_+(p) := \cos\theta_1\,\mathbf{x}(p) + \sin\theta_1\,\mathbf{n}(p) \in \tilde{M}_+$$

for each $p \in \tilde{M}_t$. The differential of the positive focal map $\tilde{\nu}_+$ is given as

$$(d\tilde{\nu}_+)_p(X) - (\cos\theta_1\,I - \sin\theta_1\,A_{\mathbf{n}})(X)$$

for each $X \in T_p\tilde{M}$. If $X \in \tilde{V}_i(p)$ $(i = 1,2,3,4)$, then we have

$$(d\tilde{\nu}_+)_p(X) = (\cos\theta_1 - \sin\theta_1\,k_i)(X)$$
$$= (\cos\theta_1 - \sin\theta_1\,\cot\theta_i)(X) = \frac{\sin(\theta_i - \theta_1)}{\sin\theta_i}\,X.$$

Note that

$$\frac{\sin(\theta_i - \theta_1)}{\sin\theta_i} = \begin{cases} 0 & \text{if } i = 1, \\ \dfrac{1}{\sqrt{2}\sin\theta_2} = \dfrac{1}{\cos\theta_1 + \sin\theta_1} & \text{if } i = 2, \\ \dfrac{1}{\sin\theta_3} = \dfrac{1}{\cos\theta_1} & \text{if } i = 3, \\ \dfrac{1}{\sqrt{2}\sin\theta_4} = \dfrac{1}{\sqrt{2}\cos\theta_2} = \dfrac{1}{\cos\theta_1 - \sin\theta_1} & \text{if } i = 4. \end{cases}$$

The positive focal map $\tilde{\nu}_+ : \tilde{M}_t \to \tilde{M}_+$ is an $SO(2) \times SO(n+2)$-equivariant submersion

$$\nu_+ : \tilde{M}_t = \frac{SO(2) \times SO(n+2)}{\mathbb{Z}_2 \times SO(n)} \longrightarrow \tilde{M}_+ = \frac{SO(2) \times SO(n+2)}{SO(2) \times SO(n)}$$

$$\cong \frac{SO(n+2)}{SO(n)} \cong V_2(\mathbb{R}^{n+2})$$

with fiber $\dfrac{SO(2) \times SO(n)}{\mathbb{Z}_2 \times SO(n)} \cong \dfrac{SO(2)}{\mathbb{Z}_2} \cong S^1$. Here $V_2(\mathbb{R}^{n+2})$ denotes the Stiefel manifold of all pairs of two orthonormal vectors in $\mathbb{R}^{n+2}$. Note that this map $\tilde{\nu}_+ : \tilde{M}_t \to \tilde{M}_+$ is *not* a Riemannian submersion.

Let $\pi : S^{2n+3}(1) \to \mathbb{C}P^{n+1}$ be the Hopf fibration with the $S^1 \cong SO(2)$-action. Since the isoparametric function $\tilde{f}$ on $S^{2n+3}(1)$ is invariant under the $S^1 \cong SO(2)$-action, a smooth function $f : \mathbb{C}P^{n+1} \to [-1,1] \subset \mathbb{R}$ can be induced by $f \circ \pi = \tilde{f}$. Then f is an isoparametric function on $\mathbb{C}P^{n+1}$.

Set $M_t := f^{-1}(t)$ for each $t \in [-1,1]$ and $M_\pm := M_{\pm 1}$. Then for each $t \in (-1,1)$ the subset M_t is the level regular hypersurface, and is a compact homogeneous isoparametric hypersurface embedded in $\mathbb{C}P^{n+1}$ and $M_t \cong \dfrac{SO(n+2)}{\mathbb{Z}_2 \times SO(n)}$. The subsets $M_\pm$ are focal manifolds which are compact minimal submanifolds embedded in $\mathbb{C}P^{n+1}$. The subset M_+ is a complex hyperquadric $Q_n(\mathbb{C}) \cong \dfrac{SO(n+2)}{SO(2) \times SO(n)}$ in $\mathbb{C}P^{n+1}$ and the subset M_- is a real projective subspace $\mathbb{R}P^{n+1} \cong \dfrac{SO(n+2)}{S(O(1) \times O(n+1))}$ in $\mathbb{C}P^{n+1}$.

Each M_t $(t \in (-1,1))$ is a homogeneous isoparametric real hypersurface in $\mathbb{C}P^{n+1}$ with *three* distinct constant principal curvatures

$$\ell_1 = -2\tan 2\theta_1, \ \ell_2 = \cot\theta_1, \ \ell_3 = -\tan\theta_1, \tag{7}$$

and the corresponding multiplicities

$$m(\ell_1) = 1, \ m(\ell_2) = n, \ m(\ell_3) = n. \tag{8}$$

It is called a *homogeneous real hypersurface of type B* ([23, 24]). The vector field $\nu := \dfrac{\mathrm{grad}f}{\|\mathrm{grad}f\|}$ is a unit normal vector field on each M_t ($t \in (-1,1)$) and A_ν denotes the shape operator of M_t in the direction of ν. For each $x \in M_t$, we express the eigenspace decomposition of $T_x M_t$ corresponding to three principal curvatures ℓ_1, ℓ_2, ℓ_3 as follows:

$$T_x M_t = V_1(x) \oplus V_2(x) \oplus V_3(x) \tag{9}$$

where $V_1(x) = \mathbb{R}\, J\nu(x)$, $V_2(x) = J(V_3(x))$.

Now we have a commutative diagram of positive focal maps and Hopf fibrations as follows:

$$
\begin{array}{ccc}
\mathbb{C}^{n+2} & = & \mathbb{C}^{n+2} \\
\cup & & \cup \\
S^{2n+3}(1) & = & S^{2n+3}(1) \\
\cup & & \cup \\
\tilde{M}_t & \xrightarrow{\ \tilde{\nu}_+\ } & \tilde{M}_+ \\
\pi \downarrow {\scriptstyle S^1} & & \pi \downarrow {\scriptstyle S^1} \\
M_t & \xrightarrow{\ \nu_+\ } & M_+ \\
\cap & & \cap \\
\mathbb{C}P^{n+1} & = & \mathbb{C}P^{n+1}
\end{array}
$$

Here

$$\tilde{M}_t = \pi^{-1}(M_t) \cong \frac{SO(2) \times SO(n+2)}{\mathbb{Z}_2 \times SO(n)} \quad (t \in (-1,1)),$$

$$\tilde{M}_+ = \pi^{-1}(M_+) \cong \frac{SO(2) \times SO(n+2)}{SO(2) \times SO(n)} \cong \frac{SO(n+2)}{SO(n)} \cong V_2(\mathbb{R}^{n+2}),$$

$$\tilde{M}_- = \pi^{-1}(M_-) \cong \frac{SO(2) \times SO(n+2)}{O(n+1)} \cong S^1 \cdot S^{n+1} \cong Q_{2,n+2}(\mathbb{R})$$

and

$$M_t = \pi(\tilde{M}_t) \cong \frac{SO(n+2)}{\mathbb{Z}_2 \times SO(n)} \quad (t \in (-1,1)),$$

$$M_+ = \pi(\tilde{M}_+) \cong \frac{SO(n+2)}{SO(2) \times SO(n)} \cong Q_n(\mathbb{C}),$$

$$M_- = \pi(\tilde{M}_-) \cong \frac{SO(n+2)}{S(O(1) \times O(n+1))} \cong \mathbb{R}P^{n+1}.$$

Under a surjective linear map $(d\pi)_p : T_p\tilde{M}_t \to T_pM_t$, we have

$$\mathrm{Ker}((d\pi)_p) = \mathbb{R}\,J\mathbf{x}(p) \subset \tilde{V}_1(p) \oplus \tilde{V}_3(p) = \mathbb{R}\,J\mathbf{x}(p) \oplus \mathbb{R}\,J\mathbf{n}(p),$$

$$(d\pi)_p(\tilde{V}_1(p) \oplus \tilde{V}_3(p)) = V_1(p),$$

$$(d\pi)_p(\tilde{V}_2(p)) = V_2(p), \quad (d\pi)_p(\tilde{V}_4(p)) = V_3(p).$$

The positive focal map $\nu_+ : M_t \to M_+$ is an $SO(n+2)$-equivariant submersion

$$\nu_+ : M_t \cong \frac{SO(n+2)}{\mathbb{Z}_2 \times SO(n)} \longrightarrow M_+ = Q_n(\mathbb{C}) \cong \frac{SO(n+2)}{SO(2) \times SO(n)}.$$

Note that it is not a Riemannian submersion and the Reeb flow of a real hypersurface M_t as an almost contact metric manifold is not isometric.

We observe that the differential of the positive focal map ν_+ has kernel $\mathrm{Ker}(d\nu_+) = V_1$. Let $\omega_{Q_n(\mathbb{C})}$ denote the Kähler form of $Q_n(\mathbb{C}) \subset \mathbb{C}P^{n+1}$. Then a relation between the pull-back form of $\omega_{Q_n(\mathbb{C})}$ by ν_+ and the restriction of the Kähler form ω of $\mathbb{C}P^{n+1}$ to M_t is given explicitly as

Lemma 4.1.

$$\nu_+^*\omega_{Q_n(\mathbb{C})} = \frac{1}{\cos(2\theta_1)}\,\omega|_{M_t}. \tag{10}$$

Here note that $\cos(2\theta_1) \neq 0$.

The vector field $J(\mathrm{grad}f)$ on $\mathbb{C}P^{n+1}$ satisfies $\omega(J(\mathrm{grad}f), \cdot) = g(\mathrm{grad}f, \cdot) = -df$, and thus it is a Hamiltonian vector field X_{-f} on $\mathbb{C}P^{n+1}$ corresponding to the Hamiltonian $-f$. For each $t \in (-1,1)$, the flow of a Hamiltonian vector field X_{-f} preserves each level set $f^{-1}(t) = M_t$ and $X_{-f}|_{M_t} = J(\mathrm{grad}f) = \|\mathrm{grad}f\|\,J\nu$ belongs to $V_1 \subset TM_t$. Note that $\|\mathrm{grad}f\|$ is constant on M_t. Hence we see that the vector field $X_{-f}|_{M_t}$ generates the S^1-action on M_t and the orbits of the S^1-action coincide with the fibers of the positive focal map $\nu_+ : M_t \to M_+$. Therefore $M_+ = Q_n(\mathbb{C})$ is regarded as a *symplectic quotient* $f^{-1}(t)/S^1$, but which is *not a Kähler quotient*, because the S^1-action on M_t is not isometric.

$$M_t\left(\cong \tfrac{SO(n+2)}{\mathbb{Z}_2 \times SO(n)}\right) = f^{-1}(t) \quad \subset \quad \mathbb{C}P^{n+1}$$

$$\nu_+ \Big\downarrow \quad S^1\left(\cong SO(2)/\mathbb{Z}_2\right)$$

$$M_+ = Q_n(\mathbb{C})\left(\cong \tfrac{SO(n+2)}{SO(2) \times SO(n)}\right) = f^{-1}(t)/S^1$$

Lemma 4.1 gives the relation between the push-forwarded symplectic form $(\nu_+)_*\omega$ on the symplectic quotient $f^{-1}(t)/S^1$ and the original Kähler form $\omega_{Q_n(\mathbb{C})}$ of $Q_n(\mathbb{C})$.

The positive focal map $\nu_+ : M_t \to M_+ = Q_n(\mathbb{C})$ can be considered also as a principle fiber bundle with structure group S^1 over $Q_n(\mathbb{C})$. We define a 1-form α_ν on M_t by

$$\alpha_\nu(X) := \omega(\nu, X) = g(J\nu, X) \tag{11}$$

for each $X \in TM_t$. The 1-form α_ν satisfies

$$(\nabla_X \alpha_\nu)(Y) = g(A_\nu(X), JY) \tag{12}$$

for each $X, Y \in TM_t$. Hence the exterior derivative of α_ν is

$$\begin{aligned}(d\alpha_\nu)(X,Y) &= (\nabla_X \alpha_\nu)(Y) - (\nabla_Y \alpha_\nu)(X)\\ &= g(A_\nu(X), JY) - g(A_\nu(Y), JX)\\ &= -\omega(A_\nu(X), Y) + \omega(A_\nu(Y), X)\end{aligned} \tag{13}$$

for each $X, Y \in TM_t$. Hence by using (7) and (9), we have the following formula:

Lemma 4.2.

$$d\alpha_\nu = -2\tan(2\theta_1)\,\omega|_{M_t}. \tag{14}$$

The principal S^1-bundle $\nu_+ : M_t \to M_+ = Q_n(\mathbb{C})$ has a connection invariant under the left group action of $SO(n+2)$ whose horizontal subspaces are defined by $(J\nu)^\perp = V_2 \oplus V_4 = V_2 \oplus J(V_2)$. Then Lemma 4.2 means that its connection form is α_ν and its connection form is $-2\tan(2\theta_1)\,\omega|_{M_t}$. Moreover, by using Lemma 4.1 we have

$$d\alpha_\nu = -2\tan(2\theta_1)\,\cos(2\theta_1)\,\nu_+^*\omega_{Q_n(\mathbb{C})} = -2\sin(2\theta_1)\,\nu_+^*\omega_{Q_n(\mathbb{C})}. \tag{15}$$

Note that $\sin(2\theta_1) \neq 0$.

Let $\varphi : L \to Q_n(\mathbb{C})$ be a Lagrangian immersion of an n-dimensional smooth manifold L^n into an n-dimensional complex hyperquadric $Q_n(\mathbb{C})$. Then we take a pull-back S^1-bundle $\hat{L}^{n+1} = \varphi^{-1}M_t$ by φ over L as

$$\begin{array}{ccccc} \hat{L}^{n+1} = \varphi^{-1}M_t & \xrightarrow{\hat{\varphi}_t} & M_t = f^{-1}(t) & \subset & \mathbb{C}P^{n+1}\\ {\scriptstyle \nu_+}\downarrow\;{\scriptstyle S^1} & & {\scriptstyle \nu_+}\downarrow\;{\scriptstyle S^1} & & \\ L^n & \xrightarrow{\varphi} & M_+ = Q_n(\mathbb{C}) & \subset & \mathbb{C}P^{n+1} \end{array}$$

Since φ is a Lagrangian immersion, i.e. $\varphi^*\omega_{Q_n(\mathbb{C})} = 0$, by Lemma 4.1 $\hat{\varphi}_t^*\omega = \cos(2\theta_1)\,\hat{\varphi}_t^*\nu_+^*\omega_{Q_n(\mathbb{C})} = \cos(2\theta_1)\,\nu_+^*\varphi^*\omega_{Q_n(\mathbb{C})} = 0$ and hence $\hat{\varphi}_t : L^n \to$

$\mathbb{C}P^{n+1}$ is also a Lagrangian immersion. Moreover, $\nu_+ : \hat{L}^{n+1} = \varphi^{-1}M_t \to L^n$ is also a principal S^1-bundle over L^n and the pull-back connection is flat.

Therefore we obtain

Theorem 4.1. *If $\varphi : L \to Q_n(\mathbb{C})$ is a Lagrangian immersion (resp. embedding) into an n-dimensional complex hyperquadric, then we have a family of Lagrangian immersions (resp. embeddings) into an $(n+1)$-dimensional complex projective space*

$$\hat{\varphi}_t : \hat{L} = \varphi^{-1}M_t \longrightarrow (M_t \subset)\,\mathbb{C}P^{n+1} \quad (t \in (-1,1)).$$

Conversely, any Lagrangian submanifold $\hat{L}$ in $\mathbb{C}P^{n+1}$ invariant under the Hamiltonian flow of the above isoparametric function f on $\mathbb{C}P^{n+1}$ is obtained in this way as $\hat{L} = \varphi^{-1}M_t$ for some ($t \in (-1,1)$ or an open submanifold of $\mathbb{R}P^{n+1} = M_{-1} = f^{-1}(-1)$ in $\mathbb{C}P^{n+1}$.

Remark 4.1. The infinitesimal Lagrangian deformation given by the above family corresponds to the closed 1-form $\hat{\varphi}_t^* \alpha_\nu$.

Remark 4.2. The same result holds also for isotropic submanifolds instead of Lagrangian submanifolds.

Remark 4.3. If we take the Gauss images of compact isoparametric hypersurfaces in $S^{n+1}(1)$ as $L^n \subset Q_n(\mathbb{C})$, then by this theorem we obtain a class of compact Lagrangian submanifolds $\nu_+^*(L^n)\,(\subset M_t)$ in $\mathbb{C}P^{n+1}$ related to isoparametric hypersurfaces. If we take compact homogeneous Lagrangian submanifolds as $L^n \subset Q_n(\mathbb{C})$, then by this theorem we obtain a class of compact Lagrangian orbits in $\mathbb{C}P^{n+1}$ which are not necessarily homogeneous in our sense.

We shall discuss the case when $L^n = S^n \subset Q_n(\mathbb{C})$ is a totally geodesic Lagrangian submanifold, which is one of real forms in $Q_n(\mathbb{C})$. It also coincides with the Gauss image of a compact isoparametric hypersurface with $g = 1$ in $S^{n+1}(1)$ (i.e. a great or small hypersphere of $S^{n+1}(1)$) in Section 4. By Theorem 4.1 we have a family of compact Lagrangian submanifolds in $\mathbb{C}P^{n+1}$ as

$$\tilde{L}^{n+1} = (\nu_+)^{-1}(S^n) \quad \subset \quad M_t = f^{-1}(t) \quad \subset \quad \mathbb{C}P^{n+1}$$

$$\nu_+ \Big\downarrow S^1 \qquad\qquad \nu_+ \Big\downarrow S^1$$

$$L^n = S^n \quad \subset \quad M_+ = Q_n(\mathbb{C}) \quad \subset \quad \mathbb{C}P^{n+1}$$

for each $t \in (-1, 1)$. In this case by direct computations we obtain

Theorem 4.2. *If $L^n = S^n \subset Q_n(\mathbb{C})$, then we have a family of compact Lagrangian submanifolds $\hat{L}_t^{n+1} = (\nu_+)^{-1}(S^n) (\subset M_t)$ $(t \in (-1, 1))$ embedded in $\mathbb{C}P^{n+1}$ with the following properties:*

(1) $\hat{L}_t^{n+1}$ *is diffeomorphic but not isometric to the Riemannian product $S^n \times S^1$:*

$$\hat{L}_t^{n+1} = \nu_+^{-1}(S^n) \cong \frac{SO(n+1)}{SO(n)} \cdot (SO(2) \times \{\mathrm{I}_n\}) \cong S^n \times S^1 .$$

(2) $\hat{L}_t^{n+1}$ *is a compact Lagrangian orbit of $SO(n+1) \times R^1$ but not homogeneous in the sense of Section 2. Here R^1 denotes the flow on $\mathbb{C}P^{n+1}$ generated by the vector field $J(\mathrm{grad} f)$.*

(3) $\hat{L}_t^{n+1}$ *is not Hamiltonian minimal.*

Remark 4.4. Note that the first statement of Theorem 5.6 in [18, p.237] was incorrect. On that line "Riemannian product" should be "warped product". More precisely, $\hat{L}_t^{n+1}$ is isometric to a warped product $S^1 \times_\rho S^n$ for a smooth function $\rho(\gamma) = \cos^2 \gamma \cos^2 \theta_1 + \sin^2 \gamma \sin^2 \theta_1$ on S^1.

In next sections we indicate that the argument of this section works also for some other homogeneous isoparametric families.

5. Construction of Lagrangian submanifolds in $Q_{n+1}(\mathbb{C})$ from a Lagrangian submanifold in $Q_n(\mathbb{C})$

Let $\{e_0, e_1, e_2, \cdots, e_{n+2}\}$ denote the standard orthonormal basis of $\mathbb{R}^{n+3}$. Set $K := SO(n+2) = \{A \in SO(n+3) \mid A(e_0) = e_0\}$, which is an orthogonal group of degree $n+2$ considered as a connected compact subgroup of $SO(n+3)$. The natural left group action of K on $Q_{n+1}(\mathbf{C}) = \tilde{G}r_2(\mathbb{R}^{n+3}) \cong \frac{SO(n+3)}{SO(2) \times SO(n+2)}$ gives the following homogeneous isoparametric family of $Q_{n+1}(\mathbb{C})$ with focal manifolds $Q_n(\mathbb{C})$ and S^{n+1}: For each $t \in (-1, 1)$,

$$
\begin{array}{ccccc}
Q_{n+1}(\mathbb{C}) & = & Q_{n+1}(\mathbb{C}) & = & Q_{n+1}(\mathbb{C}) \\
\cup & & \cup & & \cup \\
M_{-1} = S^{n+1} & \xleftarrow{\;\nu_-\;} & M_t & \xrightarrow{\;\nu_+\;} & Q_n(\mathbb{C}) = M_1
\end{array}
$$

Here $Q_n(\mathbb{C})$ is a totally geodesic complex hypersurface in $Q_{n+1}(\mathbb{C})$ and S^{n+1} is a totally geodesic Lagrangian submanifold (i.e. a real form) in $Q_{n+1}(\mathbb{C})$. Each tube M_t ($t \in (-1, 1)$) is a homogeneous isoparametric hypersurface of $Q_{n+1}(\mathbb{C})$ diffeomorphic to a compact homogeneous space $SO(n+2)/SO(n)$ The isoparametric function on $Q_{n+1}(\mathbb{C}) = \widetilde{Gr}_2(\mathbb{R}^{n+3})$ in this case is given as

$$
f(\mathbf{a} \wedge \mathbf{b}) := 2 \left(\|\mathbf{a}'\|^2 \, \|\mathbf{b}'\|^2 - (\mathbf{a}' \cdot \mathbf{b}') \right) - 1
$$

where $\mathbf{a} = a_0 e_0 + \mathbf{a}'$ denotes a decomposition of an element of $\mathbf{a} \in \mathbb{R}^{n+3}$ along the orthogonal direct sum $\mathbb{R}^{n+3} = \mathbb{R}e_0 \oplus \bigoplus_{i=1}^{n+2} \mathbb{R}e_0 = \mathbb{R}e_0 \oplus \mathbb{R}^{n+2}$. We remark that this isoparametric function f has an expression as $f = \|\mu_K\|^2 - 1$ in terms of the moment map $\mu_K : Q_{n+1}(\mathbb{C}) = \widetilde{Gr}_2(\mathbb{R}^{n+3}) \to \mathfrak{k} \cong \mathfrak{k}^*$ for the group action of K on $Q_{n+1}(\mathbb{C}) = \widetilde{Gr}_2(\mathbb{R}^{n+3})$. Using a rank 2 Riemannian symmetric pair $(U, K) = (SO(2) \times SO(n+3), SO(n+2))$ and its canonical decomposition $\mathfrak{u} = \mathfrak{k} + \mathfrak{p}$ as a symmetric Lie algebra, we can express the moment map μ_K as $\mu_K : Q_{n+1}(\mathbb{C}) = \widetilde{Gr}_2(\mathbb{R}^{n+3}) \ni \mathbf{a} \wedge \mathbf{b} \longmapsto [\mathbf{a}, \mathbf{b}] \in \mathfrak{k} \cong \mathfrak{k}^*$ (cf. [7] for more general results). Here we use the identification $\mathbb{R}^{n+3} \cong \mathfrak{p}$.

In this case we should note that the Reeb flow of the structure vector field $J\nu$ on M_t is also not isometric. Thus $\nu_+ : M_t \to Q_n(\mathbb{C}) = M_1$ is not a Riemannian submersion and it gives a symplectic reduction but not Kähler reduction of $Q_{n+1}(\mathbb{C})$. Hence we obtain

Theorem 5.1. *If $\varphi : L \to Q_n(\mathbb{C})$ is a Lagrangian immersion (resp. embedding) into an n-dimensional complex hyperquadric, then we have a family of Lagrangian immersions (resp. embeddings) into an $(n+1)$-dimensional complex hyperquadric*

$$
\hat{\varphi}_t : \hat{L} = \varphi^{-1} M_t \longrightarrow (M_t \subset) Q_{n+1}(\mathbb{C}) \quad (t \in (-1, 1)).
$$

In order to construct various Lagrangian submanifolds in $Q_{n+1}(\mathbb{C})$, we may use this method recursively:

$$
Q_{n+1}(\mathbb{C}) \supset Q_n(\mathbb{C}) \supset Q_{n-1}(\mathbb{C}) \supset \cdots \supset Q_2(\mathbb{C}) \supset Q_1(\mathbb{C}) \cong S^2.
$$

6. Construction of Lagrangian submanifolds in $\mathbb{C}P^{n+1}$: From a Lagrangian submanifold in $\mathbb{C}P^n$

Let $\{e_0, e_1, \cdots, e_{n+1}\}$ be the standard unitary basis of $\mathbb{C}P^{n+1}$. Let $S^{2n+3}(1) \subset \mathbb{C}^{n+2}$ be the unit standard hypersphere of $\mathbb{C}^{n+2} \cong \mathbb{R}^{2n+4}$ and $\pi : S^{2n+3}(1) \to \mathbb{C}P^{n+1}$ be the Hopf fibration. Put $[e_0] = \pi(e_0) \in \mathbb{C}P^{n+1}$. We know a symmetric space expression $\mathbb{C}P^{n+1} \cong SU(n+2)/S(U(1) \times U(n+1))$. We set $K := S(U(1) \times U(n+1)) = \{A \in SU(n+2) \mid A([e_0]) = [e_0]\}$. The natural left group action of K on $\mathbb{C}P^{n+1}$ gives the following homogeneous isoparametric family of $\mathbb{C}P^{n+1}$ with focal manifolds $\mathbb{C}P^n$ and $\{[e_0]\}$ (a point):

$$
\begin{array}{ccccc}
\mathbb{C}P^{n+1} & = & \mathbb{C}P^{n+1} & = & \mathbb{C}P^{n+1} \\
\cup & & \cup & & \cup \\
M_{-1} = \{[e_0]\} & \xleftarrow{\ \nu_- \ } & M_t & \xrightarrow{\ \nu_+ \ } & \mathbb{C}P^n = M_1 \\
\text{(a point)} & & & &
\end{array}
$$

Here each M_t $(t \in (-1,1))$ is a geodesic sphere $S^{2n+1} \cong \dfrac{S(U(1) \times U(n+1))}{U(n)}$ of $\mathbb{C}P^{n+1}$ with center $[e_0]$ and $\mathbb{C}P^n$ is a totally geodesic complex submanifold of $\mathbb{C}P^{n+1}$.

The corresponding isoparametric function f on $\mathbb{C}P^{n+1}$ is induced by an isoparametric function $\tilde{f} = F|_{S^{2n+3}(1)} : S^{2n+3}(1) \to \mathbb{C}P^{n+1}$ defined by

$$
F(\mathbf{z}) := -|z_0|^2 + \sum_{i=1}^{n+1} |z_i|^2 \quad (\mathbf{z} = (z_0, z_1, \cdots, z_{n+1}) \in \mathbb{C}^{n+2}).
$$

Then f is the first eigenfunction of the Laplace operator on $\mathbb{C}P^{n+1}$ and thus $J(\mathrm{grad} f)$ is a (holomorphic) Killing vector field. of $\mathbb{C}P^{n+1}$. On the other hand, $J(\mathrm{grad} f)$ is a Hamiltonian vector field corresponding to the Hamiltonian $-f$ and thus each $M_t = f^{-1}(t)$ is preserved by the flow of $J(\mathrm{grad} f)$. The flow of $J(\mathrm{grad} f)$ is generated by the natural group action of the center $U(1)$ of the isotropy subgroup $K = S(U(1) \times U(n+1)) \cong U(n+1)$ on $\mathbb{C}P^{n+1} \cong SU(n+2)/S(U(1) \times U(n+1))$.

In this case it is well-known that the flow of $J\nu$ on M_t is isometric and the induced metrics on geodesic spheres $M_t \cong S^{2n+1}$ are so-called *Berger metrics*. Under the submersion $\nu_+ : M_t \to \mathbb{C}P^n = M_1$ not only the push-forwarded symplectic form on $\mathbb{C}P^n$ is homothetic to the original Kähler form of $\mathbb{C}P^n$, but also the push-forwarded $SU(n+1)$-invariant Riemannian metric on $\mathbb{C}P^n$ is homothetic to the original Kähler metric of $\mathbb{C}P^n$. Hence $\nu_+ : M_t \to \mathbb{C}P^n = M_1$ becomes a Riemannian submersion with respect to a suitable multiple of the original Kähler metric of $\mathbb{C}P^n$. Hence the positive focal map $\nu_+ : M_t \to M_1 = \mathbb{C}P^n$ gives a symplectic and Kähler reduction of $\mathbb{C}P^{n+1}$. Therefore we obtain

Theorem 6.1. *If $\varphi : L \to \mathbb{C}P^n$ is a Lagrangian immersion (resp. embedding) into an n-dimensional complex projective space, then we have a family of Lagrangian immersions (resp. embeddings) into an $(n+1)$-dimensional complex projective space*

$$\hat{\varphi}_t : \hat{L} = \varphi^{-1}M_t \longrightarrow (M_t \subset)\mathbb{C}P^{n+1} \quad (t \in (-1,1)).$$

Conversely, any Lagrangian submanifold $\hat{L}$ in $\mathbb{C}P^{n+1}$ invariant under the Hamiltonian flow of the above isoparametric function f on $\mathbb{C}P^{n+1}$ is obtained in this way as $\hat{L} = \varphi^{-1}M_t$ for some ($t \in (-1,1)$).

Remark 6.1. In this case we have that if L is a compact homogeneous Lagrangian submanifold in $\mathbb{C}P^n$, then $\nu_+^{-1}(L)$ is a compact homogeneous Lagrangian submanifold in $\mathbb{C}P^{n+1}$. By results of [6] we see that φ is a Hamiltonian minimal Lagrangian immersion into $\mathbb{C}P^n$ if and only if $\hat{\varphi}_t$ is a Hamiltonian minimal Lagrangian immersions into $\mathbb{C}P^{n+1}$.

Acknowledgement. This paper is based on my talk at the workshop "Differential Geometry of Submanifolds and Its Related Topics"(at Saga University in Japan, on August 4–6, 2012) for the sixtieth birthday of Professor Sadahiro O'Maeda. The author sincerely appreciates Sadahiro O'Maeda for his great contribution to the submanifold theory in Japan. He also gave the author many useful comments on geometry of homogeneous real hypersurfaces in complex projective spaces in this work.

References

1. A. Amarzaya and Y. Ohnita, *Hamiltonian stability of certain minimal Lagrangian submanifolds in complex projective spaces*, Tohoku Math. J. **55** (2003), 583–610.
2. A. Amarzaya and Y. Ohnita, *Hamiltonian stability of parallel Lagrangian submanifolds embedded in complex space forms*, a preprint, http://www.sci.osaka-cu.ac.jp/ ohnita/paper/Amar-Ohnita08.pdf.
3. L. Bedulli and A. Gori, *A Hamiltonian stable minimal Lagrangian submanifolds of projective spaces with nonparallel second fundamental form*, Transf. Groups **12** (2007), 611–617.
4. L. Bedulli and A. Gori, *Homogeneous Lagrangian submanifolds*, Comm. Anal. Geom. **16** (2008), 591–615.
5. P. Dazord, *Sur la geometrie des sous-fibres et des feuilletages lagrangiens*, (French) [*On the geometry of subbundles and Lagrange foliations*] Ann. Sci. École Norm. Sup. (4) 14 (1981), no. 4, 465–480 (1982).

6. Y.-X. Dong, *Hamiltonian-minimal Lagrangian submanifolds in Kaehler manifolds with symmetries*, Nonlinear Analysis **67** (2007), 865–882.

7. H. Ma and Y. Ohnita, *On Lagrangian submanifolds in complex hyperquadrics and isoparametric hypersurfaces in spheres*. Math. Z. **261** (2009), 749–785.

8. H. Ma and Y. Ohnita, *Differential Geometry of Lagrangian Submanifolds and Hamiltonian Variational Problems*. in Harmonic Maps and Differential Geometry, Contemporary Mathematics vol. 542, Amer. Math. Soc., Providence, RI, 2011, pp. 115-134.

9. H. Ma and Y. Ohnita, *Hamiltonian stability of the Gauss images of homogeneous isoparametric hypersurfaces*, a preprint (2010), OCAMI Preprint Ser. no.10-23.

10. R. Miyaoka, *Moment maps of the spin action and the Cartan-Münzner polynomials of degree four*, Math. Ann. **355** (2013), 1067–1084.

11. H. F. Münzner, *Isoparametrische Hyperfläche in Sphären*, Math. Ann. **251** (1980), 57–71.

12. H. F. Münzner, *Isoparametrische Hyperfläche in Sphären, II*, Math. Ann. **256** (1981), 215–232.

13. K. Nomizu, *Some results in E. Cartan's theory of isoparametric families of hypersurfaces*, Bull. Amer. Math. Soc. 79 (1973), 1184-1188.

14. Y. G. Oh, *Volume minimization of Lagrangian submanifolds under Hamiltonian deformations*, Math. Z. **212** (1993), 175-192.

15. Y. Ohnita, *Stability and rigidity of special Lagrangian cones over certain minimal Legendrian orbits*, Osaka J. Math. **44** (2007), 305-334.

16. Y. Ohnita, *Geometry of Lagrangian Submanifolds and Isoparametric Hypersurfaces*, Proceedings of The Fourteenth International Workshop on Differential Geometry, **14** (2010), pp 43-67, NIMS, KMS and GRG. (OCAMI Preprint Ser. no.10-9.)

17. Y. Ohnita, *Certain Lagrangian submanifolds in Hermitian symmetric spaces and Hamiltonian stability problems*, Proceedings of The Fifteenth International Workshop on Differential Geometry and the 4th KNUGRG-OCAMI Differential Geometry Workshop, **15** (2011), pp 209-234, ed. by Y.-J. Suh, NIMS, KMS and GRG. (OCAMI Preprint Ser. no.11-14).

18. Y. Ohnita, *Certain compact homogeneous Lagrangian submanifolds in Hermitian symmetric spaces*, Proceedings of The Sixteenth International Workshop on Differential Geometry and the 5th KNUGRG-OCAMI Differential Geometry Workshop, **16** (2012), pp 225-240, ed. by Y.-J. Suh, NIMS, KMS and GRG.

19. H. Ozeki and M. Takeuchi, *On some types of isoparametric hypersurfaces in spheres* I. Tohoku Math. J.(2) **27** (1975), 515–559.

20. H. Ozeki and M. Takeuchi, *On some types of isoparametric hypersurfaces in spheres* II. Tohoku Math. J.(2) **28** (1976), 7–55.

21. B. Palmer, *Hamiltonian minimality of Hamiltonian stability of Gauss maps*, Differential Geom. Appl. **7** (1997), 51–58.

22. D. Petrecca and F. Podesta, *Construction of homogeneous Lagrangian submanifolds in CP^n and Hamiltonian stability*, Tohoku Math. J.(2) **64** (2012), 261–268.

23. R. Takagi, *Real hypersurfaces in a complex projective space with constant principal curvatures.* J. Math. Soc. Japan **27** no. 4 (1975), 43–53.

24. R. Takagi, *Real hypersurfaces in a complex projective space with constant principal curvatures* II. J. Math. Soc. Japan **27** no. 4 (1975), 507–516.

Received April 8, 2013.

Proceedings of the Workshop on
Differential Geometry of Submanifolds
and its Related Topics
Saga, August 4-6, 2012

SOME REAL HYPERSURFACES OF COMPLEX
PROJECTIVE SPACE

Tatsuyoshi HAMADA*

*Department of Applied Mathematics, Fukuoka University,
Fukuoka, 814-0180, Japan & JST, CREST, 5 Sanbancho, Chiyoda-ku,
Tokyo 102-0075, Japan
E-mail: hamada@fukuoka-u.ac.jp*

Dedicated to Professor Sadahiro Maeda for his sixtieth birthday

The study of real hypersurfaces in complex projective space has been an active
field of study over the past decade. M. Okumura classified the real hypersurfaces of type (A) in complex projective space [7]. And η-parallelism of real
hypersurfaces are discussed in M. Kimura and S. Maeda [4]. Maeda proved the
nonexistence of semi-parallel real hypersurfaces [8]. In this paper, we study real
hypersurfaces of type (A) in complex projective space with parallelism of some
symmetric tensor.

Keywords: complex projective space, real hypersurface, almost contact structure.

1. Introduction

The study of real hypersurfaces in complex projective space $P_n(\mathbf{C})$ is an
active field over the past decade. Okumura classified the real hypersurfaces
M of type (A) of complex projective space. He showed that if the almost
contact structure ϕ and the shape operator A of M are commutative, then
the real hypersurface M is locally congruent to a tube of some radius r
over a totally geodesic Kähler submanifold $P_k(\mathbf{C})$, $0 < k \leq n - 1$, where
$0 < r < \pi/\sqrt{c}$. Define $\psi = \phi A - A\phi$ of M. Then we can show that ψ is
symmetric tensor on M. We show the following theorem:

*This research was financially supported by Grant-in-Aid Scientific Research No.
24540104, Japan Society for the Promotion of Science.

Theorem 1.1. *Let M be a connected real hypersurface of $P_n(\mathbf{C})$. Then the symmetric tensor $\psi = \phi A - A\phi$ of M is parallel if and only if M is an open subset of one of*

(A_1) *a tube over a totally geodesic complex hypersurface $P_{n-1}(\mathbf{C})$ of radius*
  $r\,(0 < r < \pi/\sqrt{c})$,
(A_2) *a tube over a totally geodesic Kähler submanifold $P_k(\mathbf{C}), 0 < k < n-1$*
  of radius $r\,(0 < r < \pi/\sqrt{c})$.

2. Preliminaries

Let $P_n(\mathbf{C})$ be a complex projective space with complex dimension $n \geq 2$ with the holomorphic sectional curvature $c > 0$. M is a real hypersurface of $P_n(\mathbf{C})$ through an isometric immersion. In a neighborhood of each point, we take a unit normal local vector field N on M. The Riemannian connections $\widetilde{\nabla}$ in $P_n(\mathbf{C})$ and ∇ in M are related by the following formulas for arbitrary vector fields X and Y on M.

$$\widetilde{\nabla}_X Y = \nabla_X Y + g(AX, Y)N,$$

$$\widetilde{\nabla}_X N = -AX,$$

where g denotes the Riemannian metric of M induced from the Kähler metric G of $P_n(\mathbf{C})$ and A is the shape operator of M in $P_n(\mathbf{C})$. We denote by TM the tangent bundle of M. An eigenvector X of the shape operator A is called a *principal curvature vector*. Also an eigenvalue λ of A is called a *principal curvature*. We know that M has an almost contact metric structure induced from the Kähler structure (J, G) of $P_n(\mathbf{C})$: We define a $(1,1)$-tensor field ϕ, a vector field ξ, and a 1-form η on M by $g(\phi X, Y) = G(JX, Y)$ and $g(\xi, X) = \eta(X) = G(JX, N)$. Then we have

$$\phi^2 X = -X + \eta(X)\xi, \quad \eta(\xi) = 1, \quad \phi\xi = 0. \tag{1}$$

It follows that

$$\nabla_X \xi = \phi AX, \quad (\nabla_X \phi)Y = \eta(Y)AX - g(AX, Y)\xi. \tag{2}$$

Now we prepare without proof the following in order to prove our results (cf. [6]).

Lemma 2.1. *If ξ is a principal curvature vector, then the corresponding principal curvature α is locally constant.*

Lemma 2.2. *Assume that ξ is a principal curvature vector and the corresponding principal curvature is α. If $AX = \lambda X$ for $X \perp \xi$, then we have*

$$A\phi X = \frac{\alpha\lambda + c/2}{2\lambda - \alpha}\phi X.$$

Okumura proved the following result in [7].

Proposition 2.1. *Let M be a connected real hypersurface of complex projective space $P_n(\mathbf{C})$. Then $\phi A = A\phi$ if and only if M is an open subset of one of*

(A_1) *a tube over a totally geodesic complex hypersurface $P_{n-1}(\mathbf{C})$ of radius r $(0 < r < \pi/\sqrt{c})$,*

(A_2) *a tube over a totally geodesic Kähler submanifold $P_k(\mathbf{C})$, $0 < k < n-1$ of radius r $(0 < r < \pi/\sqrt{c})$.*

Takagi gave the complete list of homogeneous real hypersurface of $P_n(\mathbf{C})$ [11].

Proposition 2.2. *Let M be a homogeneous real hypersurface of $P_n(\mathbf{C})$ of constant holomorphic sectional curvature $c > 0$. Then M is locally congruent to the following:*

(A_1) *a tube over a totally geodesic complex hypersurface $P_{n-1}(\mathbf{C})$ of radius r $(0 < r < \pi/\sqrt{c})$*

(A_2) *a tube over totally geodesic complex projective spaces $P_k(\mathbf{C})$, $0 < k < n-1$ of radius r $(0 < r < \pi/\sqrt{c})$,*

(B) *a tube over complex quadrics Q_{n-1} and $P^n(\mathbf{R})$ of radius r $(0 < r < \pi/(2\sqrt{c}))$,*

(C) *a tube over the Segre embedding of $P_1(C) \times P_{(n-1)/2}(\mathbf{C})$ of radius r $(0 < r < \pi/(2\sqrt{c}))$, and $n(\geq 5)$ is odd,*

(D) *a tube over the Plücker embedding of the complex Grassmann manifold $G_{2,5}(\mathbf{C})$ of radisu r $(0 < r < \pi/(2\sqrt{c}))$ and $n = 9$,*

(E) *a tube over the canonical embedding of the Hermitian symmetric space $SO(10)/U(5)$ of radius r $(0 < r < \pi/(2\sqrt{c}))$ and $n = 15$.*

Kimura showed the following important result [3].

Proposition 2.3. *Let M be a real hypersurface of $P_n(\mathbf{C})$. Then M has constant principal curvatures and ξ is a principal curvature vector if and only if M is locally congruent to a homogeneous real hypersurface.*

We define the symmetric tensor $\psi = \phi A - A\phi$ on M, we can consider that Okumura's result is a case of $\psi = 0$.

3. Proof of the theorem

We may calculate the following from (2):

$$g((\nabla_X \psi)Y, Z) = \eta(AY)g(AX, Z) - \eta(Z)g(AX, AY)$$
$$+ g((\phi(\nabla_X A) - (\nabla_X A)\phi)Y, Z)$$
$$+ \eta(AZ)g(AX, Y) - \eta(Y)g(AX, AZ),$$

for any tangent vector fields X, Y and Z on M. We put $X = Y = Z = \xi$ in the above equation, because of (1), we obtain

$$g((\nabla_\xi \psi)\xi, \xi) = 2(\eta(A\xi)\eta(A\xi) - \eta(A^2\xi)).$$

By the assumption, we assert that

$$\eta(A\xi)\eta(A\xi) - \eta(A^2\xi) = 0. \tag{3}$$

We may suppose

$$A\xi = a\xi + bU, \tag{4}$$

for two functions a and b on M, where U is the unit tangent vector field orthogonal to ξ on M. We imply (4) to (3), we have

$$a^2 - (a^2 + b^2) = 0.$$

So we conclude that $b = 0$, i.e. ξ is a principal curvature vector.

We choose a principal curvature vector Y orthogonal to ξ and with the principal curvature λ. From Lemma 2.2, ϕY is also a principal curvature vector with the principal curvature $(a\lambda + c/2)/(2\lambda - a)$ and we get the following formula:

$$\psi Y = \mu \phi Y,$$

where $\mu = \lambda - (a\lambda + c/2)/(2\lambda - a)$. Using the above formula, we get

$$g((\nabla_X \psi)Y, Z) = (X\mu)g(\phi Y, Z) - \lambda \mu g(X, Y)\eta(Z)$$
$$+ \mu g(\phi \nabla_X Y, Z) - g(\phi A \nabla_X Y, Z) + g(A\phi \nabla_X Y, Z).$$

By putting $Z = \xi$ to the equation, we have

$$g((\nabla_X \psi)Y, \xi) = -\lambda \mu g(X, Y),$$

for any tangent vector field X and a principal curvature vector Y with the principal curvature λ. We may assume $X = Y$, so we observe

$$\lambda \mu = 0.$$

By Lemma 2.2,

$$\lambda \left(2\lambda^2 - 2\alpha\lambda - \frac{c}{2}\right) = 0.$$

From Lemma 2.1, we conclude that principal curvature λ is locally constant. By Propositions 2.2 and 2.3, M is homogeneous and λ does not vanish on M. We obtain

$$\mu = 0.$$

We know that these real hypersurfaces are type (A) by the Proposition 2.1.

References

1. Cecil, T. E. and Ryan, P. J., *Focal sets and real hypersurfaces in complex projective space*, Trans. Amer. Math. Soc. **269** (1982), 481–499.
2. Hamada, T., *On real hypersurfaces of a complex projective space with recurrent second fundamental tensor*, J. Ramanujan Math. Soc. **11** (1996), 103–107.
3. Kimura, M., *Real hypersurfaces and complex submanifolds in a complex projective space*, Trans. Amer. Math. Soc. **296** (1986), 137–149.
4. Kimura, M. and Maeda, S., *On real hypersurfaces of a complex projective space*, Math. Z. **202** (1989), 299–311.
5. Kobayashi, S. and Nomizu, K., *Foundations of Differential Geometry*, Vol.1 (1963) John Wiley and Sons, Inc. New York-London.
6. Maeda, Y., *On real hypersurfaces of a complex projective space*, J. Math. Soc. Japan **28** (1976), 529–540.
7. Okumra, M., *On some real hypersurfaces of a complex projective space*, Trans. Amer. Math. Soc. **212** (1975), 355–364.
8. Maeda, S., *Real hypersurfaces of complex projective spaces*, Math. Ann. **263** (1984), 473–478
9. Montiel, S., *Real hypersurfaces of a complex hyperbolic space*, J. Math. Soc. Japan **37** (1985), 515–535.
10. Niebergall, R. and Ryan, P.J., *Real hypersurfaces in complex space forms*, Tight and taut submanifolds Cecil, T.E. and Chern, S.S., Math. Sci. Res. Inst. Publ., 32 Cambridge Univ. Press, Cambridge, (1997), 233–305.
11. Takagi, R., *On homogeneous real hypersurfaces in a complex projective space*, Osaka J. Math. **10** (1973), 495–506.

Received March 31, 2013.

Proceedings of the Workshop on
Differential Geometry of Submanifolds
and its Related Topics
Saga, August 4-6, 2012

CONTACT METRIC HYPERSURFACES IN COMPLEX SPACE FORMS

Jong Taek CHO*

*Department of Mathematics, Chonnam National University,
CNU The Institute of Basic Science,
Gwangju, 500–757, Korea
E-mail: jtcho@chonnam.ac.kr*

Jun-ichi INOGUCHI[†]

*Department of Mathematical Sciences, Yamagata University,
Yamagata, 990–8560, Japan
E-mail: inoguchi@sci.kj.yamagata-u.ac.jp*

Dedicated to Professor Sadahiro Maeda for his sixtieth birthday

We study contact metric real hypersurfaces in complex projective space and complex hyperbolic space.

Keywords: contact metric manifolds, real hypersurfaces, complex space forms, (κ, μ)-spaces, Sasakian manifolds, quasi-Sasakian manifolds.

Introduction

Sasakian space forms are homogeneous contact metric manifolds and realized as specific homogeneous real hypersurfaces in non-flat complex space forms (Berndt [3], *cf.* Ejiri [12]). Recently, Adachi, Kameda and Maeda gave a classification of all contact metric hypersurfaces in non-flat complex space forms. In particular they showed that hypersurfaces (type (A_0) and type (A_1)) in the list due to Berndt [3] exhaust the Sasakian hyper-

---

*Partially supported by Basic Science Research Program through the National Research Foundation of Korea (NRF) funded by the Ministry of Education, Science and Technology (2012R1A1B3003930).
†Partially supported by Kakenhi 24540063.

surfaces in non-flat complex space forms. Moreover they showed that non-Sasakian contact metric hypersurfaces are tubes around real forms (*i.e.*, totally geodesic Lagrangian subspaces) of specific radii (type (B)).

In contact metric geometry, as a generalization of Sasakian manifolds, the notion of contact (κ, μ)-space was introduced by Blair, Koufogiorgos and Papantoniou [5]. Contact (κ, μ)-spaces provide a large class of strongly pseudo-convex CR-manifolds. Boeckx showed that every non-Sasakian contact (κ, μ)-space is homogeneous and strongly locally ϕ-symmetric [6]. Dileo and Lotta [11] showed that for non-Sasakian contact metric manifolds of dimension greater than 3, the (κ, μ)-condition is equivalent to the local CR-symmetry. Boeckx and the first named author of the present paper showed that Sasakian manifolds and non-Sasakian contact $(\kappa, 2)$-spaces are characterized as strongly pseudo-convex CR-manifolds with parallel Tanaka-Webster curvature and torsion [7]. For bi-Legendre foliations of contact (κ, μ)-spaces, we refer to Cappelletti Montano [8].

It is natural to expect that these contact metric hypersurfaces are contact (κ, μ)-spaces. In this paper we determine contact (κ, μ)-hypersurfaces in complex projective space and complex hyperbolic space.

1. Contact metric manifolds

Let M be a manifold of odd dimension $m = 2n - 1$. Then M is said to be an *almost contact manifold* if its structure group $GL_m\mathbb{R}$ of the linear frame bundle is reducible to $U_{n-1} \times \{1\}$. This is equivalent to existence of an endomorphism field ϕ, a vector field ξ and a 1-form η satisfying

$$\phi^2 = -I + \eta \otimes \xi, \quad \eta(\xi) = 1. \tag{1}$$

From these conditions one can deduce that $\phi\xi = 0$ and $\eta \circ \phi = 0$.

Moreover, since $U_{n-1} \times \{1\} \subset SO_{2n-1}$, M admits a Riemannian metric g satisfying

$$g(\phi X, \phi Y) = g(X, Y) - \eta(X)\eta(Y)$$

for all X, $Y \in \mathfrak{X}(M)$. Here $\mathfrak{X}(M)$ denotes the Lie algebra of all smooth vector fields on M. Such a metric is called an *associated metric* of the almost contact manifold $M = (M; \phi, \xi, \eta)$. With respect to the associated metric g, η is metrically dual to ξ, that is $g(X, \xi) = \eta(X)$ for all $X \in \mathfrak{X}(M)$. A structure (ϕ, ξ, η, g) on M is called an *almost contact metric structure*, and a manifold M equipped with an almost contact metric structure is said to be an *almost contact metric manifold*.

A plane section Π at a point p of $(M; \phi, \xi, \eta, g)$ is said to be a ϕ-*section* if it is invariant under ϕ_p. The sectional curvature function of ϕ-sections are called the ϕ-*sectional curvature*.

On an almost contact metric manifold M, we define an endomorphism field h by $h = (\mathcal{L}_\xi \phi)/2$. Here $\mathcal{L}_\xi$ denotes the Lie differentiation by ξ.

The *fundamental 2-form* Φ of $(M; \phi, \xi, \eta, g)$ is defined by

$$\Phi(X, Y) = g(X, \phi Y), \quad X, Y \in \mathfrak{X}(M).$$

An almost contact metric manifold M is said to be a *contact metric manifold* if $\Phi = d\eta$. On a contact metric manifold, η is a *contact form*, i.e., $(d\eta)^{n-1} \wedge \eta \neq 0$. Thus every contact metric manifold is orientable. More precisely, the volume element dv_g induced from the associated metric g is related to the contact form η by the following formula ([4, Theorem 4.6]):

$$dv_g = \frac{(-1)^{n-1}}{2^{n-1}(n-1)!} \eta \wedge (d\eta)^{n-1}.$$

An almost contact metric manifold M is said to be *normal* if

$$[\phi X, \phi Y] + \phi^2[X, Y] - \phi[\phi X, Y] - \phi[X, \phi Y] + 2d\eta(X, Y)\xi = 0$$

for any $X, Y \in \mathfrak{X}(M)$.

Definition 1.1. An almost contact metric manifold is said to be a *quasi-Sasakian manifold* if it is normal and $d\Phi = 0$.

Definition 1.2. A contact metric manifold is said to be a *Sasakian manifold* if it is normal. In particular, Sasakian manifolds of constant ϕ-sectional curvature are called *Sasakian space forms*.

Sasakian manifolds are quasi-Sasakian since $\Phi = d\eta$. Note that Sasakian manifolds satisfy

$$R(X, Y)\xi = \eta(Y)X - \eta(X)Y.$$

According to [5], a contact metric manifold M is said to be a *contact* (κ, μ)-*space* if there exist a pair $\{\kappa, \mu\}$ of real numbers such that

$$R(X, Y)\xi = (\kappa I + \mu h)(\eta(Y)X - \eta(X)Y)$$

for all $X, Y \in \mathfrak{X}(M)$.

It should be remarked that on contact (κ, μ)-spaces, $\kappa \leq 1$ holds. If $\kappa = 1$, then the structure is Sasakian.

A *pseudo-homothetic deformation* of a contact metric structure (ϕ, ξ, η, g) is a new structure $(\phi, \xi/a, a\eta, ag + a(a-1)\eta \otimes \eta)$ with positive

constant a. The pseudo-homothetic deformation of a contact (κ, μ)-space is
a contact $(\widetilde{\kappa}, \widetilde{\mu})$-space with $\widetilde{\kappa} = (\kappa + a^2 - 1)/a^2$ and $\widetilde{\mu} = (\mu + 2a - 2)/a^2$.

The *Boeckx invariant* $\mathcal{I}$ of a *non-Sasakian* contact (κ, μ)-space is defined by $\mathcal{I} = (1 - \mu/2)/\sqrt{1 - \kappa}$. Two non-Sasakian contact (κ, μ)-spaces are related by pseudo-homothetic deformation if and only if their Boeckx invariants agree (see [6]).

For more informations on contact metric geometry, we refer to Blair's monograph [4].

2. Real hypersurfaces

Let M be a real hypersurface of a Kähler manifold $\widetilde{M}_n$ of complex dimension n. Since we are working in local theory, we may assume that M is orientable. Take a unit normal vector filed N of M in $\widetilde{M}_n$. Then the Levi-Civita connections $\widetilde{\nabla}$ of $\widetilde{M}_n$ and ∇ of M are related by the following *Gauss formula* and *Weingarten formula*:

$$\widetilde{\nabla}_X Y = \nabla_X Y + g(AX, Y)N, \quad \widetilde{\nabla}_X N = -AX, \quad X \in \mathfrak{X}(M).$$

Here g is the Riemannian metric of M induced by the Kähler metric $\tilde{g}$ of the ambient space $\widetilde{M}_n$. The endomorphism field A is called the *shape operator* of M derived from N.

An eigenvector X of the shape operator A is called a *principal curvature vector*. The corresponding eigenvalue λ of A is called a *principal curvature*. Define a vector field ξ on M by $\xi = -\epsilon JN$ with $\epsilon = \pm 1$. We call ϵ the *sign* of M relative to N. The vector field ξ is called the *structure vector field*. Next, define the 1-form η and the endomorphism field ϕ by

$$\eta(X) = g(\xi, X) = \tilde{g}(JX, N), \quad g(\phi X, Y) = \tilde{g}(JX, Y), \quad X, Y \in \mathfrak{X}(M),$$

respectively. Then one can see that (ϕ, ξ, η, g) is an almost contact metric structure on M, that is, it satisfies (1).

It follows that

$$(\nabla_X \phi)Y = \epsilon(\eta(Y)AX - g(AX, Y)\xi), \quad \nabla_X \xi = \epsilon \phi AX. \tag{2}$$

A real hypersurface is said to be *Hopf* if its structure vector field ξ is a principal curvature vector field. The next lemma is well known.

Lemma 2.1. ([17], *cf.* [19]) *If M is a Hopf hypersurface in a non-flat complex space form. Then the principal curvature α corresponding to ξ is a constant.*

Proposition 2.1. ([21]) *Let M be an orientable real hypersurface in a Kähler manifold $\widetilde{M}_n$. Then M satisfies $d\eta = k\Phi$ for some constant k if and only if $A\phi + \phi A = -2\epsilon k\phi$. In this case M is Hopf.*

3. Real hypersurfaces in complex space forms

Now let us assume that the ambient space is a *complex space form $\widetilde{M}_n(c)$* of constant holomorphic sectional curvature c. As is well known, a complete and simply connected complex space form is a *complex projective space* $\mathbb{C}P_n(c)$, a *complex Euclidean space* $\mathbb{C}^n$ or a *complex hyperbolic space* $\mathbb{C}H_n(c)$, according as $c > 0$, $c = 0$ or $c < 0$.

Takagi classified the homogeneous real hypersurfaces of $\mathbb{C}P_n(c)$ into six types [26]. Cecil and Ryan extensively studied Hopf hypersurfaces, which are realized as tubes over certain Kähler submanifolds in $\mathbb{C}P_n(c)$ [9]. Kimura proved the equivalence of extrinsic homogeneity and constancy of principal curvatures in the class of all Hopf hypersurfaces in $\mathbb{C}P_n(c)$ [15]. As a result we know the following classification table.

Theorem 3.1. ([26],[15]) *Let M be a Hopf hypersurface of $\mathbb{C}P_n(c)$. Then M has constant principal curvatures if and only if M is locally holomorphically congruent to one of the following real hypersurfaces:*

(A_1) *a geodesic hypersphere of radius r, where $0 < r < \pi/\sqrt{c}$,*
(A_2) *a tube of radius r over a totally geodesic $\mathbb{C}P_\ell(c)(1 \leq \ell \leq n-2)$ via Segre imbedding, where $0 < r < \pi/(\sqrt{c})$,*
(B) *a tube of radius r over a complex quadric Q^{n-1}, where $0 < r < \pi/(2\sqrt{c})$,*
(C) *a tube of radius r over a $\mathbb{C}P_1(c) \times \mathbb{C}P_{(n-1)/2}(c)$, where $0 < r < \pi/(2\sqrt{c})$ and $n \geq 5$ is odd,*
(D) *a tube of radius r over the Plücker imbedding of complex Grassmannian $Gr_2(\mathbb{C}^5) \subset \mathbb{C}P_9(c)$, where $0 < r < \pi/(2\sqrt{c})$.*
(E) *a tube of radius r over a Hermitian symmetric space $SO_{10}/U_5 \subset \mathbb{C}P_{15}(c)$, where $0 < r < \pi/(2\sqrt{c})$.*

It should be remarked that a tube of radius r over a complex quadric Q^{n-1} is realized also as a tube of radius $\pi/(2\sqrt{c}) - r$ over a totally geodesic Lagrangian real projective space $\mathbb{R}P^n(c/4)$ (see [9]).

The tube around $\mathbb{C}P_\ell(c)$ ($1 \leq \ell < n-2$) of radius $r = \pi/(2\sqrt{c})$ is the *quadric* in the sense of Tanaka [27] and Yamaguchi [29].

Corresponding table for $\mathbb{C}H_n(c)$ was obtained by Berndt.

Theorem 3.2. ([2]) *Let M be a Hopf hypersurface of $\mathbb{C}H_n(c)$. Then M has constant principal curvatures if and only if M is locally holomorphically congruent to one of the following real hypersurfaces:*

(A_0) *a horosphere,*
(A_1) *a geodesic hypersphere ($A_{1,0}$) or a tube over a complex hyperbolic hyperplane $\mathbb{C}H_{n-1}(c)$ ($A_{1,1}$),*
(A_2) *a tube over a totally geodesic $\mathbb{C}H_\ell(c)$ ($1 \le \ell \le n-2$),*
(B) *a tube over a totally geodesic Lagrangian real hyperbolic space $\mathbb{R}H^n(c/4)$.*

We call simply type (A) for real hypersurfaces of type (A_1), (A_2) in $\mathbb{C}P_n(c)$ and ones of type (A_0), (A_1) or (A_2) in $\mathbb{C}H_n(c)$.

The *Gauss equation* for a real hypersurface $M \subset \widetilde{M}_n(c)$ is given by

$$R(X,Y)Z = \frac{c}{4}\{g(Y,Z)X - g(X,Z)Y$$
$$+ g(\phi Y, Z)\phi X - g(\phi X, Z)\phi Y - 2g(\phi X, Y)\phi Z\}$$
$$+ g(AY,Z)AX - g(AX,Z)AY.$$

In particular we have

$$R(X,Y)\xi = \frac{c}{4}\{\eta(Y)X - \eta(X)Y\} + g(AY,\xi)AX - g(AX,\xi)AY. \tag{3}$$

The operator h is computed as follows

$$2hX = \pounds_\xi(\phi X) - \phi(\pounds_\xi X) = [\xi, \phi X] - \phi[\xi, X]$$
$$= \nabla_\xi(\phi X) - \nabla_{\phi X}\xi - \phi(\nabla_\xi X) + \phi(\nabla_X \xi)$$
$$= (\nabla_\xi \phi)X - \nabla_{\phi X}\xi + \phi(\nabla_X \xi).$$

It should be remarked that h is not self-adjoint in general. Indeed, we can show that h is self-adjoint if and only if M is a Hopf hypersurface.

Inserting (2) into the last equation, we have (*cf.* [10, (31)])

$$hX = \frac{\epsilon}{2}\{\eta(X)A\xi - (\phi A\phi)X - AX\}.$$

When M is a Hopf hypersurface satisfying $A\xi = \alpha\xi$, we have

$$AX = -2\epsilon hX + \alpha\eta(X)\zeta - (\phi A\phi)X \tag{4}$$

If we assume, in addition, that M satisfies $A\phi + \phi A = \gamma\phi$ with nonzero constant γ. Then we have

$$A\phi X = \gamma\phi X - \phi AX$$
$$\phi A\phi X = \gamma\phi^2 X - \phi^2 AX = \gamma\{-X + \eta(X)\xi\} + AX - \eta(AX)\xi.$$

Combining this with (4), we get

$$AX = -\epsilon h X + \frac{\gamma}{2} X + \left(\alpha - \frac{\gamma}{2}\right) \eta(X)\xi. \tag{5}$$

Inserting this into (3) we arrive at the formula:

$$R(X,Y)\xi = \left\{\left(\frac{c}{4} + \frac{\gamma}{2}\alpha\right) I - \epsilon \alpha h\right\} \{\eta(Y)X - \eta(X)Y\}. \tag{6}$$

Theorem 3.3. *Let $M \subset \widetilde{M}_n(c)$ be an orientable real hypersurface in a non-flat complex space form with sign ϵ. If M satisfies the condition $A\phi + \phi A = \gamma\phi$ for some nonzero constant γ. Then M satisfies*

$$R(X,Y)\xi = (\kappa I + \mu h)\{\eta(Y)X - \eta(X)Y\}$$

with $\kappa = c/4 + \gamma\alpha/2$ and $\mu = -\epsilon\alpha$.

Corollary 3.1. *Let $M \subset \widetilde{M}_n(c)$ be an orientable real hypersurface in a non-flat complex space form with sign ϵ. If M is contact metric, that is $d\eta = \Phi$, then M is a contact (κ, μ)-space with $\kappa = c/4 - \epsilon\alpha$ and $\mu = -\epsilon\alpha$.*

Real hypersurfaces with $A\phi + \phi A = \gamma\phi$ are classified by Adachi, Kameda and Maeda (see also Suh [25, Lemma 3.1] for the case $c < 0$ and $n > 2$):

Lemma 3.1. ([1]) *Let $M \subset \widetilde{M}_n(c)$ be a orientable real hypersurface with $n \geq 2$ and $c \neq 0$. Then M satisfies $\phi A + A\phi = \gamma\phi$ for some nonzero constant γ if and only if M is of type (A_0), (A_1) or (B).*

Let M be a real hypersurface of type (A_1) in $\mathbb{C}P_n(c)$, type (A_0) in $\mathbb{C}H_n(c)$ or type (A_1) in $\mathbb{C}H_n(c)$. Then M is Hopf and has two distinct principal curvatures α and λ. It is easy to see that M satisfies $A\phi + \phi A = 2\lambda\phi$. Berndt showed that these real hypersurfaces are Sasakian space forms up to homothety [3] (see also Ejiri [12] for (A_1) case). Adachi, Kameda and Maeda classified all Sasakian hypersurfaces in $\mathbb{C}P_n(c)$ and $\mathbb{C}H_n(c)$.

Theorem 3.4. ([1]) *Let $M \subset \widetilde{M}_n(c)$ be an oriented real hypersurface with $n \geq 2$ and $c \neq 0$. Then the following conditions are mutually equivalent:*

(1) *M is a Sasakian manifold;*
(2) *M is a Sasakian space form of ϕ-sectional curvature $H = c + 1$;*
(3) *M is locally holomorphically congruent to one of the following homogeneous real hypersurfaces:*

(A_1) *a geodesic hypersphere of radius $r = (2/\sqrt{c})\tan^{-1}(\sqrt{c}/2)$ in $\mathbb{C}P_n(c)$ with sign $\epsilon = -1$ $(H > 1)$.*

$(A_{1,0})$ *a geodesic hypersphere of radius* $r = (2/\sqrt{-c})\tanh^{-1}(\sqrt{-c}/2)$ *in* $\mathbb{C}H_n(c)$ *with sign* $\epsilon = -1$, *where* $-4 < c < 0$ *and* $-3 < H < 1$.;

(A_0) *a horosphere in* $\mathbb{C}H_n(-4)$ *with sign* $\epsilon = -1$ ($H = -3$).

$(A_{1,1})$ *a tube around totally geodesic* $\mathbb{C}H_{n-1}(c)$ *of radius* $r = (2/\sqrt{-c})\coth^{-1}(\sqrt{-c}/2)$ *in* $\mathbb{C}H_n(c)$ *with sign* $\epsilon = -1$, *where* $c < -4$, ($H < -3$).

Non-Sasakian contact metric hypersurfaces are classified as follows:

Theorem 3.5. ([1]) *Let* $M \subset \widetilde{M}_n(c)$ *with* $n \geq 2$ *and* $c \neq 0$ *be a real hypersurface. If* M *is contact metric but not Sasakian, then* M *is locally holomorphically congruent to one of the following homogeneous real hypersurfaces of type* (B):

- *a tube of* $r = (2/\sqrt{c})\tan^{-1}\{(\sqrt{c+4} - \sqrt{c})/2\} < \pi/(2\sqrt{c})$ *around the complex quadric* $Q^{n-1} \subset \mathbb{C}P_n(c)$ *with sign* $\epsilon = 1$;
- *a tube of radius* $r = (1/\sqrt{-c})\tanh^{-1}(\sqrt{-c}/2)$ *around the totally geodesic Lagrangian real hyperbolic space* $\mathbb{R}H^n(c/4) \subset \mathbb{C}H_n(c)$ *with sign* $\epsilon = -1$ *and* $-4 < c < 0$.

These tubes have three distinct principal curvatures unless $c = -3$. *In case* $c = -3$, *the tubes have two distinct principal curvature* $3/2$ *and* 2.

It should be remarked that Vernon [28] classified real hypersurfaces in $\mathbb{C}H_n(-4)$ satisfying $d\eta = k\Phi$ with nonzero constant k.

Corollary 3.1 implies that these hypersurfaces of type (B) are contact (κ, μ)-spaces. We compute the values of κ and μ of these hypersurfaces.

Example 3.1. Let M be a tube of radius $r = (2/\sqrt{c})\tan^{-1}\{(\sqrt{c+4} - \sqrt{c})/2\} < \pi/(2\sqrt{c})$ around the complex quadric $Q^{n-1} \subset \mathbb{C}P_n(c)$ with sign $\epsilon = 1$. Then M is a contact metric Hopf hypersurface. The principal curvature α corresponding to ξ is $\alpha = c/2$. By Corollary 3.1, M is a non-Sasakian (κ, μ)-space with $\kappa = -c/4$, $\mu = -\sqrt{c}/2$. The Boeckx invariant is $\mathcal{I} = \sqrt{1 + c/4} > 1$.

Example 3.2. Let M be a tube of radius $r = (1/\sqrt{-c})\tanh^{-1}(\sqrt{-c}/2)$ around the totally geodesic Lagrangian real hyperbolic space $\mathbb{R}H^n(c/4) \subset \mathbb{C}H_n(c)$ with sign $\epsilon = -1$ and $-4 < c < 0$. Then M is a contact metric Hopf hypersurface. The principal curvature corresponding to ξ is $\alpha = -c/2$. Hence M is a contact (κ, μ)-space with $\kappa = 3c/4 < 0$, $\mu = -c/2 > 0$. The Boeckx invariant is $0 < \mathcal{I} = (c+4)/(2\sqrt{4-3c}) < 1$.

Homogeneous real hypersurfaces of type (A) are characterized as follows:

Proposition 3.1. ([13], [20], [22], [18], [23], [16]) *Let $M \subset \widetilde{M}_n(c)$ $(n \geq 2)$ be a real hypersurface in a non-flat complex space form. Then the following conditions are mutually equivalent:*

- *M satisfies $A\phi = \phi A$;*
- *M is locally holomorphically congruent to a type (A) hypersurface;*
- *ξ is a Killing vector field;*
- *$h = 0$.*

In these cases, M is a quasi-Sasakian manifold.

Type (A_2) hypersurfaces are characterized as the only quasi-Sasakian hypersurfaces in $\mathbb{C}P_n(c)$ and $\mathbb{C}H_n(c)$ which are non-homothetic Sasakian (see also Olszak [24]).

Remark 3.1. Let $M^n(c)$ be an n-dimensional Riemannian space form of curvature c. Then its unit tangent bundle $T_1 M^n(c)$ equipped with standard contact metric structure (for details, see Blair [4]) is a contact (κ, μ)-space with $\kappa = (2 - c)c$ and $\mu = -2c$. The Boeckx invariant for $T_1 M^n(c)$ with $c \neq 1$ is given by $\mathcal{I} = (1 + c)/|1 - c|$. In particular, the unit tangent bundles of Euclidean space $\mathbb{R}^n$ and real hyperbolic space $\mathbb{R}H^n(-1)$ are contact $(0, 0)$-space with $\mathcal{I} = 1$ and contact $(-3, 2)$-space with $\mathcal{I} = 0$, respectively. It would be interesting to find contact (κ, μ)-real hypersurfaces with $\mathcal{I} < 0$ in Kähler manifolds of non-constant holomorphic sectional curvature (*e.g.*, Hermitian symmetric spaces of rank greater than 1).

References

1. T. Adachi, M. Kameda and S. Maeda, Real hypersurfaces which are contact in a nonflat complex space form, Hokkaido Math. J. **40** (2011), 205–217.
2. J. Berndt, Real hypersurfaces with constant principal curvatures in a complex hyperbolic space, J. Reine Angew. Math. **395** (1989), 132–141.
3. J. Berndt, Real hypersurfaces with constant principal curvatures in complex space forms, *Geometry and Topology of Submanifolds*, III, (M. Boyom, J. M. Morvan and L. Verstraelen eds), World Scientific, pp. 10–19, 1990.
4. D. E. Blair, *Riemannian Geometry of Contact and Symplectic Manifolds*, 2nd. ed., Progress in Math. **203**, Birkhäuser, Boston, Basel, Berlin, 2010.
5. D. E. Blair, Th. Koufogiorgos and B. J. Papantoniou, Contact metric manifolds satisfying a nullity condition, Israel J. Math. **91** (1995), 189–214.
6. E. Boeckx, A full classification of contact metric (κ, μ)-spaces, Illinois J. Math. **44** (2000), 212–219.

7. E. Boeckx and J. T. Cho, Pseudo-Hermitian symmetries, Israel J. Math. **166** (2008), 125–145.
8. B. Cappelletti Montano, The foliated structure of contact metric (κ, μ)-spaces, Ilinois J. Math. **53** (2009), no. 4, 1157–1172.
9. T. Cecil and P.J. Ryan, Focal sets and real hypersurfaces in complex projective spaces, Trans. Amer. Math. Soc. **269** (1982), 481–499.
10. J. T. Cho, Geometry of CR-manifolds of contact type, *Proceedings of the Eighth International Workshop on Differential Geometry* **8** (2004), 137–155.
11. G. Dileo and A. Lotta, A classification of spherical symmetric CR manifolds, Bull. Austral. Math. Soc. **80** (2009), 251–274.
12. N. Ejiri, A generalization of minimal cones, Trans. Amer. Math. Soc. **276** (1983), 347–360.
13. S. Kanemaki, Quasi-Sasakian manifolds, Tohoku Math. J. **29** (1977), 227–233.
14. Y. H. Kim and S. Maeda, Real hypersurfaces in a nonflat complex space form and their almost contact metric structures, Coll. Math. **124** (2011), no. 1, 117–131.
15. M. Kimura, Real hypersurfaces and complex submanifolds in complex projective space, Trans. Amer. Math. Soc. **296** (1986), 137–149.
16. S. Maeda and S. Udagawa, Real hypersurfaces of a compex projective space in terms of holomorphic distribution, Tsukuba J. Math. **14** (1990), no. 1, 39–52.
17. Y. Maeda, On real hypersurfaces of a complex projective space, J. Math. Soc. Japan **28**(1976), no. 3, 529–540.
18. S. Montiel and A. Romero, On some real hypersurfaces of a complex hyperbolic space, Geom. Dedicata **20**(1986), 245–261.
19. R. Niebergall and P. J. Ryan, Real hypersurfaces in complex space forms, *Tight and taut submanifolds*, (T. E. Cecil and S. S. Chern eds.), Math. Sci. Res. Inst. Publ., **32** Cambridge Univ. Press, Cambridge, pp. 233–305, 1997.
20. M. Okumura, Certain almost contact hypersurfaces in Kaehlerian manifolds of constant holomorphic sectional curvature, Tohoku Math. J. (2) **16** (1964), 270–284.
21. M. Okumura, Contact hypersurfaces in certain Kaehlerian manifolds, Tohoku Math. J. (2) **18** (1966), no. 1, 74–102.
22. M. Okumura, On some real hypersurfaces of a complex projective space, Trans. Amer. Math. Soc. **212** (1975), 355–364.
23. M. Okumura, Compact real hypersurfaces of a complex projective space, J. Differ. Geom. **12** (1977), 595–598.
24. Z. Olszak, Curvature properties of quasi-Sasakian manifolds, Tensor N. S. **38** (1982), 19–28.
25. Y. J. Suh, On real hypersurfaces of a complex space form with η-parallel Ricci tensor, Tsukuba J. Math. **14** (1990), no. 1, 27–37.
26. R. Takagi, On homogeneous real hypersurfaces in a complex projective space, Osaka J. Math. **10**(1973), 495–506.
27. N. Tanaka, On the pseudo-conformal geometry of hypersurfaces of the space n complex variables, J. Math. Soc. Japan **14** (1962), 397–429.

28. M. H. Vernon, Contact hypersurfaces of a complex hyperbolic space, Tohoku Math. J. (2) **39** (1987), 215–222.
29. K. Yamaguchi, Non-degenerate real hypersurfaces in complex manifolds admitting large groups of pseudo-conformal transformations I, II, Nagoya Math. J. **62** (1976), 55–96, **69** (1978), 9–31.

Received January 30, 2013.
Revised February 13, 2013.

Proceedings of the Workshop on
Differential Geometry of Submanifolds
and its Related Topics
Saga, August 4-6, 2012

NON-HOMOGENEOUS η-EINSTEIN REAL HYPERSURFACES IN A 2-DIMENSIONAL NONFLAT COMPLEX SPACE FORM

Kazuhiro OKUMURA

Asahikawa National College of Technology,
Shunkodai 2-2 Asahikawa 071-8142, Japan
E-mail: okumura@asahikawa-nct.ac.jp

Dedicated to Professor Sadahiro Maeda for his sixtieth birthday

We investigate geometric properties of 3-dimensional real hypersurfaces with $A\xi = 0$ in a complex 2-dimensional nonflat complex space form from the viewpoints of their shape operators, Ricci tensors and *-Ricci tensors.

Keywords: complex space forms, real hypersurfaces, shape operator, Ricci tensor, *-Ricci tensor, η-Einstein, *-Einstein, η-parallel.

1. Introduction

We denote by $\widetilde{M}_n(c)$ $(n \geq 2)$ an n-dimensional nonflat complex space form of constant holomorphic sectional curvature $c(\neq 0)$. Namely, $\widetilde{M}_n(c)$ is congruent to either a complex projective space of constant holomorphic sectional curvature $c(> 0)$ or a complex hyperbolic space of constant holomorphic sectional curvature $c(< 0)$.

In the theory of real hypersurfaces in $\widetilde{M}_n(c)$, Hopf hypersurfaces and homogeneous real hypersurfaces play a remarkable role. The former are real hypersurfaces such that the characteristic vector ξ is a principal vector at its each point. The latter are real hypersurfaces which are expressed as orbits of some subgroup of the isometry group $I(\widetilde{M}_n(c))$ of $\widetilde{M}_n(c)$.

In this paper, we classify real hypersurfaces M^3 in $\widetilde{M}_2(c)$ satisfying $A\xi = 0$, where A is the shape operator of M^3 and ξ is the characteristic vector field of M^3. These real hypersurfaces include both a homogeneous real hypersurface and a non-homogeneous real hypersurface in $\widetilde{M}_2(c)$. One is a geodesic sphere of radius $\pi/(2\sqrt{c}\,)$ in $\mathbb{C}P^2(c)$ and the other is a non-homogeneous Hopf hypersurface in $\widetilde{M}_2(c)$ having three distinct principal curvature 0, λ, $c/(4\lambda)$, where λ is a smooth function on M^3. In particular, the latter is a tube of radius $\pi/(2\sqrt{c}\,)$ over a non-totally geodesic complex curve which dose not have the principal curvatures $\pm\sqrt{c}/2$ in $\mathbb{C}P^2(c)$ or a Hopf hypersurface in $\mathbb{C}H^2(c)$ constructed from two contact curves in a 3-dimensional sphere S^3 (for detail [5], [6]).

In addition, we study properties of real hypersurfaces M^3 with $A\xi = 0$ in $\widetilde{M}_2(c)$ in terms of the shape operator, Ricci tensor and *-Ricci tensor. Non-homogeneous real hypersurfaces with $A\xi = 0$ in $\widetilde{M}_2(c)$ are interesting. For example, these real hypersurfaces are counter examples to results of [4] and [16]. In Section 5, we introduce the classification of η-Einstein real hypersurfaces, the classification of real hypersurfaces with η-parallel Ricci tensor and the classification of *-Einstein real hypersurfaces in $\widetilde{M}_n(c)$.

The author would like to thank Professor Sadahiro Maeda for his valuable suggestion and encouragement during the preparation of this paper.

2. Fundamental materials, Hopf hypersurfaces and homogeneous real hyperusurfaces in $\widetilde{M}_n(c)$

Let M^{2n-1} be a real hypersurface with a unit local vector field $\mathcal{N}$ of a nonflat complex space form $\widetilde{M}_n(c)$ through an isometric immersion $\iota_M : M^{2n-1} \longrightarrow \widetilde{M}_n(c)$. The Riemannian connections $\widetilde{\nabla}$ of $\widetilde{M}_n(c)$ and ∇ of M^{2n-1} are related by

$$\widetilde{\nabla}_X Y = \nabla_X Y + g(AX, Y)\mathcal{N}, \tag{1}$$

$$\widetilde{\nabla}_X \mathcal{N} = -AX \tag{2}$$

for vector fields X and Y tangent to M^{2n-1}, where g denotes the induced metric from the standard Riemannian metric of $\widetilde{M}_n(c)$ and A is the shape operator of M^{2n-1} in $\widetilde{M}_n(c)$. (1) is called *Gauss's formula*, and (2) is called *Weingarten's formula*. It is known that M^{2n-1} admits an *almost contact metric structure* (ϕ, ξ, η, g) induced from the Kähler structure J of $\widetilde{M}_n(c)$. The characteristic vector field ξ of M^{2n-1} is defined as $\xi = -J\mathcal{N}$ and this structure satisfies

$$\phi^2 = -I + \eta \otimes \xi, \ \eta(X) = g(X, \xi), \ \eta(\xi) = 1, \ \phi\xi = 0, \ \eta(\phi X) = 0,$$

$$g(\phi X, Y) = -g(X, \phi Y), \text{ and } \ g(\phi X, \phi Y) = g(X, Y) - \eta(X)\eta(Y) \tag{3}$$

for vector fields X and Y tangent to M^{2n-1}, where I denotes the identity map of the tangent bundle TM of M^{2n-1}.

It follows from the fact that $\widetilde{\nabla}J = 0$ and Equations (1) and (2) that

$$\nabla_X \xi = \phi AX. \tag{4}$$

Let R be the curvature tensor of M^{2n-1}. We have the equation of Gauss given by:

$$\begin{aligned}
g(R(X,Y)Z,W) = &(c/4)\{g(Y,Z)g(X,W) - g(X,Z)g(Y,W) \\
&+ g(\phi Y, Z)g(\phi X, W) - g(\phi X, Z)g(\phi Y, W) \\
&- 2g(\phi X, Y)g(\phi Z, W)\} + g(AY,Z)g(AX,W) \\
&- g(AX,Z)g(AY,W).
\end{aligned} \tag{5}$$

We usually call M^{2n-1} a *Hopf hypersurface* if the characteristic vector ξ is a principal curvature vector at each point of M^{2n-1}. It is known that every tube of sufficiently small constant radius around each Kähler submanifold of $\widetilde{M}_n(c)$ is a Hopf hypersurface. This fact tells us that the notion of Hopf hypersurface is natural in the theory of real hypersurfaces in $\widetilde{M}_n(c)$ (see [15]).

The following lemma clarifies a fundamental property which is a useful tool in the theory of Hopf hypersurfaces in $\widetilde{M}_n(c)$.

Lemma 2.1. *For a Hopf hypersurface M^{2n-1} with principal curvature δ corresponding to the characteristic vector field ξ in $\widetilde{M}_n(c)$, we have the following:*

(1) *δ is locally constant on M^{2n-1};*
(2) *If X is a tangent vector of M^{2n-1} perpendicular to ξ with $AX = \lambda X$, then $(2\lambda - \delta)A\phi X = (\delta\lambda + (c/2))\phi X$.*

Remark 2.1. When $c < 0$, in Lemma 2.1(2) there exists a case that both of equations $2\lambda - \delta = 0$ and $\delta\lambda + (c/2) = 0$ hold. In fact, for example we take a horosphere in $\mathbb{C}H^n(c)$. It is known that this real hypersurface has two distinct constant principal curvatures $\lambda = \sqrt{|c|}/2$, $\delta = \sqrt{|c|}$ or $\lambda = -\sqrt{|c|}/2$, $\delta = -\sqrt{|c|}$. Hence, when $c < 0$, we must consider two cases $2\lambda - \delta = 0$ and $2\lambda - \delta \neq 0$.

In $\mathbb{C}P^n(c)$ ($n \geq 2$), a Hopf hypersurface all of whose principal curvatures are constant is locally congruent to a homogeneous real hypersurface. Moreover, these real hypersurfaces are one of the following (see [9], [15], [17]):

(A$_1$) A geodesic sphere of radius r, where $0 < r < \pi/\sqrt{c}$;

(A_2) A tube of radius r around a totally geodesic $\mathbb{C}P^{\ell}(c)$ ($1 \leqq \ell \leqq n-2$), where $0 < r < \pi/\sqrt{c}$;

(B) A tube of radius r around a complex hyperquadric $\mathbb{C}Q^{n-1}$, where $0 < r < \pi/(2\sqrt{c}\,)$;

(C) A tube of radius r around a $\mathbb{C}P^1(c) \times \mathbb{C}P^{(n-1)/2}(c)$, where $0 < r < \pi/(2\sqrt{c}\,)$ and n ($\geqq 5$) is odd;

(D) A tube of radius r around a complex Grassmann $\mathbb{C}G_{2,5}$, where $0 < r < \pi/(2\sqrt{c}\,)$ and $n = 9$;

(E) A tube of radius r around a Hermitian symmetric space $SO(10)/U(5)$, where $0 < r < \pi/(2\sqrt{c}\,)$ and $n = 15$.

These real hypersurfaces are said to be of types (A_1), (A_2), (B), (C), (D) and (E). Summing up real hypersurfaces of types (A_1) and (A_2), we call them hypersurfaces of type (A). The numbers of distinct principal curvatures of these real hypersurfaces are $2, 3, 3, 5, 5, 5$, respectively. The principal curvatures of these real hypersurfaces in $\mathbb{C}P^n(c)$ are given as follows (cf. [15]):

	(A_1)	(A_2)	(B)	(C, D, E)
λ_1	$\frac{\sqrt{c}}{2}\cot(\frac{\sqrt{c}}{2}r)$	$\frac{\sqrt{c}}{2}\cot(\frac{\sqrt{c}}{2}r)$	$\frac{\sqrt{c}}{2}\cot(\frac{\sqrt{c}}{2}r-\frac{\pi}{4})$	$\frac{\sqrt{c}}{2}\cot(\frac{\sqrt{c}}{2}r-\frac{\pi}{4})$
λ_2	—	$-\frac{\sqrt{c}}{2}\tan(\frac{\sqrt{c}}{2}r)$	$\frac{\sqrt{c}}{2}\cot(\frac{\sqrt{c}}{2}r+\frac{\pi}{4})$	$\frac{\sqrt{c}}{2}\cot(\frac{\sqrt{c}}{2}r+\frac{\pi}{4})$
λ_3	—	—		$\frac{\sqrt{c}}{2}\cot(\frac{\sqrt{c}}{2}r)$
λ_4	—	—		$-\frac{\sqrt{c}}{2}\tan(\frac{\sqrt{c}}{2}r)$
δ	$\sqrt{c}\cot(\sqrt{c}\,r)$	$\sqrt{c}\cot(\sqrt{c}\,r)$	$\sqrt{c}\cot(\sqrt{c}\,r)$	$\sqrt{c}\cot(\sqrt{c}\,r)$

The multiplicities of these principal curvatures are given as follows (cf. [15]):

	(A_1)	(A_2)	(B)	(C)	(D)	(E)
$m(\lambda_1)$	$2n-2$	$2n-2\ell-2$	$n-1$	2	4	6
$m(\lambda_2)$	—	2ℓ	$n-1$	2	4	6
$m(\lambda_3)$	—	—	—	$n-3$	4	8
$m(\lambda_4)$	—	—	—	$n-3$	4	8
$m(\delta)$	1	1	1	1	1	1

Remark 2.2. A geodesic sphere $G(r)$ of radius r ($0 < r < \pi/\sqrt{c}$) in $\mathbb{C}P^n(c)$ is congruent to a tube of radius $(\pi/\sqrt{c})-r$ around a totally geodesic $\mathbb{C}P^{n-1}(c)$ of $\mathbb{C}P^n(c)$. In fact, $\lim_{r \to \pi/\sqrt{c}} G(r) = \mathbb{C}P^{n-1}(c)$.

In $\mathbb{C}H^n(c)$ ($n \geq 2$), a Hopf hypersurface all of whose principal curvatures are constant is locally congruent to one of the following (see [15]):

- (A_0) A horosphere in $\mathbb{C}H^n(c)$;
- $(A_{1,0})$ A geodesic sphere of radius r, where $0 < r < \infty$;
- $(A_{1,1})$ A tube of radius r around a totally geodesic $\mathbb{C}H^{n-1}(c)$, where $0 < r < \infty$;
- (A_2) A tube of radius r around a totally geodesic $\mathbb{C}H^\ell(c)$ ($1 \leqq \ell \leqq n-2$), where $0 < r < \infty$;
- (B) A tube of radius r around a totally real totally geodesic $\mathbb{R}H^n(c/4)$, where $0 < r < \infty$.

These real hypersurfaces are said to be of types (A_0), $(A_{1,0})$, $(A_{1,1})$, (A_2) and (B). Here, type (A_1) means either type $(A_{1,0})$ or type $(A_{1,1})$. Summing up real hypersurfaces of types (A_0), (A_1) and (A_2), we call them hypersurfaces of type (A). A real hypersurface of type (B) with radius $r = (1/\sqrt{|c|})\log_e(2+\sqrt{3})$ has two distinct constant principal curvatures $\lambda_1 = \delta = \sqrt{3|c|}/2$ and $\lambda_2 = \sqrt{|c|}/(2\sqrt{3})$ (see the table of principal curvatures in the case of $c < 0$). Except for this real hypersurface, the numbers of distinct principal curvatures of Hopf hypersurfaces with constant principal curvatures are $2, 2, 2, 3, 3$, respectively. The principal curvatures of these real hypersurfaces in $\mathbb{C}H^n(c)$ are given as follows (cf. [15]):

	(A_0)	$(A_{1,0})$	$(A_{1,1})$	(A_2)	(B)
λ_1	$\frac{\sqrt{|c|}}{2}$	$\frac{\sqrt{|c|}}{2}\coth(\frac{\sqrt{|c|}}{2}r)$	$\frac{\sqrt{|c|}}{2}\tanh(\frac{\sqrt{|c|}}{2}r)$	$\frac{\sqrt{|c|}}{2}\coth(\frac{\sqrt{|c|}}{2}r)$	$\frac{\sqrt{|c|}}{2}\coth(\frac{\sqrt{|c|}}{2}r)$
λ_2	—	—	—	$\frac{\sqrt{|c|}}{2}\tanh(\frac{\sqrt{|c|}}{2}r)$	$\frac{\sqrt{|c|}}{2}\tanh(\frac{\sqrt{|c|}}{2}r)$
δ	$\sqrt{|c|}$	$\sqrt{|c|}\coth(\sqrt{|c|}r)$	$\sqrt{|c|}\coth(\sqrt{|c|}r)$	$\sqrt{|c|}\coth(\sqrt{|c|}r)$	$\sqrt{|c|}\tanh(\sqrt{|c|}r)$

The multiplicities of these principal curvatures are given as follows (cf. [15]):

	(A_0)	$(A_{1,0})$	$(A_{1,1})$	(A_2)	(B)
$m(\lambda_1)$	$2n-2$	$2n-2$	$2n-2$	$2n-2\ell-2$	$n-1$
$m(\lambda_2)$	—	—	—	2ℓ	$n-1$
$m(\delta)$	1	1	1	1	1

Note that there exist many homogeneous non-Hopf hypersurfaces in $\mathbb{C}H^n(c)$. J. Berndt and H. Tamaru [1] classified homogeneous real hypersurfaces in $\mathbb{C}H^n(c)$.

Theorem 2.1 (J. Berndt and H. Tamaru, 2007). *Let M^{2n-1} be a homogeneous real hypersurface in $\mathbb{C}H^n(c)$ $(n \geqq 2)$. Then M^{2n-1} is locally congruent to one of the following real hypersurfaces:*

(1) *A horosphere in $\mathbb{C}H^n(c)$;*
(2) *A tube of radius r around a totally geodesic $\mathbb{C}H^\ell(c)$ $(0 \leqq \ell \leqq n-1)$, where $0 < r < \infty$;*
(3) *A tube of radius r around a totally real totally geodesic $\mathbb{R}H^n(c/4)$, where $0 < r < \infty$;*
(4) *The minimal ruled real hypersurface S determined by a horocycle in a totally geodesic $\mathbb{R}H^2(c/4)$, or an equidistant hypersurface to S;*
(5) *A tube of radius r around the normally homogeneous submanifold F_k of $\mathbb{C}H^n(c)$ with real normal bundle of rank k $(2 \leqq k \leqq n-1)$, where $0 < r < \infty$;*
(6) *A tube of radius r around the normally homogeneous submanifold $F_{k,\varphi}$ of $\mathbb{C}H^n(c)$ with normal bundle of rank $2k$ $(2 \leqq 2k \leqq 2[(n-1)/2])$ and constant Kähler angle $\varphi(\in (0, \pi/2))$, where $0 < r < \infty$.*

Note that homogeneous real hypersurfaces $(4), (5)$ and (6) of Theorem 2.1 are non-Hopf hypersurfaces in $\mathbb{C}H^n(c)$.

3. Ruled real hypersurfaces in $\widetilde{M}_n(c)$

Next we recall ruled real hypersurfaces in a nonflat complex space form, which are typical examples of non-Hopf hypersurfaces. A real hypersurface M^{2n-1} is called a *ruled real hypersurface* of a nonflat complex space form $\widetilde{M}_n(c)$ $(n \geqq 2)$ if the holomorphic distribution T^0 defined by $T^0(x) = \{X \in T_x M \mid X \perp \xi\}$ for $x \in M^{2n-1}$ is integrable and each of its maximal integral manifolds is a totally geodesic complex hypersurface $M_{n-1}(c)$ of $\widetilde{M}_n(c)$. A ruled real hypersurface is constructed in the following manner. Given an arbitrary regular real smooth curve γ in $\widetilde{M}_n(c)$ which is defined on an interval I we have at each point $\gamma(t)$ $(t \in I)$ a totally geodesic complex hypersurface $M_{n-1}^{(t)}(c)$ that is orthogonal to the plane spanned by $\{\dot{\gamma}(t), J\dot{\gamma}(t)\}$. Then we see that $M^{2n-1} = \bigcup_{t \in I} M_{n-1}^{(t)}(c)$ is a ruled real hypersurface in $\widetilde{M}_n(c)$. The following is a well-known characterization of ruled real hypersurfaces in terms of the shape operator A.

Lemma 3.1. *For a real hypersurface M^{2n-1} in a nonflat complex space form $\widetilde{M}_n(c)$ $(n \geqq 2)$, the following conditions are mutually equivalent:*

(1) M^{2n-1} *is a ruled real hypersurface;*
(2) *The shape operator A of M^{2n-1} satisfies the following equalities on the open dense subset $M_1 = \{x \in M^{2n-1} | \nu(x) \neq 0\}$ with a unit vector field U orthogonal to ξ : $A\xi = \mu\xi + \nu U$, $AU = \nu\xi$, $AX = 0$ for an arbitrary tangent vector X orthogonal to ξ and U, where μ, ν are differentiable functions on M_1 by $\mu = g(A\xi, \xi)$ and $\nu = \|A\xi - \mu\xi\|$;*
(3) *The shape operator A of M^{2n-1} satisfies $g(Av, w) = 0$ for arbitrary tangent vectors $v, w \in T_xM$ orthogonal to ξ_x at each point $x \in M^{2n-1}$.*

We treat a ruled real hypersurface locally, because generally this hypersurface has singularities. When we study ruled real hypersurfaces, we usually omit points where ξ is principal and suppose that ν does not vanish everywhere, namely a ruled hypersurface M^{2n-1} is usually supposed $M_1 = M^{2n-1}$.

4. Conditions on the shape operator

It is well known that there are no *totally umbilic* real hypersurfaces in a nonflat complex space form $\widetilde{M}_n(c)$. From this fact, we consider a weaker condition of totally umbilic. A real hypersurface M^{2n-1} of $\widetilde{M}_n(c)$ $(n \geq 2)$ is called *totally η-umbilic* if its shape operator A is of the form $A = \alpha I + \beta\eta \otimes \xi$ for some smooth functions α and β on M^{2n-1}. This definition is equivalent to saying that $Au = \alpha u$ for each vector u on M^{2n-1} which is orthogonal to the characteristic vector ξ of M^{2n-1}, where α is a smooth function on M^{2n-1}. The following classification theorem of totally η-umbilic real hypersurfaces M^{2n-1} shows that every totally η-umbilic real hypersurface is a member of Hopf hypersurfaces with constant principal curvatures, so that these two functions α and β are automatically constant on M^{2n-1} (see [15]):

Theorem 4.1. *Let M^{2n-1} be a totally η-umbilic real hypersurface of a nonflat complex space form $\widetilde{M}_n(c)$ $(n \geqq 2)$ with shape operator $A = \alpha I + \beta\eta \otimes \xi$. Then M is locally congruent to one of the following:*

(P) *A geodesic sphere $G(r)$ of radius r $(0 < r < \pi/\sqrt{c}\,)$ in $\mathbb{C}P^n(c)$, where $\alpha = (\sqrt{c}/2)\cot(\sqrt{c}\,r/2)$ and $\beta = -(\sqrt{c}/2)\tan(\sqrt{c}\,r/2)$;*
(H$_i$) *A horosphere in $\mathbb{C}H^n(c)$, where $\alpha = \beta = \sqrt{|c|}/2$;*
(H$_{ii}$) *A geodesic sphere $G(r)$ of radius r $(0 < r < \infty)$ in $\mathbb{C}H^n(c)$, where $\alpha = (\sqrt{|c|}/2)\coth(\sqrt{|c|}\,r/2)$ and $\beta = (\sqrt{|c|}/2)\tanh(\sqrt{|c|}\,r/2)$;*

(H$_{iii}$) *A tube of radius r ($0 < r < \infty$) around a totally geodesic complex hyperplane $\mathbb{C}H^{n-1}(c)$ in $\mathbb{C}H^n(c)$, where $\alpha = (\sqrt{|c|}/2)\tanh(\sqrt{|c|}\,r/2)$ and $\beta = (\sqrt{|c|}/2)\coth(\sqrt{|c|}\,r/2)$.*

In $\widetilde{M}_n(c)$, real hypersurfaces do not admit the parallelism of the shape operator A. So we study a weaker condition in the following. The shape operator A of M^{2n-1} in $\widetilde{M}_n(c)$ is called *η-parallel* if

$$g((\nabla_X A)Y, Z) = 0 \tag{6}$$

for all vectors X, Y and Z orthogonal to ξ on M^{2n-1}. M. Kimura and S. Maeda [10] classified Hopf hypersurfaces having η-parallel shape operator A in $\mathbb{C}P^n(n)$ and Y. J. Suh [16] classified these real hypersurfaces of $\mathbb{C}H^n(c)$.

Theorem 4.2 (M. Kimura and S. Maeda, 1989). *Let M^{2n-1} be a connected Hopf hypersurface in $\mathbb{C}P^n(c)\,(n \geq 2)$. Then M^{2n-1} has η-parallel shape operator A if and only if M^{2n-1} is locally congruent to one of the following:*

(i) *A geodesic sphere $G(r)$ of radius r, where $0 < r < \pi/\sqrt{c}$;*
(ii) *A tube of radius r around a totally geodesic $\mathbb{C}P^\ell(c)$ ($1 \leq \ell \leq n - 2$), where $0 < r < \pi/\sqrt{c}$;*
(iii) *A tube of radius r around a complex hyperquadric $\mathbb{C}Q^{n-1}$, where $0 < r < \pi/(2\sqrt{c}\,)$.*

Theorem 4.3 (Y. J. Suh, 1990). *Let M^{2n-1} be a connected Hopf hypersurface in $\mathbb{C}H^n(c)$ ($n \geq 2$). Then M^{2n-1} has η-parallel shape operator A if and only if M^{2n-1} is locally congruent to one of the following:*

(i) *A horosphere in $\mathbb{C}H^n(c)$;*
(ii) *A geodesic sphere $G(r)$ of radius r ($0 < r < \infty$);*
(iii) *A tube of radius r around a totally geodesic $\mathbb{C}H^{n-1}(c)$, where $0 < r < \infty$;*
(iv) *A tube of radius r around a totally geodesic $\mathbb{C}H^\ell(c)$ ($1 \leq \ell \leq n - 2$), where $0 < r < \infty$;*
(v) *A tube of radius r around a totally real totally geodesic $\mathbb{R}H^n(c/4)$, where $0 < r < \infty$.*

In addition, there are non-Hopf hyperusrfaces with η-parallel shape operator A in $\widetilde{M}_n(c)$ (see [10]):

Proposition 4.1 (M. Kimura and S. Maeda, 1989). *The shape operator of any ruled real hypersurface M^{2n-1} in $\widetilde{M}_n(c)$ ($n \geq 2$) is η-parallel.*

5. Conditions on the Ricci tensor and the *-Ricci tensor

The Ricci tensor S of an arbitrary real hypersurface M^{2n-1} in $\widetilde{M}_n(c)(n \geq 2)$ is defined by:

$$g(SX, Y) = \text{Trace of } (Z \mapsto R(Z, X)Y)$$

for all vectors X, Y and Z on M^{2n-1}, which is expressed as:

$$SX = \frac{c}{4}((2n+1)X - 3\eta(X)\xi) + (\text{Trace } A)AX - A^2X. \qquad (7)$$

It is well known that there are also no Einstein real hypersurfaces in a nonflat complex space form $\widetilde{M}_n(c)$. So, many researchers have focused on real hypersurfaces satisfying a weaker condition. A real hypersurface M^{2n-1} in $\widetilde{M}_n(c)$ is called η-Einstein if its Ricci tensor S is of the form

$$SX = aX + b\eta(X)\xi, \qquad (8)$$

where a, b are smooth functions on M^{2n-1}. Recently, the classification of η-Einstein real hypersurfaces was completed (cf. [3], [6], [8], [11], [13]).

Theorem 5.1 (M. Kon, 1979, T. E Cecil and P. J. Ryan, 1982).
Let M^{2n-1} be a connected η-Einstein real hypersurface in $\mathbb{C}P^n(c)$ $(n \geq 3)$. Then the functions a and b of Equation (8) are constant on M^{2n-1}. Moreover, M^{2n-1} is locally congruent to one of the following:

 (i) *A geodesic sphere $G(r)$ of radius r, where $0 < r < \pi/\sqrt{c}$;*
 (ii) *A tube of radius r around a totally geodesic $\mathbb{C}P^\ell(c)$ $(1 \leq \ell \leq n-2)$, where $0 < r < \pi/\sqrt{c}$ and $\cot^2(\sqrt{c}\,r/2) = \ell/(n-\ell-1)$;*
(iii) *A tube of radius r around a complex hyperquadric $\mathbb{C}Q^{n-1}$, where $0 < r < \pi/(2\sqrt{c}\,)$ and $\cot^2(\sqrt{c}\,r) = n-2$.*

Theorem 5.2 (H. S. Kim and P. J. Ryan, 2008). *Let M^3 be a connected η-Einstein real hypersurface in $\mathbb{C}P^2(c)$. Then the functions a and b of Equation (8) are constant on M^3. Moreover, M^3 is locally congruent to one of the following:*

 (i) *A geodesic sphere $G(r)$ of radius r, where $0 < r < \pi/\sqrt{c}$;*
 (ii) *A non-homogeneous real hypersurface with $A\xi - 0$ in $\mathbb{C}P^2(c)$.*

Theorem 5.3 (S. Montiel, 1985). *Let M^{2n-1} be a connected η-Einstein real hypersurface in $\mathbb{C}H^n(c)(n \geq 3)$. Then the functions a and b of Equation (8) are constant on M^{2n-1}. Moreover, M^{2n-1} is locally congruent to one of the following:*

 (i) *A horosphere in $\mathbb{C}H^n(c)$;*

(ii) *A geodesic sphere $G(r)$ of radius r, where $0 < r < \infty$;*

(iii) *A tube of radius r around a totally geodesic $\mathbb{C}H^{n-1}(c)$, where $0 < r < \infty$.*

Theorem 5.4 (T. A. Ivey and P. J. Ryan, 2009). *Let M^3 be a connected η-Einstein real hypersurface in $\mathbb{C}H^2(c)$. Then the functions a and b of Equation (8) are constant on M^3. Moreover, M^3 is locally congruent to one of the following:*

(i) *A horosphere in $\mathbb{C}H^2(c)$;*

(ii) *A geodesic sphere $G(r)$ of radius r, where $0 < r < \infty$;*

(iii) *A tube of radius r around a totally geodesic $\mathbb{C}H^1(c)$, where $0 < r < \infty$;*

(iv) *A non-homogeneous real hypersurface with $A\xi = 0$ in $\mathbb{C}H^2(c)$.*

Putting $X = \xi$ in Equation (8), we find

$$S\xi = \alpha\xi \tag{9}$$

where, α is a smooth function. Then we can get the following problem:

Problem 5.1. *Classify real hypersurfaces satisfying $S\xi = \alpha\xi$, where α is a smooth function.*

Obviously, η-Einstein real hypersurfaces in $\widetilde{M}_n(c)$ satisfy Equation (9). It is easy to find that ruled real hypersurfaces in $\widetilde{M}_n(c)$ also satisfy Equation (9). In fact,

$$S\xi = \left(\frac{c}{4}(2n-2) - \nu^2\right)\xi, \tag{10}$$

where, $\nu = \|A\xi - \mu\xi\|$, $\mu = g(A\xi, \xi)$.

In $\widetilde{M}_2(c)$, Mayuko. Kon [12] investigated Problem 5.1:

Theorem 5.5 (M. Kon, 2011). *Let M^3 be a compact real hypersurface in $\widetilde{M}_2(c)$. If the Ricci tensor S satisfies $S\xi = \alpha\xi$, α is a smooth function, then M^3 is a Hopf hypersurface.*

Next, we investigate the Ricci parallel condition. It is well known that there are no real hypersurfaces with parallel Ricci tensor (i. e., $\nabla S = 0$) in $\widetilde{M}_n(c)$. We prepare the following equation on an arbitrary real hypersurface M^{2n-1} in $\widetilde{M}_n(c)$:

$$(\nabla_X S)Y = -\frac{3c}{4}\{g(\phi AX, Y)\xi + \eta(Y)\phi X\} + (X(\text{Trace } A))AY$$
$$+ ((\text{Trace } A)I - A)(\nabla_X A)Y - (\nabla_X A)AY. \tag{11}$$

A Ricci tensor S of M^{2n-1} in $\widetilde{M}_n(c)$ is called *η-parallel* if

$$g((\nabla_X S)Y, Z) = 0 \tag{12}$$

for all vectors X, Y and Z orthogonal to ξ on M^{2n-1}. In [16], Y. J. Suh investigated the classification problem of Hopf hypersurfaces with η-parallel Ricci tensor in $\widetilde{M}_n(c)$. In recent years, S. Maeda [14] modified classification results of [16]:

Theorem 5.6 (Y. J. Suh, 1990, S. Maeda, 2013). *Let M^{2n-1} be a connected Hopf hypersurface in $\mathbb{C}P^n(c)$ $(n \geq 2)$. Suppose that M^{2n-1} has η-parallel Ricci tensor. Then M^{2n-1} is locally congruent to one of the following:*

(i) *A geodesic sphere $G(r)$ of radius r, where $0 < r < \pi/\sqrt{c}$;*
(ii) *A tube of radius r around a totally geodesic $\mathbb{C}P^\ell(c)$ $(1 \leq \ell \leq n-2)$, where $0 < r < \pi/\sqrt{c}$;*
(iii) *A tube of radius r around a complex hyperquadric $\mathbb{C}Q^{n-1}$, where $0 < r < \pi/(2\sqrt{c})$;*
(iv) *A non-homogeneous real hypersurface with $A\xi = 0$ in $\mathbb{C}P^2(c)$.*

Theorem 5.7 (Y. J. Suh, 1990, S. Maeda, 2013). *Let M^{2n-1} be a connected Hopf hypersurface in $\mathbb{C}H^n(c)$ $(n \geq 2)$. Suppose that M^{2n-1} has η-parallel Ricci tensor. Then M^{2n-1} is locally congruent to one of the following:*

(i) *A horosphere in $\mathbb{C}H^n(c)$;*
(ii) *A geodesic sphere $G(r)$ of radius r, where $0 < r < \infty$;*
(iii) *A tube of radius r around a totally geodesic hypersurface $\mathbb{C}H^{n-1}$, where $0 < r < \infty$;*
(iv) *A tube of radius r around a totally geodesic $\mathbb{C}H^\ell(c)$ $(1 \leq \ell \leq n-2)$, where $0 < r < \infty$;*
(v) *A tube of radius r around a complex hyperquadric $\mathbb{R}H^n(c/4)$, where $0 < r < \pi/\infty$;*
(vi) *A non-homogeneous real hypersurface with $A\xi = 0$ in $\mathbb{C}H^2(c)$.*

The **- Ricci tensor* S^* of an any real hypersurface M^{2n-1} in $\widetilde{M}_n(c)$ is defined by:

$$g(S^*X, Y) = \frac{1}{2} \text{ Trace of } (Z \mapsto R(X, \phi Y)\phi Z)$$

for all vectors X, Y and Z on M^{2n-1}, which is expressed as:

$$S^*X = \frac{cn}{2}(X - \eta(X)\xi) - \phi A\phi AX. \tag{13}$$

A real hypersurface M^{2n-1} in $\widetilde{M}_n(c)$ is called *-Einstein if

$$g(S^*X, Y) = \frac{\rho^*}{2(n-1)} g(X, Y) \qquad (14)$$

for all vectors X and Y orthogonal to ξ on M^{2n-1}, where ρ^* is *-scalar curvature i. e., $\rho^* = \text{Trace } S^*$.

In [4], T. Hamada studied the classification problem of *-Einstein real hypersurfaces in $\widetilde{M}_n(c)$. Recently, T. A. Ivey and P. J. Ryan [7] refined classification results of [4].

Theorem 5.8 (T. Hamada, 2002, T.A. Ivey and P.J. Ryan, 2011).
*Let M^{2n-1} be a *-Einstein real hypersurface in $\mathbb{C}P^n(c)$ $(n \geq 2)$. Then M^{2n-1} is locally congruent to one of the following:*

(i) *A geodesic sphere $G(r)$ of radius r, where $0 < r < \pi/\sqrt{c}$;*
(ii) *A tube of radius $\pi/(2\sqrt{c}\,)$ around a totally geodesic $\mathbb{C}P^\ell(c)$ $(1 \leqq \ell \leqq n - 2)$;*
(iii) *A tube of radius r around a complex hyperquadric $\mathbb{C}Q^{n-1}$, where $0 < r < \pi/(2\sqrt{c}\,)$;*
(iv) *A non-homogeneous real hypersurface with $A\xi = 0$ in $\mathbb{C}P^2(c)$.*

Theorem 5.9 (T. Hamada, 2002, T.A. Ivey and P.J. Ryan, 2011).
*Let M^{2n-1} be a *-Einstein real hypersurface in $\mathbb{C}H^n(c)$ $(n \geq 2)$. Then M^{2n-1} is locally congruent to one of the following:*

(i) *A horosphere in $\mathbb{C}H^2(c)$;*
(ii) *A geodesic sphere $G(r)$ of radius r, where $0 < r < \infty$;*
(iii) *A tube of radius r around a totally geodesic $\mathbb{C}H^{n-1}(c)$, where $0 < r < \infty$;*
(iv) *A tube of radius r around a totally real totally geodesic $\mathbb{R}H^n(c/4)$, where $0 < r < \infty$;*
(v) *A non-homogeneous real hypersurface with $A\xi = 0$ in $\mathbb{C}H^2(c)$.*

6. Main theorem

In this section, we shall classify 3-dimensional real hypersurfaces M^3 with $A\xi = 0$ in a 2-dimensional nonflat complex space form $\widetilde{M}_2(c)$.

Theorem 6.1. *Let M^3 be a 3-dimensional real hypersruface in $\widetilde{M}_2(c)$. Suppose that $A\xi = 0$ on M^3. Then a certain open dense subset $\mathcal{U}$ on M^3 is locally congruent to either a geodesic sphere $G(\pi/(2\sqrt{c}\,))$ in $\mathbb{C}P^2(c)$ or a non-homogeneous real hypersurface in $\widetilde{M}_2(c)$ having three distinct principal* *curvatures 0, λ and $c/(4\lambda)$, where λ is a smooth function on M^3. Moreover, these two real hypersurfaces satisfy the following properties on the Ricci tensor S and the *-Ricci tensor S^*:*

(1) $SX = \dfrac{3}{2}cX - c\eta(X)\xi$ *for each vector X on M^3, that is, M^3 is η-Einstein;*

(2) $g((\nabla_X S)Y, Z) = 0$ *for all vectors X, Y and Z orthogonal to ξ on M^3, that is, the Ricci tensor S is η-parallel;*

(3) $g(S^* X, Y) = \dfrac{7}{4}cg(X, Y)$ *for all vectors X, Y orthogonal to ξ on M^3, that is, M^3 is *-Einstein.*

Proof. Since $A\xi = 0$, we have $\phi A\xi = 0$. This equation implies that ξ is principal and its principal curvature $\delta = 0$. We take an orthogonal frame field $\{e, \phi e, \xi\}$ such that

$$Ae = \lambda e.$$

It follows from Lemma 2.1 (2) that we obtain

$$2\lambda A\phi e = \frac{c}{2}\phi e. \tag{15}$$

We suppose $\lambda = 0$. From (15), we have $c = 0$, which contradicts $c \neq 0$. So, we may assume that $\lambda \neq 0$. Then, we can get

$$A\phi e = \frac{c}{4\lambda}\phi e.$$

We consider the case of $\lambda \neq c/(4\lambda)$ at some point p on M^3. It follows from the continuity of λ that $\lambda \neq c/(4\lambda)$ on some sufficiently small neighborhood $\mathcal{U}$ of the point p. If λ is constant on M^3, then $c/(4\lambda)$ also is constant on M^3. Thus, our real hypersurface has distinct three constant principal curvatures 0, λ and $c/(4\lambda)$ $(\lambda \neq c/(4\lambda))$ on $\mathcal{U}$. However, this case cannot occur (see the table of the principal curvatures in Section 2). Hence, λ is a non-constant smooth function on M^3. Therefore, M^3 is locally congruent to a non-homogeneous real hypersurface in $\widetilde{M}_2(c)$ having three distinct principal curvatures 0, λ and $c/(4\lambda)$.

Next, we investigate the case $\lambda = c/(4\lambda)$ at some point p on M^3. Then we can see $\lambda = c/(4\lambda)$ on some sufficiently small neighborhood $\mathcal{V}$ of the point p. In fact, we suppose that there exists no neighborhood $\mathcal{V}$ of p such that $\lambda = c/(4\lambda)$ on $\mathcal{V}$. Then there exists a sequence $\{p_n\}$ on M^3 such that

$$\lim_{n \to \infty} p_n = p \quad \text{and} \quad \left(\lambda - \frac{c}{4\lambda}\right)(p_n) \neq 0 \text{ for each } n.$$

On the other hand, for each point p_n there exists a neighborhood $\mathcal{V}_{p_n}$ of the point p_n satisfying

$$\left(\lambda - \frac{c}{4\lambda}\right)(q) \neq 0 \text{ for each } q \in \mathcal{V}_{p_n}.$$

It follows from the continuity of the principal curvature λ, we obtain $(\lambda - c/(4\lambda))(p) \neq 0$, which is contradiction. Therefore $\lambda^2 = (c/4)$ on some sufficiently small neighborhood $\mathcal{V}$ of the point p. Namely, our real hypersurface has locally two constant principal curvatures 0, $\sqrt{c}/2$ or 0, $-\sqrt{c}/2$. In this case, M^3 is locally congruent to a geodesic sphere $G(\pi/(2\sqrt{c}))$ in $\mathbb{C}P^2(c)$ (see the table of the principal curvatures in Section 2 and Remark 2.2).

Finally, from equations (7), (11) and (13) we obtain immediately properties (1), (2) and (3) of Theorem 6.1. $\square$

We can characterize these two real hypersurfaces from the viewpoint of the η-parallelism of the shape operator A in the following:

Proposition 6.1. *Let M^3 be a 3-dimensional real hypersurface in $\widetilde{M}_2(c)$. Suppose that $A\xi = 0$ on M^3. Then the following hold:*

(1) *If M^3 has η-parallel shape operator if and only if a certain open dense $\mathcal{U}$ of M is locally congruent to a geodesic sphere $G(\pi/(2\sqrt{c}))$ in $\mathbb{C}P^2(c)$.*
(2) *If M^3 does not have η-parallel shape operator if and only if a certain open dense $\mathcal{U}$ of M is locally congruent to a non-homogeneous real hypersurface in $\widetilde{M}_2(c)$ having three distinct principal curvature 0, λ and $c/(4\lambda)$, where λ is a smooth function on M^3.*

References

1. J. Berndt and H. Tamaru, *Cohomogeneity one actions on noncompact symmetric spaces of rank one*, Trans. Amer. Math. Soc. **359** (2007), 3425–3439.
2. D. E. Blair, *Riemannian geometry of contact and symplectic manifolds*, Progress in Mathematics **203**, 2010.
3. T. E. Cecil and P. J. Ryan, *Focal sets real hypersurfaces in complex projective space*, Trans. Amer. Math. Soc. **269** (1982), 481–499.
4. T. Hamada, *Real hypersurfaces of complex space forms in terms of Ricci *-tensor*, Tokyo J. Math. **25** (2002), 473–483.
5. T. Ivey, *A d'Alembert formula for Hopf hypersurfaces*, Results in Math. **60** (2011), 293–309.
6. T. A. Ivey and P. J. Ryan, *Hopf hypersurfaces of small Hopf principal curvature in $\mathbb{C}H^2$*, Geom. Dedicata. **141** (2009), 147–161.
7. T. A. Ivey and P. J. Ryan, *The *-Ricci tensor for hypersurfaces in $\mathbb{C}P^n$ and $\mathbb{C}H^n$*, Tokyo J. Math. **34** (2011), no. 2, 445–471.

8. H. S. Kim and P. J. Ryan, *A classification of pseudo-Einstein hypersurfaces in* $\mathbb{C}P^2$, Differential Geom. Appl. **26** (2008), 106–112.

9. M. Kimura, *Real hypersurfaces and complex submanifolds in complex projective space*, Trans. Amer. Math. Soc. **296** (1986), 137–149.

10. M. Kimura and S. Maeda, *On real hypersurfacesof a complex projective space*, Math. Z. **202** (1989), 299–311.

11. Masahiro Kon, *Pseudo-Einstein real hypersurfaces in complex space forms*, J. Differential Geom. **14** (1979), 339–354.

12. Mayuko Kon, *On 3-dimensional compact real hypersurfaces of a complex space form*, Proceedings of The Fifteenth International Workshop on Diff. Geom. **15** (2011) 47–56.

13. S. Montiel, *Real hypersurfaces of a complex hyperbolic space* J. Math. Soc. Japan **37** (1985), 515–535.

14. S. Maeda, *Hopf hypersurfaces with η-parallel Ricci tensors in a nonflat complex space form*, to appear in Sci. Math. Japonicae.

15. R. Niebergall and P. J. Ryan, *Real hypersurfaces in complex space form*, in: Tight and Taut submanifolds, T.E. Cecil and S.S. Chern (eds.), Cambridge Univ. Press, 1998, 233–305.

16. Y. J. Suh, *On real hypersurfaces of a complex space form with η-parallel Ricci tensor*, Tsukuba J. Math. **14** (1990), 27–37.

17. R. Takagi, *On homogeneous real hypersurfaces in a complex projective space*, Osaka J. Math. **10**(1973), 495-506.

Received March 31, 2013.
Revised March 27, 2013.

Proceedings of the Workshop on
Differential Geometry of Submanifolds
and its Related Topics
Saga, August 4-6, 2012

SECTIONAL CURVATURES OF RULED REAL HYPERSURFACES IN A NONFLAT COMPLEX SPACE FORM

Hiromasa TANABE

Department of Science, Matsue College of Technology,
Matsue, Shimane 690-8518, Japan
E-mail: h-tanabe@matsue-ct.jp

Sadahiro MAEDA

Department of Mathematics, Saga University,
1 Honzyo, Saga 840-8502, Japan
E-mail: smaeda@ms.saga-u.ac.jp

In this note we give sharp inequalities for the sectional curvatures of ruled real hypersurfaces in a nonflat complex space form $\widetilde{M}_n(c)$ $(n \geq 2)$.

Keywords: nonflat complex space forms, ruled real hypersurfaces, sectional curvatures.

1. Introduction

Let $\widetilde{M}_n(c)$ be an n-dimensional nonflat complex space form, that is to say, $\widetilde{M}_n(c)$ is either a complex projective space $\mathbb{C}P^n(c)$ of constant holomorphic sectional curvature $c(> 0)$ or a complex hyperbolic space $\mathbb{C}H^n(c)$ of constant holomorphic sectional curvature $c(< 0)$. The class of ruled real hypersurfaces in $\widetilde{M}_n(c)$ is one of fundamental objects in the geometry of real hypersurfaces in a nonflat complex space form. For example, they are typical examples of non-Hopf hypersurfaces (for Hopf hypersurfaces, see [5], for instance).

Adachi, Kim and the second author ([4]) investigated the sectional curvatures of some Hopf hypersurfaces in $\widetilde{M}_n(c)$. On the other hand, in classical differential geometry, it is well-known that every ruled surface in 3-dimensional Euclidean space $\mathbb{R}^3$ has nonpositive Gaussian curvature. In this note, we compute the sectional curvature of every ruled real hypersurface in a nonflat complex space form $\widetilde{M}_n(c)\,(n \geq 2)$.

2. Ruled real hypersurfaces in $\widetilde{M}_n(c)$

Let M^{2n-1} be a real hypersurface isometrically immersed into an n-dimensional nonflat complex space form $\widetilde{M}_n(c)\,(n \geq 2)$ of constant holomorphic sectional curvature c and $\mathcal{N}$ be a unit normal local vector field on M^{2n-1} in $\widetilde{M}_n(c)$. Then the Riemannian connections $\widetilde{\nabla}$ of $\widetilde{M}_n(c)$ and ∇ of M^{2n-1} are related by $\widetilde{\nabla}_X Y = \nabla_X Y + g(AX, Y)\mathcal{N}$ and $\widetilde{\nabla}_X \mathcal{N} = -AX$ for vector fields X and Y tangent to M^{2n-1}, where g denotes the Riemannian metric on M^{2n-1} induced from the standard metric on $\widetilde{M}_n(c)$, and A is the shape operator of M in $\widetilde{M}_n(c)$. It is known that M^{2n-1} admits an almost contact metric structure induced by the Kähler structure (J, g) of the ambient space $\widetilde{M}_n(c)$. That is, we have a quadruple (ϕ, ξ, η, g) defined by

$$\xi = -J\mathcal{N}, \quad \eta(X) = g(X, \xi) = g(JX, \mathcal{N}) \quad \text{and} \quad \phi X = JX - \eta(X)\mathcal{N}$$

for each tangent vector $X \in TM^{2n-1}$. Then the structure satisfies

$$\phi^2 X = -X + \eta(X)\xi, \quad g(\phi X, \phi Y) = g(X, Y) - \eta(X)\eta(Y), \qquad (2.1)$$
$$\eta(\xi) = 1, \quad \phi\xi = 0 \quad \text{and} \quad \eta(\phi X) = 0$$

for all vectors $X, Y \in TM^{2n-1}$. We call ξ *the characteristic vector field* on M^{2n-1}.

Now, we recall the definition of ruled real hypersurfaces in a nonflat complex space form. A real hypersurface M^{2n-1} in $\widetilde{M}_n(c)\,(n \geq 2)$ is said to be *ruled* if the *holomorphic distribution* T^0 defined by $T^0(x) = \{X \in T_x M^{2n-1} | X \perp \xi\}$ for $x \in M^{2n-1}$ is integrable and each of its leaves (i.e., maximal integral manifolds) is a totally geodesic complex hypersurface $\widetilde{M}_{n-1}(c)$ of $\widetilde{M}_n(c)$. A ruled real hypersurface in $\widetilde{M}_n(c)$ is constructed in the following way. Given an arbitrary regular real curve $\gamma : I \to \widetilde{M}_n(c)$ defined on some open interval $I (\subset \mathbb{R})$, we attach at each point $\gamma(s)\,(s \in I)$ a totally geodesic complex hypersurface $\widetilde{M}_{n-1}^{(s)}(c)$ that is orthogonal to the plane spanned by $\dot{\gamma}(s)$ and $J\dot{\gamma}(s)$. Then we see that $M^{2n-1} = \bigcup_{s \in I} \widetilde{M}_{n-1}^{(s)}(c)$ is a ruled real hypersurface in $\widetilde{M}_n(c)$.

For a real hypersurface M^{2n-1} we define two functions $\mu, \nu : M^{2n-1} \to \mathbb{R}$ by $\mu = g(A\xi, \xi)$ and $\nu = \|A\xi - \mu\xi\|$, where A denotes the shape operator A of M^{2n-1}. The following lemma gives a characterization of ruled real hypersurfaces by the property of their shape operators.

Lemma. *Let M^{2n-1} be a real hypersurface in a nonflat complex space form $\widetilde{M}_n(c)\,(n \geq 2)$ through an isometric immersion. Then the following three conditions are mutually equivalent.*

(1) *M^{2n-1} is a ruled real hypersurface.*
(2) *The set $M_1 = \{x \in M^{2n-1} | \nu(x) \neq 0\}$ is an open dense subset of M^{2n-1} and there exists a unit vector field U on M_1 such that it is orthogonal to ξ and satisfies that*

$$A\xi = \mu\xi + \nu U, \quad AU = \nu\xi \quad and \quad AX = 0$$

   for an arbitrary tangent vector X orthogonal to both ξ and U.
(3) *The shape operator A of M^{2n-1} satisfies $g(AX, Y) = 0$ for any tangent vectors $X, Y \in T_x M^{2n-1}$ orthogonal to ξ at each point $x \in M^{2n-1}$.*

3. Main result

Let K be the sectional curvature of a real hypersurface M^{2n-1} in a nonflat complex space form $\widetilde{M}_n(c)\,(n \geq 2)$. That is, we define K by $K(X, Y) = g(R(X, Y)Y, X)$ for orthonormal tangent vectors X, Y on M^{2n-1} with the curvature tensor R of M^{2n-1}. It is known that the equation of Gauss can be written as

$$\begin{aligned}
g(R(X,Y)Z, W) =& \frac{c}{4}\{g(Y, Z)g(X, W) - g(X, Z)g(Y, W)\\
&+ g(\phi Y, Z)g(\phi X, W) - g(\phi X, Z)g(\phi Y, W)\\
&- 2g(\phi X, Y)g(\phi Z, W)\}\\
&+ g(AY, Z)g(AX, W) - g(AX, Z)g(AY, W)
\end{aligned}$$

for vector fields X, Y, Z, W on M^{2n-1}, so that

$$K(X, Y) = \frac{c}{4}\left\{1 + 3g(\phi X, Y)^2\right\} + g(AX, X)g(AY, Y) - g(AX, Y)^2. \quad (3.1)$$

We shall prove

Theorem. *Let M^{2n-1} be a ruled real hypersurface in a nonflat complex space form $\widetilde{M}_n(c)\,(n \geq 2)$. We define a function $\nu : M^{2n-1} \to \mathbb{R}$ by $\nu = \|A\xi - \mu\xi\|$, where $\mu = g(A\xi, \xi)$ and A is the shape operator of M^{2n-1} in $\widetilde{M}_n(c)$. Then the sectional curvature K of M^{2n-1} satisfies the following:*

(1) *When $c > 0$, we have $\dfrac{c}{4} - \nu^2 \leq K \leq c$;*
(2) *When $c < 0$, we have $\min\left\{c, \dfrac{c}{4} - \nu^2\right\} \leq K \leq \dfrac{c}{4}$.*

Remark 3.1. These estimates are sharp in the sense that at each point of M^{2n-1} we can take a pair (X, Y) of orthonormal tangent vectors satisfying $K(X, Y) = k$ for given k with $K_{\min} \leq k \leq K_{\max}$.

Remark 3.2. By definition, every leaf of the holomorphic distribution T^0 of a ruled real hypersurface M^{2n-1} in $\widetilde{M}_n(c)$ is a totally geodesic complex hypersurface $\widetilde{M}_{n-1}(c)$ of $\widetilde{M}_n(c)$. Thus we see for orthonormal vectors $X, Y \in T^0(x)$ at each point $x \in M$ that $|c/4| \leq |K(X, Y)| \leq |c|$ when $n \geq 3$ and that $|K(X, Y)| = |c|$ when $n = 2$.

Proof of theorem. For an arbitrary 2-plane P in the tangent space $T_x M^{2n-1}$ at a point $x \in M^{2n-1}$, there exist an orthonormal pair of vectors $X, Y \in T^0(x)$ and a real number t such that vectors $\cos t \cdot \xi + \sin t \cdot X$ and Y form the orthonormal basis of the plane P. So, we only have to evaluate $K(\cos t \cdot \xi + \sin t \cdot X, Y)$. By using (2.1), (3.1) and (2), (3) of Lemma, we obtain

$$K(\cos t \cdot \xi + \sin t \cdot X, Y) = \frac{c}{4}\left\{1 + 3\sin^2 t\, g(\phi X, Y)^2\right\} - \nu^2 \cos^2 t\, g(Y, U)^2 \tag{3.2}$$

in the following way:

$$
\begin{aligned}
K(&\cos t \cdot \xi + \sin t \cdot X, Y) \\
&= \frac{c}{4}\left\{1 + 3g(\phi(\cos t \cdot \xi + \sin t \cdot X), Y)^2\right\} \\
&\quad + g(A(\cos t \cdot \xi + \sin t \cdot X), (\cos t \cdot \xi + \sin t \cdot X))g(AY, Y) \\
&\quad - g(A(\cos t \cdot \xi + \sin t \cdot X), Y)^2 \\
&= \frac{c}{4}\left\{1 + 3\sin^2 t\, g(\phi X, Y)^2\right\} - \cos^2 t\, g(A\xi, Y)^2 \\
&= \frac{c}{4}\left\{1 + 3\sin^2 t\, g(\phi X, Y)^2\right\} - \cos^2 t\, g(\mu\xi + \nu U, Y)^2 \\
&= \frac{c}{4}\left\{1 + 3\sin^2 t\, g(\phi X, Y)^2\right\} - \nu^2 \cos^2 t\, g(Y, U)^2.
\end{aligned}
$$

When $c > 0$, we find that the expression of (3.2) takes its maximum value c at $t = \pi/2$ and $X = \phi Y$, that is, for example, in the case that a plane section P is a ϕ-section spanned by U and ϕU. Also, it takes the minimum value $(c/4) - \nu(x)^2$ at $t = 0, Y - U$, that is, in the case that the plane P is spanned by U and ξ.

When $c < 0$, it follows from (3.2) that

$$K(\cos t \cdot \xi + \sin t \cdot X, Y) \geq \frac{c}{4}(1 + 3\sin^2 t) - \nu^2 \cos^2 t$$

$$= \frac{c}{4} - \nu^2 + \left(\frac{3c}{4} + \nu^2\right)\sin^2 t.$$

So, if $(3c/4) + \nu(x)^2 \geq 0$, the sectional curvature K takes its minimum value $(c/4) - \nu(x)^2$ at $t = 0, Y = U, X = \phi U$. If $(3c/4) + \nu(x)^2 < 0$, its minimum value is $(c/4) - \nu(x)^2 + \{(3c/4) + \nu(x)^2\} = c$ (at $t = \pi/2, Y = U, X = \phi U$). Hence we have $\min \{c, (c/4) - \nu(x)^2\} \leq K \leq c/4$. Obviously, the maximum value of K is $c/4$ from (3.2). Thus the proof of Theorem is finished. $\square$

Remark 3.3. For a ruled real hypersurface M^{2n-1} in a nonflat complex space form $\widetilde{M}_n(c)$ $(n \geq 2)$, it is known that every integral curve of ϕU is a geodesic on M^{2n-1} and the function ν satisfies the differential equation

$$\phi U \nu = \nu^2 + (c/4)$$

([2, 3]). Therefore, along an arbitrary integral curve γ of ϕU the function ν is given as follows:

(1) When $c > 0$,

$$\nu(\gamma(s))^2 = (c/4) \tan^2 \left((\sqrt{c}/2)s + a \right)$$

with some constant $a = a_\gamma$;

(2) When $c < 0$,

$$\nu(\gamma(s))^2 = \begin{cases} (|c|/4) \tanh^2 \left((\sqrt{|c|}/2)s + a \right) & \text{if} \quad \nu(\gamma(0)) < \sqrt{|c|}/2, \\ |c|/4 & \text{if} \quad \nu(\gamma(0)) = \sqrt{|c|}/2, \\ (|c|/4) \coth^2 \left((\sqrt{|c|}/2)s + a \right) & \text{if} \quad \nu(\gamma(0)) > \sqrt{|c|}/2 \end{cases}$$

with some constant $a = a_\gamma$.

As an immediate consequence of these facts, we obtain

Corollary. *Let M^{2n-1} be a ruled real hypersurface in a nonflat complex space form $\widetilde{M}_n(c)$ $(n \geq 2)$. Then, on an integral curve $\gamma(s)$ of ϕU, which is a geodesic on M^{2n-1}, the sectional curvature K of M^{2n-1} satisfies the following inequalities:*

(1) *When $c > 0$, we have $-\infty < K \leq c$;*
(2) *When $c < 0$, we have*

$$\begin{cases} c \leq K \leq \dfrac{c}{4} & \text{if} \quad \nu(\gamma(0)) \leq \sqrt{|c|}/2, \\ -\infty < K \leq \dfrac{c}{4} & \text{if} \quad \nu(\gamma(0)) > \sqrt{|c|}/2. \end{cases}$$

In [1], Adachi, Bao and the second author studied congruence classes of minimal ruled real hypersurfaces in a nonflat complex space form with respect to the action of its isometry group. They showed that those in a complex projective space $\mathbb{C}P^n(c)$ $(n \geq 2)$ are congruent to each other and that those in a complex hyperbolic space $\mathbb{C}H^n(c)$ $(n \geq 2)$ are classified into three classes, which are called axial, parabolic and elliptic. By virtue of their work, corresponding to the classification of minimal ruled real hypersurfaces in $\mathbb{C}H^n(c)$, we obtain the following results: (1) On minimal ruled real hypersurfaces of axial type, the function ν satisfies $\nu < \sqrt{|c|}/2$, so that we have $c \leq K \leq c/4$; (2) On those of parabolic type, the function ν satisfies $\nu = \sqrt{|c|}/2$, so that $c \leq K \leq c/4$; (3) On those of elliptic type, the function ν satisfies $\nu > \sqrt{|c|}/2$, so that $-\infty < K \leq c/4$.

References

1. T. Adachi, T. Bao and S. Maeda, *Congruence classes of minimal ruled real hypersurfaces in a nonflat complex space form*, to appear in Hokkaido Math. J.
2. M. Kimura, *Sectional curvatures of holomorphic planes on a real hypersurface in $P^n(\mathbb{C})$*, Math. Ann. **276** (1987), 487–497.
3. S. Maeda, T. Adachi and Y. H. Kim, *A characterization of the homogeneous minimal ruled real hypersurface in a complex hyperbolic space*, J. Math. Soc. Japan **61** (2009), 315–325.
4. S. Maeda, T. Adachi and Y. H. Kim, *Characterizations of geodesic hyperspheres in a nonflat complex space form*, Glasgow Math. J. **55** (2013), 217–227.
5. R. Niebergall and P.J. Ryan, *Real hypersurfaces in complex space forms*, Tight and taut submanifolds (T.E. Cecil and S.S. Chern, eds.), Cambridge Univ. Press, 1998, 233–305.

Received January 7, 2013.
Revised March 2, 2013.

Proceedings of the Workshop on
Differential Geometry of Submanifolds
and its Related Topics
Saga, August 4-6, 2012

TOTALLY GEODESIC KÄHLER IMMERSIONS INTO A COMPLEX SPACE FORM, AND A NON-EXISTENCE THEOREM FOR HESSIAN METRICS OF POSITIVE CONSTANT HESSIAN SECTIONAL CURVATURE

Tomonori NODA

Department of Mathematics, Osaka Dental University,
8-1 Kuzuhahanazono-cho, Hirakata-shi, Osaka 573-1121, Japan
E-mail: noda-t@cc.osaka-dent.ac.jp

Nobutaka BOUMUKI

Department of Mathematics, Tokyo University of Science,
1-3 Kagurazaka, Shinjuku-ku, Tokyo 162-8601, Japan
E-mail: boumuki@rs.tus.ac.jp

Dedicated to Professor Sadahiro Maeda for his sixtieth birthday

The main purpose of this paper is to demonstrate that there exists no simply connected Hessian manifold (M, ∇, g) of constant Hessian sectional curvature c in the case where (i) ∇ is complete and (ii) $c > 0$. We achieve the purpose by considering totally geodesic Kähler immersions into a complex space form.

Keywords: Hessian manifold, Hessian metric of positive constant Hessian sectional curvature, complete affine connection, Kähler immersion, complex space form, Kähler metric of negative constant holomorphic sectional curvature.

1. Introduction and the main result

A smooth manifold M with a flat, torsion-free affine connection ∇ is said to be *Hessian*, if there exists a Riemannian metric g on M such that ∇g is a symmetric tensor field of type $(0,3)$, where ∇g is defined as follows:

$$(\nabla g)(Y, Z; X) := X(g(Y, Z)) - g(\nabla_X Y, Z) - g(Y, \nabla_X Z)$$

for $X, Y, Z \in \mathfrak{X}(M)$. Such a metric g is called a *Hessian metric* on (M, ∇). A Hessian manifold (M, ∇, g) is said to be of *constant Hessian sectional curvature* c, if

$$(\nabla\gamma)(Y, Z; X) = \frac{c}{2}\{g(X, Y)Z + g(X, Z)Y\} \text{ for all } X, Y, Z \in \mathfrak{X}(M),$$

where $\gamma := \nabla^g - \nabla$ and ∇^g denotes the Levi-Civita connection on (M, g).

The main purpose of this paper is to demonstrate the following non-existence theorem:

Theorem 1.1. *Let M be a simply connected manifold with a flat, torsion-free affine connection ∇. Suppose that ∇ is complete. Then, (M, ∇) admits no Hessian metrics of positive constant Hessian sectional curvature.*

Theorem 1.1 implies

Corollary 1.1. *There exists no simply connected Hessian manifold (M, ∇, g) of constant Hessian sectional curvature c in the case where (i) ∇ is complete and (ii) $c > 0$.*

Remark that Corollary 1.1 is no longer true if one omits either the condition (i) or (ii) (see Section 4).

This paper is organized as follows:

§2 Totally geodesic Kähler immersions into a complex space form
   We study totally geodesic Kähler immersions into a complex space form and conclude Proposition 2.1, which implies that a complex Euclidean space $(\mathbb{C}^n, J_0)$ admits no Kähler metrics of negative constant holomorphic sectional curvature. Proposition 2.1 will play an important role in Subsection 3.2.

§3 Proof of Theorem 1.1
   This section consists of two subsections. In Subsection 3.1 we prepare two Propositions 3.1 and 3.2 for the proof of Theorem 1.1. In Subsection 3.2 we devote ourselves to proving Theorem 1.1.

§4 Remarks
   By referring to the book of Shima [8], we see that Corollary 1.1 is no longer true if one omits either the condition (i) or (ii).

Notation

Throughout this paper, we use the following notation (n1)—(n4):

(n1) $\mathfrak{X}(M)$: the set of smooth vector fields on M,
(n2) T_pM : the tangent space of M at p,
(n3) TM : the tangent bundle over M,
(n4) $f|_A$: the restriction of a mapping f to a subset A.

Acknowledgments

The authors would like to express gratitude to Professors Hitoshi Furuhata and Takashi Kurose for nice advice.

2. Totally geodesic Kähler immersions into a complex space form

By $(\mathbb{M}^n, J_M, g_M)$ we denote an n-dimensional complex space form, that is, $(\mathbb{M}^n, J_M, g_M)$ is one of the following (1), (2) and (3):

(1) A complex Euclidean space $\mathbb{C}^n$ equipped with the standard complex structure J_0 and the standard metric $\langle\,,\,\rangle$.

(2) A complex projective space $\mathbb{P}^n$ equipped with the standard complex structure J_P (see Kobayashi-Nomizu [6, p.134] for J_P) and the Fubini-Study metric g_P.

(3) An open unit ball $\mathbb{B}^n$ (centered at $0 \in \mathbb{C}^n$) equipped with $J_B := J_0|_{\mathbb{B}^n}$ and the Bergman metric g_B.

In this section we first deal with a rigidity problem (cf. Lemma 2.1) and an extension problem (cf. Lemma 2.2) about totally geodesic Kähler immersions into $(\mathbb{M}^n, J_M, g_M)$, and afterwards obtain Proposition 2.1 from Lemma 2.2. Throughout this paper, we mean a holomorphic isometric immersion by a *Kähler immersion*.

We are going to prove two Lemmas 2.1 and 2.2, and Proposition 2.1. Before proving them, we need to remark the following:

Remark 2.1.

(i) Calabi [1] has already asserted more excellent statements than our statements on Lemmas 2.1 and 2.2.

(ii) Proposition 2.1 is immediate from the result of Kobayashi [4].

Here, the authors are pointed out by Professors Hitoshi Furuhata and Takashi Kurose that Proposition 2.1 is a special case of a more general statement in Kobayashi [4].

From now on, let us prove

Lemma 2.1. *Let (V, J, g) be a connected Kähler manifold, and let $f_i : (V, J, g) \to (\mathbb{M}^n, J_M, g_M)$ be totally geodesic Kähler immersions $(i = 1, 2)$. Then, there exists a holomorphic isometry ϕ of $(\mathbb{M}^n, J_M, g_M)$ satisfying $\phi \circ f_1 = f_2$.*

Proof. Denote by G the holomorphic isometry group of $(\mathbb{M}^n, J_M, g_M)$. Fix any $o \in V$. Since G is transitive on $\mathbb{M}^n$, there exists a $\psi \in G$ satisfying $\psi(f_1(o)) = f_2(o)$. For the reason we assume that $f_1(o) = f_2(o)$ from the beginning; and set $\bar{o} := f_1(o) = f_2(o)$. First, let us show that

$$\text{there exists an } A \in U(n)_{\bar{o}} \text{ such that } A \circ (df_1)_o = (df_2)_o, \qquad (1)$$

where $U(n)_{\bar{o}}$ denotes the unitary group of the vector space $(T_{\bar{o}}\mathbb{M}^n, J_M, g_M)$. Take an arbitrary orthonormal basis $\{e_a, J(e_a)\}_{a=1}^k$ of (T_oV, g), where $k := \dim_{\mathbb{C}} V$. In this case, $\{df_1(e_a), J_M(df_1(e_a))\}_{a=1}^k$ must be orthonormal in $(T_{\bar{o}}\mathbb{M}^n, g_M)$ because f_1 is a Kähler immersion. Hence, we can choose $n - k$ vectors $\bar{v}_b \in T_{\bar{o}}\mathbb{M}^n$ such that $\{df_1(e_a), J_M(df_1(e_a))\}_{a=1}^k \bigcup \{\bar{v}_b, J_M(\bar{v}_b)\}_{b=1}^{n-k}$ is an orthonormal basis of $(T_{\bar{o}}\mathbb{M}^n, g_M)$. Similarly, one can choose $\bar{w}_b \in T_{\bar{o}}\mathbb{M}^n$ such that $\{df_2(e_a), J_M(df_2(e_a))\}_{a=1}^k \bigcup \{\bar{w}_b, J_M(\bar{w}_b)\}_{b=1}^{n-k}$ becomes an orthonormal basis of $(T_{\bar{o}}\mathbb{M}^n, g_M)$. Consequently one has (1), if we define a linear isomorphism $A : T_{\bar{o}}\mathbb{M}^n \to T_{\bar{o}}\mathbb{M}^n$ by

$$A(df_1(e_a)) := df_2(e_a), \ A(J_M(df_1(e_a))) := J_M(df_2(e_a)), \ A(\bar{v}_b) := \bar{w}_b,$$
$$A(J_M(\bar{v}_b)) := J_M(\bar{w}_b).$$

The linear isotropy subgroup at $\bar{o} \in \mathbb{M}^n$ of G may be considered as the unitary group $U(n)$. So, it follows from (1) that there exists a $\phi \in G$ satisfying $\phi(\bar{o}) = \bar{o}$ and $(d\phi)_{\bar{o}} = A$. Then

$$\phi(f_1(o)) = \bar{o} = f_2(o), \ d(\phi \circ f_1)_o = A \circ (df_1)_o = (df_2)_o.$$

Since V is connected, the rest of the proof is to verify that the following subset $V' \subset V$ is non-empty, open and closed:

$$V' := \{\, p \in V \mid \phi(f_1(p)) = f_2(p), \ d(\phi \circ f_1)_p = (df_2)_p \,\}.$$

It is natural that V' is closed and $o \in V'$. Thus we may only verify that V' is open. Take any $p \in V'$ and set $\bar{p} := \phi(f_1(p)) = f_2(p)$. Let O and $\bar{O}$ be regular neighborhoods of $p \in V$ and $\bar{p} \in \mathbb{M}^n$ such that $\phi(f_1(O)), f_2(O) \subset \bar{O}$. Since $\phi \circ f_1$ and f_2 map each geodesic into a geodesic, they can be expressed in O by

$$\phi \circ f_1 = \mathrm{Exp}_{\bar{p}} \circ d(\phi \circ f_1)_p \circ \mathrm{Exp}_p^{-1}, \ f_2 = \mathrm{Exp}_{\bar{p}} \cup (df_2)_p \cap \mathrm{Exp}_p^{-1},$$

respectively. This yields $\phi \circ f_1 \equiv f_2$ on O, and $p \in O \subset V'$. Hence V' is open in V. $\qquad \square$

By virtue of Lemma 2.1 we can deduce

Lemma 2.2. *Let (N, J, g) be a simply connected, Kähler manifold with a base point $x \in N$. Suppose that for every $p \in N$, there exist a convex neighborhood V_p of p and a totally geodesic Kähler immersion $f_p : (V_p, J, g) \to (\mathbb{M}^n, J_M, g_M)$. Then, there exists a totally geodesic Kähler immersion $\widetilde{f} : (N, J, g) \to (\mathbb{M}^n, J_M, g_M)$ such that $\widetilde{f}|_{V_x} = f_x$.*

Proof. Take a curve $\tau(t) : [0, 1] \to N$ joining x to any $y \in N$. Then, there exists a finite sequence of convex neighborhoods $V_1, V_2, \ldots, V_m$ such that $\tau([0, 1]) \subset \bigcup_{i=1}^{m} V_i$, $x \in V_1$, $y \in V_m$ and $V_i \cap V_{i+1} \neq \emptyset$ for $1 \le i \le m - 1$. Let us extend the immersion $f_1 : (V_1, J, g) \to (\mathbb{M}^n, J_M, g_M)$ to a totally geodesic Kähler immersion $\widetilde{f} : (\bigcup_{i=1}^{m} V_i, J, g) \to (\mathbb{M}^n, J_M, g_M)$. On the one hand, the intersection $V_i \cap V_{i+1}$ is convex, and so it is connected. On the other hand, both f_i and f_{i+1} are totally geodesic Kähler immersions of $(V_i \cap V_{i+1}, J, g)$ into $(\mathbb{M}^n, J_M, g_M)$. Therefore Lemma 2.1 assures that there exists a holomorphic isometry $\phi_{i,i+1}$ of $(\mathbb{M}^n, J_M, g_M)$ satisfying $\phi_{i,i+1} \circ f_{i+1} \equiv f_i$ on $V_i \cap V_{i+1}$. Hence the immersion $f_i : V_i \to \mathbb{M}^n$ can be extended to $\widetilde{f}_{i,i+1} : V_i \cup V_{i+1} \to \mathbb{M}^n$ as follows:

$$\widetilde{f}_{i,i+1}(z) := \begin{cases} f_i(z) & \text{if } z \in V_i, \\ \phi_{i,i+1}(f_{i+1}(z)) & \text{if } z \in V_{i+1}. \end{cases}$$

By induction on i, we can extend $f_1 : V_1 \to \mathbb{M}^n$ to a totally geodesic Kähler immersion $\widetilde{f} : \bigcup_{i=1}^{m} V_i \to \mathbb{M}^n$. Since N is simply connected, it follows from the monodromy principle that the extension, as indicated above, does not depend on the choice of the curve $\tau(t)$. □

Now, we are in a position to demonstrate

Proposition 2.1. *A complex Euclidean space $(\mathbb{C}^n, J_0)$ admits no Kähler metrics of negative constant holomorphic sectional curvature.*

Remark 2.2. (i) Proposition 2.1 does not hold, if one replaces J_0 with another complex structure $\hat{J}$. Indeed; let us consider a diffeomorphism $f : \mathbb{C}^1 = \mathbb{R}^2 \to H^2$, $(x, y) \mapsto (x, e^y)$. Then one can get a Kähler metric $\hat{g}$ of negative constant holomorphic sectional curvature on $(\mathbb{C}^1, \hat{J})$ by setting

$$\hat{g} := f^* g_H, \ \hat{J} := df^{-1} \circ J_H \circ df.$$

Here $J_H := J_0|_{H^2}$, we denote by H^2 the upper half plane and by g_H the Poincaré metric on H^2. (ii) Needless to say, $(\mathbb{C}^n, J_0)$ can admit Kähler metrics of non-negative constant holomorphic sectional curvature.

Proof of Proposition 2.1. We use proof by contradiction. Suppose that there exists a Kähler metric g of constant holomorphic sectional curvature $\bar{c} < 0$ on $(\mathbb{C}^n, J_0)$. First, let us confirm that for every $p \in \mathbb{C}^n$,

there exist open subsets $V \subset \mathbb{C}^n$ with $p \in V$, $U \subset \mathbb{B}^n$ and a Kähler isomorphism $f : (V, J_0, g') \to (U, J_B, g_B)$, $\qquad(2)$

where $g' := \lambda \cdot g$ and $\lambda := -\bar{c}(n+1)/2$. Note that the above f is a totally geodesic Kähler immersion of (V, J_0, g') into $(\mathbb{B}^n, J_B, g_B)$, and that both metrics g' and g_B are of constant holomorphic sectional curvature $-2/(n+1)$. Now, fix any $\bar{p} \in \mathbb{B}^n$. Let F be a linear isomorphism of $T_p\mathbb{C}^n$ onto $T_{\bar{p}}\mathbb{B}^n$ satisfying

$$F \circ (J_0)_p = (J_B)_{\bar{p}} \circ F, \ g'_p(u, v) = (g_B)_{\bar{p}}\big(F(u), F(v)\big)$$

for all $u, v \in T_p\mathbb{C}^n$. We denote by $T^{g'}$ and T^{g_B} (resp. $R^{g'}$ and R^{g_B}) the torsion (resp. curvature) tensors with respect to the Levi-Civita connections $\nabla^{g'}$ and ∇^{g_B}, respectively. Proposition 7.6 [6, p.169] assures that F maps $R_p^{g'}$ into $R_{\bar{p}}^{g_B}$; and therefore Theorem 7.4 [5, p.261–262] implies that there exist open subsets $V \subset \mathbb{C}^n$, $U \subset \mathbb{B}^n$ and an affine isomorphism f of $(V, \nabla^{g'})$ onto (U, ∇^{g_B}) such that $p \in V$, $\bar{p} \in U$ and

$$f(p) = \bar{p}, \ (df)_p = F$$

because $T^{g'} = T^{g_B} = 0$ and $\nabla^{g'} R^{g'} = \nabla^{g_B} R^{g_B} = 0$. From now on, let us show that $f : V \to U$ is Kähler. One may assume that V is convex by taking a small subset of V (if necessary). Let q be any point of V, and let τ be a curve in V from p to q. Set $\bar{q} := f(q)$ and $\bar{\tau} := f \circ \tau$. Since the parallel displacement along $\bar{\tau}$ corresponds to that of along τ under f and since $\nabla^{g'} J_0 = \nabla^{g'} g' = 0$ and $\nabla^{g_B} J_B = \nabla^{g_B} g_B = 0$, the affine isomorphism f maps the complex structure and the metric tensor on V into those of U. This means that $f : V \to U$ is Kähler, and thus (2) follows. Lemma 2.2, together with (2), enables us to have a totally geodesic Kähler immersion $\widetilde{f} : (\mathbb{C}^n, J_0, g') \to (\mathbb{B}^n, J_B, g_B)$ such that $\widetilde{f}|_V = f$. In particular, $\widetilde{f}$ is a holomorphic mapping of the whole $(\mathbb{C}^n, J_0)$ into the bounded domain $(\mathbb{B}^n, J_B)$; and therefore it must be a constant mapping, which is a contradiction. $\square$

3. Proof of Theorem 1.1

3.1. *Preliminary to the proof of Theorem 1.1*

Our aim in this subsection is to prepare two Propositions 3.1 and 3.2 for the proof of Theorem 1.1.

Proposition 3.1. *Let M be a simply connected manifold of $\dim_{\mathbb{R}} M = n$, and let ∇ be a flat, torsion-free affine connection on M. Suppose that ∇ is complete. Then, there exists an affine isomorphism f of $(\mathbb{R}^n, D)$ onto (M, ∇). Here D denotes the Levi-Civita connection on $(\mathbb{R}^n, \langle\,,\,\rangle)$.*

Proof. Fix any $x \in \mathbb{R}^n$ and $p \in M$. Any linear isomorphism $F : T_x \mathbb{R}^n \to T_p M$ satisfies

$$F\big(T_x^D(u,v)\big) = 0 = T_p^{\nabla}\big(F(u), F(v)\big),$$
$$F\big(R_x^D(u,v)w\big) = 0 = R_p^{\nabla}\big(F(u), F(v)\big)F(w)$$

for all $u, v, w \in T_x \mathbb{R}^n$, where T^D and T^{∇} (resp. R^D and R^{∇}) denote the torsion (resp. curvature) tensors with respect to D and ∇, respectively. Hence, Theorem 7.8 [5, p.265] implies that there exists a unique affine isomorphism f of $(\mathbb{R}^n, D)$ onto (M, ∇) satisfying $f(x) = p$ and $(df)_x = F$, because both $\mathbb{R}^n$ and M are simply connected and both D and ∇ are complete. □

Shima [8] allows us to assert the following statement, where D is the same notation as that in Proposition 3.1:

Proposition 3.2. *Let h be a Riemannian metric on $\mathbb{R}^n$. Then, the following* (i) *and* (ii) *are equivalent:*

(i) $(\mathbb{R}^n, D, h)$ *is a Hessian manifold of constant Hessian sectional curvature c;*

(ii) $(\mathbb{C}^n, J_0, h^T)$ *is a Kähler manifold of constant holomorphic sectional curvature $-c$ (see Shima [8, (2.3), p.20] for h^T).*

Here we identify $T\mathbb{R}^n$ with $\mathbb{C}^n$ in the usual way.

Proof. Refer to Proposition 2.6 and Corollary 3.1 in Shima [8, p.20–21, p.44], and read the statements on page [8, p.20] for J_0. □

3.2. *Proof of Theorem* 1.1

Now, we are in a position to prove Theorem 1.1.

Proof of Theorem 1.1. Let us use proof by contradiction. Suppose that there exists a Hessian metric g of positive constant Hessian sectional curvature on (M, ∇). By Proposition 3.1 there exists an affine isomorphism f of $(\mathbb{R}^n, D)$ onto (M, ∇), where $n = \dim_{\mathbb{R}} M$. Hence, $h := f^* g$ becomes a

Hessian metric of positive constant Hessian sectional curvature on $(\mathbb{R}^n, D)$. Proposition 3.2 assures that h^T is a Kähler metric of negative constant holomorphic sectional curvature on $(\mathbb{C}^n, J_0)$. That contradicts Proposition 2.1. $\square$

4. Remarks

By referring to the book of Shima [8], we see that Corollary 1.1 is no longer true if one omits either the condition (i) or (ii),

$$(i) \; \nabla \text{ is complete, (ii) } c > 0.$$

Example 4.1 implies that Corollary 1.1 is no longer true if one omits the condition (i):

Example 4.1 (cf. Shima [8, p.47]). *The following* $(\Omega, D, g = Dd\varphi)$ *is a simply connected Hessian manifold of constant Hessian sectional curvature* $c > 0$ *and* D *is not complete:*

$$\Omega := \left\{ (x^1, \ldots, x^n) \in \mathbb{R}^n \; \middle| \; x^n > \frac{1}{2} \sum_{j=1}^{n-1} (x^j)^2 \right\},$$

$$\varphi := -\frac{1}{c} \log \left(x^n - \frac{1}{2} \sum_{j=1}^{n-1} (x^j)^2 \right).$$

Two Examples 4.2 and 4.3 imply that Corollary 1.1 is no longer true if one omits the condition (ii):

Example 4.2 (cf. Shima [8, p.46]). *The following* $(\Omega, D, g = Dd\varphi)$ *is a simply connected Hessian manifold of constant Hessian sectional curvature* $c = 0$ *and* D *is complete:*

$$\Omega := \mathbb{R}^n, \; \varphi := \frac{1}{2} \sum_{i=1}^{n} (x^i)^2.$$

Example 4.3 (cf. Shima [8, p.49]). *The following* $(\Omega, D, g = Dd\varphi)$ *is a simply connected Hessian manifold of constant Hessian sectional curvature* $c < 0$ *and* D *is complete:*

$$\Omega := \mathbb{R}^n, \; \varphi := -\frac{1}{c} \log \left(\sum_{i=1}^{n} e^{x^i} + 1 \right).$$

Appendix: A topic related to our paper

Furuhata-Kurose [3] deals with a classification problem about simply connected Hessian manifolds (M, ∇, g) of non-positive constant Hessian sectional curvature, where ∇ are not necessary complete (see Furuhata [2] and Kurose [7] also). It seems that by the paper [3] and our paper, one can classify simply connected Hessian manifolds (M, ∇, g) of constant Hessian sectional curvature in the case where ∇ are complete.

References

1. E. Calabi, *Isometric imbedding of complex manifolds*, Ann. of Math. (2) **58** (1953), 1–23.
2. H. Furuhata, *Spaces of nonpositive constant Hessian curvature*, http://www.math.sci.hokudai.ac.jp/~furuhata/research/doc/0911.pdf
3. H. Furuhata and T. Kurose, *Hessian manifolds of nonpositive constant Hessian sectional curvature*, Tohoku Math. J. (2) (to appear).
4. S. Kobayashi, *Invariant distances on complex manifolds and holomorphic mappings*, J. Math. Soc. Japan **19** (1967), 460–480.
5. S. Kobayashi and K. Nomizu, *Foundations of differential geometry I*, Interscience Publishers, New York-London, 1963.
6. S. Kobayashi and K. Nomizu, *Foundations of differential geometry II*, Interscience Publishers, New York-London-Sydney, 1969.
7. T. Kurose, *The classification of Hessian manifolds of constant Hessian sectional curvature* (in Japanese), RIMS Kokyuroku **1623**, (2009), 22–29.
8. H. Shima, *The geometry of Hessian structures*, World Scientific Publishing, Singapore, 2007.

Received January 8, 2013.

Proceedings of the Workshop on
Differential Geometry of Submanifolds
and its Related Topics
Saga, August 4-6, 2012

ARCHIMEDEAN THEOREMS AND W-CURVES

Dong-Soo KIM

Department of Mathematics, Chonnam National University,
Kwangju 500-757, Korea
E-mail: dosokim@chonnam.ac.kr

Young Ho KIM

Department of Mathematics, Kyungpook National University,
Taegu 702-701, Korea
E-mail: yhkim@knu.ac.kr

Dedicated to Professor Sadahiro Maeda for his sixtieth birthday

We introduce some properties of W-curves in Euclidean space with the
chord property and some properties on parabolas, spheres and paraboloids.
Archimedes found interesting geometric properties regarding parabolas and
paraboloids. We set up the converse problems for those obtained by
Archimedes.

Keywords: W-curve, parabola, paraboloid.

1. Introduction

In Euclidean 3-space, straight lines, circles and helices are the most natural
curves that can be explained in terms of the curvature and torsion. Generalizing such curves in an arbitrary dimensional Euclidean space, they are the
so-called W-curves whose Frenet curvatures are constant. More precisely,
let $c = c(t)$ be a Frenet curve of order d, that is, $c'(t) \wedge \cdots \wedge c^{(d)} \neq 0$ along
c. If c is a W-curve of order d, then all the curvatures $\kappa_1, \kappa_2, \cdots, \kappa_{d-1}$ are
constant along c. If it is compact, it is parameterized up to congruence by

$$c(t) = (\cos a_1 t, \sin a_1 t, \cdots, \cos a_d t, \sin a_d t)$$

for some non-zero real numbers $a_1, \cdots, a_d$. In case that it is of non-compact,

$$c(t) = (\cos a_1 t, \sin a_1 t, \cdots, \cos a_{d-1} t, \sin a_{d-1} t, bt + e).$$

for some non-zero real numbers $a_1, \cdots, a_{d-1}, b$. What can other geometric properties describe such curves?

In a plane, next to straight lines and circles, the most simple and interesting curves in a plane are parabolas and ellipses. They are also part of conic sections and can be treated in Affine Differential Geometry with constant affine curvature.

Some geometric properties of ellipses were studied by the present authors in terms of the curvature and the support function ([2]).

As was described in [1], a circle is characterized by the fact that the chord joining any two points on it meets the circle at the same angle, which is called the chord property.

This paper is a survey of some results on the W-curves, parabolas and paraboloids based on the present authors' works.

2. Chord property and W-curves

It is well-known that a circle is characterized as a closed plane curve such that the chord joining any two points on it meets the curve at the same angle at the two points. It can be interpreted as follows:

Chord Property. *Let $X = X(s)$ be a unit speed closed curve in the Euclidean plane $\mathbb{E}^2$ and $T(s) = X'(s)$ be its unit tangent vector field. Then $X = X(s)$ is a circle if and only if it satisfies Condition:*
(C): $\langle X(t) - X(s), T(t) - T(s) \rangle = 0$ holds identically.

Actually, one can show that a unit speed plane curve $X(s)$ satisfies Condition (C) if and only if it is either a circle or a straight line.

In views of the chord property, it is natural to ask the following question:

Question. "Which Euclidean space curves satisfy Condition (C)?"

B.-Y. Chen and the present authors proved

Theorem 2.1 ([1]). *A unit speed smooth curve $X = X(s)$ in the Euclidean m-space $\mathbb{E}^m$ is a W-curve if and only if the chord joining any two points on it meets the curve at the same angle.*

Theorem 2.2 ([1]). *A unit speed smooth curve $X = X(s)$ in the Euclidean 3-space $\mathbb{E}^3$ is a W-curve if and only if the difference of the values of cosine of the two angles between the curve and the chord joining any two given points on the curve depends only on the arc-length of the curve between the two points.*

They studied space curves satisfying a condition more general than (C), namely:

(A): $\langle X(s) - X(t), T(s) - T(t) \rangle$ *depends only on $s - t$, where $X = X(s)$ is a unit speed curve in Euclidean m-space $\mathbb{E}^m$ and $T(s) = X'(s)$.*

In the Euclidean 3-space $\mathbb{E}^3$, the chord property (C) and condition (A) are equivalent.

Theorem 2.3 ([1]). *A unit speed smooth curve $X = X(s)$ in the Euclidean 3-space $\mathbb{E}^3$ is a W-curve if and only if $X = X(s)$ satisfies Condition (A).*

3. Parabolas

Theorem 3.1 ([8]). *A curve X of class C^3 given by a function $f : \mathbb{R} \to \mathbb{R}$ is a parabola if it satisfies the following condition.*
(C_1) *For any two numbers x_1 and x_2, the pair of tangent lines to X at $x = x_1$ and at $x = x_2$ meet at $x = (x_1 + x_2)/2$.*

Proof. First, suppose that X satisfies (C_1). Then the tangent lines given by

$$
\begin{aligned}
y - f(x_1) &= f'(x_1)(x - x_1), \\
y - f(x_2) &= f'(x_2)(x - x_2)
\end{aligned}
\tag{1}
$$

have the point of intersection at $x = (x_1 + x_2)/2$. Hence we get

$$
2\{f(x_1) - f(x_2)\} = (x_1 - x_2)\{f'(x_1) + f'(x_2)\}.
\tag{2}
$$

Differentiating (2) with respect to x_1, we obtain

$$
f'(x_1) - f'(x_2) = (x_1 - x_2)f''(x_1).
\tag{3}
$$

Once more, we differentiate (3) with respect to x_1. Then we have

$$
(x_1 - x_2)f'''(x_1) = 0,
\tag{4}
$$

which shows that $f(x)$ is a quadratic polynomial. $\square$

In [8], it was shown that a curve X given by $y = f(x)$ is a parabola if it satisfies (C_1) and $f(x)$ is analytic. The present authors obtained the following theorems for parabolas ([5]).

Theorem 3.2. *A curve X of class C^2 given by $y = f(x)$ is a parabola if it satisfies the following: For any chord AB on X with $A = (x_1, y_1), B = (x_2, y_2)$, the tangent line to X.*

Theorem 3.3. *A convex curve X that for any point p on L there are two tangent lines of X through p which are perpendicular to each other, and the chord connecting the points of tangency passes through F.*

4. Surface property, spheres and paraboloids

Let $S^2(a)$ be a sphere of radius a in the Euclidean space $\mathbb{E}^3$. By an elementary calculus, one can see that if two parallel planes with distance h intersect $S^2(a)$, then the surface area of the section of $S^2(a)$ between the planes is $2\pi ah$. We call this property Condition (S) which is called the surface property ([3]).

(S): Let M be a surface in $\mathbb{E}^3$. For any two parallel planes with distance h both of which intersect the sphere, the surface area of the section of M between the planes is $2\pi ah$.

In fact, Archimedes proved Condition (S) for a sphere ([6], p.78). For a differential geometric proof, see Archimedes' Theorem ([7], pp.116-118).

Naturally, we can raise the following question: Are there any other surfaces which satisfy Condition (S)?

As an answer to this question, we have

Theorem 4.1 ([3]). *Let M be a closed and convex surface in the 3-dimensional Euclidean space $\mathbb{E}^3$. If M satisfies Condition (S), then it is a sphere $S^2(a)$ of radius a.*

To prove the theorem, they set up the co-area formula associated with Gaussian curvature of M at a point $p \in M$.

Let M be a closed and convex surface in the 3-dimensional Euclidean space $\mathbb{E}^3$. Then the Gaussian curvature K is non-negative. For a fixed point $p \in M$ and for a sufficiently small $h > 0$, consider a plane π_1 parallel to the tangent plane π_1 of M at p with distance h which intersects M. We denote by $M_p(h)$ the surface area between the two planes π_1 and π_2.

Without loss of generality, we may assume that the point p is the origin of the Cartesian coordinate system (x, y, z) of $\mathbb{E}^3$ and the tangent plane of M at p is $z = 0$, and $M = \text{graph(f)}$ for a non-negative convex function $f : R^2 \to R$. Then we have

$$M_p(h) = \iint_{f(X)<h} \sqrt{1 + |\nabla f|^2} dx dy, \tag{5}$$

where ∇f denotes the gradient vector of f.

It was proved in [3] that if the Gaussian curvature $K(p)$ of M at p is positive, then we have

$$M_p'(0) = \frac{2\pi}{\sqrt{K(p)}}. \tag{6}$$

Since M is closed, there exists a point p where $K(p) > 0$. Hence we see that $U = \{p \in M | K(p) > 0\}$ is nonempty. Together with Condition (C), (6) implies that at every point $p \in U$, $K(p) = \frac{1}{a}$. Thus continuity of K shows that $U = M$, and hence we have $K = \frac{1}{a}$ on M. This completes the proof.

A paraboloid in $\mathbb{E}^3$ has an interesting volume property: Consider a section of a paraboloid of rotation cut off by a plane not necessarily perpendicular to its axis. Let p be the point of contact of the tangent plane parallel to the base. The line through p, parallel to the axis of the paraboloid meets the base at a point v. Archimedes shows that the volume of the section is $3/2$ times the volume of the cone with the same base and vertex p ([7], Chapter 7 and Appendix A).

Proposition 4.1. *The volume of such a section of a paraboloid of rotation in the 3-dimensional Euclidean space $\mathbb{E}^3$ is proportional to $\overline{pv}^2$, where the ratio depends only on the paraboloid.*

For the converse problem, the authors proved

Theorem 4.2 ([3]). *Let M be a smooth convex hypersurface in the $(n+1)$-dimensional Euclidean space $\mathbb{E}^{n+1}$. Then M is an elliptic paraboloid if and only if there exists a line L for which M satisfies Condition:*
For any point p on M and any hyperplane section of M parallel to the tangent plane of M at p, let v denote the point where the line through p parallel to L meets the hyperplane. Then the volume of the section is a times $\overline{pv}^{(n+2)/2}$ for some constant a which depends only on the hypersurface M.

Let $S^n(a)$ be a hypersphere with radius a in the Euclidean space $\mathbb{E}^{n+1}$. Consider two parallel hyperplanes π_1 and π_2 intersecting the sphere with distance $t > 0$, where π_1 is tangent space of $S^n(a)$ at p.

We denote by $A_p(t), V_p(t)$ and $S_p(t)$ the n-dimensional area of the section in π_2 enclosed by $\pi_2 \cap S^n(a)$, the $(n+1)$-dimensional volume of the region bounded by the sphere and the plane π_2 and the n-dimensional surface area of the region of $S^n(a)$ between the two planes π_1 and π_2, respectively.

Then, for a sufficiently small $t > 0$, we can have the following properties of the sphere $S^n(a)$.

(A) The n-dimensional area $A_p(t)$ of the section is independent of the point p.

(V) The n-dimensional volume $V_p(t)$ of the region is independent of the point p.

(S) The n-dimensional surface area $S_p(t)$ of the region is independent of the point p.

For the converse problem, the authors provide the following:

Theorem 4.3 ([4]). *Let M be a complete and convex hypersurface in the $(n+1)$-dimensional Euclidean space $\mathbb{E}^{n+1}$. Suppose that M satisfies one of the following conditions.*

(1) *The n-dimensional area $A_p(t)$ of the section is independent of the point $p \in M$.*

(2) *The $(n+1)$-dimensional volume $V_p(t)$ of the region is independent of the point $p \in M$.*

(3) *The n-dimensional surface area $S_p(t)$ of the region is independent of the point $p \in M$.*

Then the hypersurface M is a round hypersphere $S^n(a)$.

References

1. Chen, B.-Y., Kim, D.-S. and Kim, Y. H., *New characterizations of W-curves*, Publ. Math. Debrecen **69** (2006), 457–472.
2. Kim, D.-S. and Kim, Y. H., *A characterization of ellipse*, Amer. Math. Monthly **114** (2007), 65–69.
3. Kim, D.-S. and Kim, Y. H., *Some characterizations of spheres and elliptic paraboloids*, Lin. Alg. Appl. (2012) 437, 113–120.
4. Kim, D.-S. and Kim, Y. H., *Some characterizations of spheres and elliptic paraboloids II*, Lin. Alg. Appl. (2013) 438, 1358–1364.
5. Kim, D.-S., Kim, Y. H. and Park, J. H., *Some characterization of parabolas*, to appear in Kyungpook Math. J.
6. Pressley, A., *Elementary differential geometry*, Undergraduate Mathematics Series, Springer-Verlag London, Ltd., London, 2001.

7. Stein, S., Archimedes. *What did he do besides cry Eureka?,* Mathematical Association of America, Washington, DC, 1999.
8. Stenlund, M., *On the tangent lines of a parabola, College Math. J.* **32** (2001), no. 3, 194–196.

Received January 16, 2013.

Proceedings of the Workshop on
Differential Geometry of Submanifolds
and its Related Topics
Saga, August 4-6, 2012

ON THE CONSTRUCTION OF COHOMOGENEITY ONE SPECIAL LAGRANGIAN SUBMANIFOLDS IN THE COTANGENT BUNDLE OF THE SPHERE

Kaname HASHIMOTO

Osaka City University Advanced Mathematical Institute,
3-3-138 Sugimoto, Sumiyoshi-ku, Osaka, 558-8585 Japan
E-mail: h-kaname@sci.osaka-cu.ac.jp

Dedicated to Professor Sadahiro Maeda for his sixtieth birthday

It is known that there exist $SO(n + 1)$-invariant complete Ricci-flat Kähler metrics on the cotangent bundle T^*S^n of the standard sphere S^n, so called the *Stenzel metrics*. We are studying cohomogeneity one special Lagrangian submanifolds in the cotangent bundle of the sphere with respect to the Stenzel metric ([10]). In this paper we discuss a description of cohomogeneity one special Lagrangian submanifold invariant under the group action induced by isotropy representations of Riemannian symmetric spaces of rank 2.

Keywords: special Lagrangian submanifold, minimal submanifold, isoparametric hypersurface.

1. Introduction

A compact hypersurface N in the unit standard sphere S^{n+1} is *homogeneous* if it is obtained as an orbit of a compact connected subgroup of $SO(n + 1)$. It is well known that any homogeneous hypersurface in S^{n+1} can be obtained as a principal orbit of the isotropy representation of a Riemannian symmetric space of rank 2 ([2]). A homogeneous hypersurface N in S^{n+1} is a hypersurface with constant principal curvatures, which is called *isoparametric* ([12]). Then the number g of distinct principal curvatures must be 1, 2, 3, 4 or 6 ([12], [7] and see [8], [9] for general isoparametric hypersurfaces). Denote by (m_1, m_2) the multiplicities of its principal curvatures. The isotropy representation of a Riemannian symmetric space G/K of rank 2 induces a group action of K on S^{n+1} and thus T^*S^{n+1} in a natural way.

In the cases of $g = 1, 2$ such group actions of $SO(p) \times SO(n + 1 - p)$ $(1 \leq p \leq n)$ are induced on T^*S^{n+1}. We classified cohomogeneity one special Lagrangian submanifolds in T^*S^{n+1} under the group actions ([5]).

In this paper we shall discuss the construction of cohomogeneity one special Lagrangian submanifolds in the case when $g = 3, 4$ and G/K is of classical type. We refer Halgason's textbook ([6]) for the general theory

Table 1.   Homogeneous hypersurfaces in spheres

g	(G, K)	m_1, m_2	g	(G, K)	m_1, m_2
3	$SU(3)/SO(3)$	1, 1	4	$SO(5) \times SO(5)/SO(5)$	2, 2
3	$SU(3) \times SU(3)/SU(3)$	2, 2	4	$SO(2 + m)/SO(2) \times SO(m)$	$1, m - 2$
3	$SU(6)/Sp(3)$	4, 4	4	$SU(2 + m)/S(U(2) \times U(m))$	$2, 2m - 3$
3	E_6/F_4	8, 8	4	$Sp(2 + m)/Sp(2) \times Sp(m)$	$4, 4m - 5$
6	$G_2/SO(4)$	1, 1	4	$SO(10)/U(5)$	4, 5
6	$G_2 \times G_2/G_2$	2, 2	4	$E_6/U(1) \times Spin(10)$	6, 9

of Riemannian symmetric space. Let M be a simply connected semisimple Riemannian symmetric space. If G is the identity component of the full group of isometrics of M, then G acts transitively on M and we can write $M = G/K$, where K is the isotropy subgroup of G at a point $p \in M$. Since S is simply connected and G is connected, K is also connected. If $\mathfrak{g} = \mathfrak{k} \oplus \mathfrak{p}$ is the canonical decomposition of $\mathfrak{g}$ associated to the symmetric pair (G, K), then the isotropy representation of K on T_pM is equivalent to the adjoint representation of K on $\mathfrak{p}$. The isotropy representation of G/K at p is a Lie group homomorphism $\mathrm{Ad}_{\mathfrak{p}} : K \to SO(\mathfrak{p})$. So a K-orbit through $X \in \mathfrak{p}$ is denoted by $\mathrm{Ad}_{\mathfrak{p}}(K)X$.

Let G/K be an $(n + 1)$-dimensional rank 2 symmetric space of compact type. Define an $\mathrm{Ad}_{\mathfrak{p}}(K)$-invariant inner product of $\mathfrak{p}$ from the Killing from of $\mathfrak{g}$. Then the vector space $\mathfrak{p}$ can be identified with $\mathbb{R}^{n+1}$ with respect to the inner product. Let $\mathfrak{a}$ be a maximal abelian subspace of $\mathfrak{p}$. Since for each $X \in \mathfrak{p}$ there is an element $k \in K$ such that $\mathrm{Ad}_{\mathfrak{p}}(K)X \in \mathfrak{a}$, every orbit in $\mathfrak{p}$ under K meets $\mathfrak{p}$. The unit hypersphere in $\mathfrak{p}$ is denoted by S^n. Since the action of K on $\mathfrak{p}$ is an orthogonal representation, an orbit $\mathrm{Ad}_{\mathfrak{p}}(K)X$ is a submanifold of the hypersphere S^n in $\mathfrak{p}$. For a *regular* element $H \in \mathfrak{a} \cap S^n$, we obtain a homogeneous hypersurface $N = \mathrm{Ad}_{\mathfrak{p}}(K)H \subset S^n \subset \mathfrak{p} \cong \mathbb{R}^{n+1}$. Conversely, every homogeneous hypersurfaces in a sphere is obtained in this way ([2]).

We briefly recall the Stenzel metric on T^*S^n. We denote the cotangent bundle of the n-sphere $S^n \cong SO(n + 1)/SO(n)$ by $T^*S^n = \left\{ (x, \xi) \in \mathbb{R}^{n+1} \times \mathbb{R}^{n+1} \mid \|x\| = 1, \langle x, \xi \rangle = 0 \right\}$. We identify the tangent bundle and the cotangent bundle of S^n by the Riemannian metric on S^n. Since any unit cotangent vector of S^n can be translated to another one, the Lie group $SO(n+1)$ acts on T^*S^n with cohomogeneity one by $g\cdot(x,\xi)=(gx,g\xi)$ for $g\in SO(n+1)$. Let Q^n be a complex quadric in $\mathbb{C}^{n+1}$ defined by

$$Q^n=\left\{z=(z_1,\ldots,z_{n+1})\in\mathbb{C}^{n+1}\;\middle|\;\sum_{i=1}^{n+1}z_i^2=1\right\}.$$

The Lie group $SO(n+1,\mathbb{C})$ acts on Q^n transitively, hence $Q^n\cong SO(n+1,\mathbb{C})/SO(n,\mathbb{C})$. According to Szöke ([11]), we can identify T^*S^n with Q^n through the following diffeomorphism:

$$\Phi:T^*S^n\ni(x,\xi)\longmapsto x\cosh(\|\xi\|)+\sqrt{-1}\,\frac{\xi}{\|\xi\|}\sinh(\|\xi\|)\in Q^n.$$

The diffeomorphism Φ is equivariant under the action of $SO(n+1)$. Thus we frequently identify T^*S^n with Q^n. Then consider a holomorphic n-form Ω_{Stz} given by

$$\frac{1}{2}d(z_1^2+z_2^2+\cdots+z_{n+1}^2-1)\wedge\Omega_{Stz}=dz_1\wedge\ldots\wedge dz_{n+1}.$$

The *Stenzel metric* is a complete Ricci-flat Kähler metric on Q^n defined by $\omega_{Stz}:=\sqrt{-1}\partial\bar{\partial}\,u(r^2)$, where $r^2=\|z\|^2=\sum_{i=0}^{n+1}z_i\bar{z}_i$ and u is a smooth real-valued function satisfying the differential equation

$$\frac{d}{dt}(U'(t))^n=c\,n\,(\sinh t)^{n-1}\quad(c>0)$$

where $U(t)=u(\cosh t)$. The Kähler form ω_{Stz} is exact, that is, $\omega_{Stz}=d\alpha_{Stz}$ where $\alpha_{Stz}:=-\mathrm{Im}(\bar{\partial}u(r^2))$.

Let K be a compact connected Lie subgroup of $SO(n+1)$ with Lie algebra $\mathfrak{k}$. Then the group action of K on Q^n is Hamiltonian with respect to ω_{Stz} and its moment map $\mu:Q^n\to\mathfrak{k}^*$ is given by

$$\langle\mu(z),X\rangle=\alpha_{Stz}(Xz)=u'(\|z\|^2)\langle Jz,Xz\rangle\quad(z\in Q^n,X\in\mathfrak{k}).\quad(1)$$

Choose a subset Σ of T^*S^n such that every K-orbit in T^*S^n meets Σ. In general assume that K has the Hamiltonian group action on a symplectic manifold M. We define the center of $\mathfrak{k}^*$ to be $Z(\mathfrak{k}^*)=\{X\in\mathfrak{k}^*\mid\mathrm{Ad}^*(k)X=X\ (\forall k\in K)\}$. It is easy to see that the inverse image $\mu^{-1}(c)$ of $c\in\mathfrak{k}^*$ is invariant under the group action of K if and only if $c\in Z(\mathfrak{k}^*)$.

Proposition 1.1. *Let L be a connected isotropic submanifold, i.e., $\omega|_L\equiv 0$, of M invariant under the action of K. Then $L\subset\mu^{-1}(c)$ for some $c\in Z(\mathfrak{k}^*)$.*

Proposition 1.2. *Let L be a connected submanifold of M invariant under the action of K. Suppose that the action of K on L is of cohomogeneity one (possibly transitive). Then L is an isotropic submanifold, i.e., $\omega|_L \equiv 0$, if and only if $L \subset \mu^{-1}(c)$ for some $c \in Z(\mathfrak{k}^*)$.*

The notion of special Lagrangian submanifolds in a Ricci-flat Kähler manifold was introduced and investigated by Harvey and Lawson ([1], see also [4]) In [5] we studied in detail and classified cohomogeneity one special Lagrangian submanifolds in T^*S^{n+1} under the group action of $SO(p) \times SO(n + 1 - p)$ $(1 \le p \le n)$. The case of $n = 3$ were investigated in detail by Ionel and Min-Oo ([3]). In this paper, by generalizing the arguments of [5], we provide a construction of cohomogeneity one special Lagrangian submanifolds in T^*S^n under the group action induced by the isotropy representation of a Riemannian symmetric space G/K of rank 2.

For the group action of K induced by the isotropy representation of G/K, the moment map formula (1) becomes

$$\mu(Z) = -u'(\|Z\|^2)\sqrt{-1}[Z, \overline{Z}] = -2u'(\|Z\|^2)[X, Y] \in \mathfrak{k} \cong \mathfrak{k}^* \qquad (2)$$

for each $Z = X + \sqrt{-1}Y \in Q^n \subset \mathfrak{p}^\mathbb{C} \cong \mathbb{C}^{n+1}$ with $X, Y \in \mathfrak{p} \cong \mathbb{R}^{n+1}$.

Now we consider only the case where the inverse image $\mu^{-1}(0)$ of $0 \in \mathfrak{k}^*$. In the same way as [2], the orbit space of K-action on $\mu^{-1}(0)$ can be explicitly parametrized by a complex coordinate $\tau = t + \sqrt{-1}\xi_1 \in \mathbb{C}$.

2. $g = 3$

2.1. $(G, K) = (SU(3), SO(3))$

We consider the case of $(G, K) = (SU(3), SO(3))$. We denote by $\mathfrak{g}$ and $\mathfrak{k}$ the Lie algebras of G and K respectively. The canonical decomposition of $\mathfrak{g}$ is given by $\mathfrak{g} = \mathfrak{k} \oplus \mathfrak{p}$, where

$$\mathfrak{k} = \mathfrak{so}(3) \quad \text{and} \quad \mathfrak{p} = \left\{ \sqrt{-1}X \in M_3(\mathbb{R}) \mid {}^tX = X, \operatorname{Tr}X = 0 \right\}.$$

Then the isotropy representation of K is defined by $\operatorname{Ad}_\mathfrak{p}(k)X = kX^tk$ for $k \in SO(3)$ and $X \in \mathfrak{p}$. We define an inner product on $\mathfrak{p}$ by $\langle X, Y \rangle = -\operatorname{Tr}(XY)$ for $X, Y \in \mathfrak{p}$. Let

$$\mathfrak{a} = \left\{ \sqrt{-1} \begin{pmatrix} a_1 & & \\ & a_2 & \\ & & a_3 \end{pmatrix} \; \middle| \; \begin{array}{l} a_1 + a_2 + a_3 = 0, \\ a_1, a_2, a_3 \in \mathbb{R} \end{array} \right\}.$$

Then $\mathfrak{a}$ is a maximal abelian subspace of $\mathfrak{p}$. The group action of $K = SO(3)$ is naturally induced on the complex quadric Q^4 in $\mathfrak{p}^\mathbb{C} = \{Z \in M_3(\mathbb{C}) \mid {}^tZ = Z, \operatorname{Tr}Z = 0\}$.

Theorem 2.1. *Let τ be a regular curve in the complex plane $\mathbb{C}$. Define a curve σ in $\mu^{-1}(0) \cap \Phi(\Sigma)$ by*

$$\sigma(s) = \frac{1}{\sqrt{6}} \begin{pmatrix} 2\cos\tau(s) \\ -\cos\tau(s) + \sqrt{3}\sin\tau(s) \\ -\cos\tau(s) - \sqrt{3}\sin\tau(s) \end{pmatrix} \in \mathfrak{a}^{\mathbb{C}}.$$

Then the K-orbit $L = K \cdot \sigma$ through a curve σ is a cohomogeneity one Lagrangian submanifold under the group action of K in Q^4. Conversely, such a cohomogeneity one Lagrangian submanifold in Q^4 is obtained in this way. Moreover, L is a special Lagrangian submanifold of phase θ if and only if τ satisfies

$$\operatorname{Im}\left(e^{\sqrt{-1}\theta} \tau'(s)(3\cos^2\tau(s) - \sin^2\tau(s))(\sin\tau(s)) \right) = 0. \tag{3}$$

2.2. $(G, K) = (SU(3) \times SU(3), SU(3))$

We consider the case of $(G, K) = (SU(3) \times SU(3), SU(3))$. The canonical decomposition of $\mathfrak{g}$ is given by $\mathfrak{g} = \mathfrak{k} \oplus \mathfrak{p}$, where

$$\mathfrak{k} = \{(X, X) \mid X \in \mathfrak{su}(3)\} \quad \text{and} \quad \mathfrak{p} = \{(X, -X) \mid X \in \mathfrak{su}(3)\} \cong \mathfrak{su}(3).$$

Since $\mathfrak{p}$ is linearly isomorphic to $\mathfrak{su}(3)$, we identify them. Then the linearly isotropy representation of K is defined by $\operatorname{Ad}_{\mathfrak{p}}(k)X = kX^t k$ for $k \in SU(3)$ and $X \in \mathfrak{p}$. We define an inner product on $\mathfrak{p}$ by $\langle X, Y \rangle = -\operatorname{Tr}(XY)$ for $X, Y \in \mathfrak{p} = \mathfrak{su}(3)$. Let

$$\mathfrak{a} = \left\{ \sqrt{-1} \begin{pmatrix} a_1 & & \\ & a_2 & \\ & & a_3 \end{pmatrix} \;\middle|\; \begin{array}{l} a_1 + a_2 + a_3 = 0, \\ a_1, a_2, a_3 \in \mathbb{R} \end{array} \right\}.$$

Then $\mathfrak{a}$ is a maximal abelian subspace of $\mathfrak{p}$. The group action of $K = SU(3)$ is naturally induced on Q^7 in $\mathfrak{p}^{\mathbb{C}} = \mathfrak{sl}(3, \mathbb{C})$.

Theorem 2.2. *Let τ be a regular curve in the complex plane $\mathbb{C}$. Define a curve σ in $\mu^{-1}(0) \cap \Phi(\Sigma)$ by*

$$\sigma(s) = \frac{1}{\sqrt{6}} \begin{pmatrix} 2\cos\tau(s) \\ -\cos\tau(s) + \sqrt{3}\sin\tau(s) \\ -\cos\tau(s) - \sqrt{3}\sin\tau(s) \end{pmatrix} \in \mathfrak{a}^{\mathbb{C}}.$$

Then the K-orbit $L = K \cdot \sigma$ through a curve σ is a cohomogeneity one Lagrangian submanifold under the group action of K in Q^7. Conversely, such a cohomogeneity one Lagrangian submanifold in Q^7 is obtained in this way. Moreover, L is a special Lagrangian submanifold of phase θ if and only if τ satisfies

$$\mathrm{Im}\left(e^{\sqrt{-1}\theta}\tau'(s)(3\cos^2\tau(s)-\sin^2\tau(s))^2(\sin\tau(s))^2\right)=0. \qquad (4)$$

2.3. $(G,K)=(SU(6),Sp(3))$

We consider the case of $(G,K)=(SU(6),Sp(3))$. The canonical decomposition of $\mathfrak{g}$ is given by $\mathfrak{g}=\mathfrak{k}\oplus\mathfrak{p}$, where

$$\mathfrak{k}=\mathfrak{su}(6)\quad\text{and}\quad\mathfrak{p}=\left\{\begin{pmatrix}X&Y\\\overline{Y}&-\overline{X}\end{pmatrix}\mid X\in\mathfrak{su}(n),Y\in\mathfrak{o}(n,\mathbb{C})\right\}.$$

Then

$$\mathfrak{p}^{\mathbb{C}}=\left\{\begin{pmatrix}V_{11}&V_{12}\\V_{21}&V_{22}\end{pmatrix}\mid V_{11},V_{22}\in\mathfrak{sl}(n,\mathbb{C}),V_{12},V_{21}\in\mathfrak{o}(n,\mathbb{C})\right\}.$$

We define an inner product on $\mathfrak{p}$ by $\langle X,Y\rangle=-\mathrm{Tr}(XY)$ for $X,Y\in\mathfrak{p}$. Then the isotropy representation of K is defined by $\mathrm{Ad}_{\mathfrak{p}}(k)X=kX^tk$ for $k\in Sp(3)$ and $X\in\mathfrak{p}$. Let

$$\mathfrak{a}=\left\{\sqrt{-1}\begin{pmatrix}H&\\&H\end{pmatrix}\mid H=\begin{pmatrix}a_1&&\\&a_2&\\&&a_3\end{pmatrix},\begin{array}{l}a_1+a_2+a_3=0,\\a_1,a_2,a_3\in\mathbb{R}\end{array}\right\}.$$

Then $\mathfrak{a}$ is a maximal abelian subspace of $\mathfrak{p}$. The group action of $K=Sp(3)$ is naturally induced on Q^{14} in $\mathfrak{p}^{\mathbb{C}}$.

Theorem 2.3. *Let τ be a regular curve in the complex plane $\mathbb{C}$. Define a curve σ in $\mu^{-1}(0)\cap\Phi(\Sigma)$ by $\sigma(s)=\sqrt{-1}\begin{pmatrix}H&O\\O&H\end{pmatrix}\in\mathfrak{a}^{\mathbb{C}}$, where*

$$H=\frac{1}{\sqrt{6}}\begin{pmatrix}2\cos\tau(s)&&\\&-\cos\tau(s)+\sqrt{3}\sin\tau(s)&\\&&-\cos\tau(s)-\sqrt{3}\sin\tau(s)\end{pmatrix}.$$

Then the K-orbit $L=K\cdot\sigma$ through a curve σ is a cohomogeneity one Lagrangian submanifold under the group action of K in Q^{14}. Conversely, such a cohomogeneity one Lagrangian submanifold in Q^{14} is obtained in this way. Moreover, L is a special Lagrangian submanifold of phase θ if and only if τ satisfies

$$\mathrm{Im}\left(e^{\sqrt{-1}\theta}\tau'(s)(3\cos^2\tau(s)-\sin^2\tau(s))^4(\sin\tau(s))^4\right)=0. \qquad (5)$$

3. $g = 4$

3.1. $(G, K) = (SO(m + 2), SO(2) \times SO(m))$

We consider the case of $(G, K) = (SO(m + 2), SO(2) \times SO(m))$. We denote by $\mathfrak{g}$ and $\mathfrak{k}$ the Lie algebras of G and K respectively. The canonical decomposition of $\mathfrak{g}$ is given by $\mathfrak{g} = \mathfrak{k} \oplus \mathfrak{p}$, where

$$\mathfrak{k} = \left\{ \begin{pmatrix} A & O \\ O & B \end{pmatrix} \middle| \begin{array}{l} A \in \mathfrak{o}(2), \\ B \in \mathfrak{o}(m) \end{array} \right\} \quad \text{and} \quad \mathfrak{p} = \left\{ \begin{pmatrix} O & X \\ -{}^t X & O \end{pmatrix} \middle| X \in M_{2,m}(\mathbb{R}) \right\}.$$

Since $\mathfrak{p}$ is linearly isomorphic to $M_{2,m}(\mathbb{R})$, we identify them. We define an inner product by $\langle X, Y \rangle = \mathrm{Tr}(X^t Y)$ for $X, Y \in M_{2,m}(\mathbb{R})$. Then the isotropy representation of K is defined by $\mathrm{Ad}_{\mathfrak{p}}(k)X = k_1 X k_2^{-1}$ for $k = \begin{pmatrix} k_1 & O \\ O & k_2 \end{pmatrix} \in SO(2) \times SO(m + 1)$ and $X \in \mathfrak{p}$. We take a maximal abelian subspace of $\mathfrak{p}$ as

$$\mathfrak{a} = \left\{ \begin{pmatrix} O & H \\ -{}^t H & O \end{pmatrix} \middle| H = \begin{pmatrix} a_1 & 0 & 0 \cdots 0 \\ 0 & a_2 & 0 \cdots 0 \end{pmatrix} \in M_{2,m}(\mathbb{R}) \right\}.$$

The group action of $K = SO(2) \times SO(m)$ is naturally induced on Q^{2m-1} in $\mathfrak{p}^{\mathbb{C}} \cong M_{2,m}(\mathbb{C})$.

Theorem 3.1. *Let τ be a regular curve in the complex plane $\mathbb{C}$. Define a curve σ in $\mu^{-1}(0) \cap \Phi(\Sigma)$ by $\sigma(s) = \begin{pmatrix} O & H \\ -{}^t H & O \end{pmatrix} \in \mathfrak{a}^{\mathbb{C}}$, where*

$$H = \begin{pmatrix} \cos \tau(s) & 0 & 0 \cdots 0 \\ 0 & \sin \tau(s) & 0 \cdots 0 \end{pmatrix}.$$

Then the K-orbit $L = K \cdot \sigma$ through σ is a cohomogeneity one Lagrangian submanifold under the group action of K in Q^{2m-1}. Conversely, such a cohomogeneity one Lagrangian submanifold in Q^{2m-1} is obtained in this way. Moreover, L is a special Lagrangian submanifold of phase θ if and only if τ satisfies

$$\mathrm{Im}\left(e^{\sqrt{-1}\theta} \tau'(s) \cos 2\tau(s)(\sin 2\tau(s))^{m-2} \right) = 0. \tag{6}$$

3.2. $(G, K) = (SU(m + 2), S(U(2) \times U(m)))$

We consider the case of $(G, K) = (SU(m+2), S(U(2) \times U(m)))$. The canonical decomposition of $\mathfrak{g}$ is given by $\mathfrak{g} = \mathfrak{k} \oplus \mathfrak{p}$, where

$$\mathfrak{k} = \left\{ \begin{pmatrix} A & O \\ O & B \end{pmatrix} \middle| \begin{array}{l} A \in \mathfrak{u}(2), \\ B \in \mathfrak{u}(m) \end{array} \right\} \quad \text{and} \quad \mathfrak{p} = \left\{ \begin{pmatrix} O & X \\ -{}^t \overline{X} & O \end{pmatrix} \middle| X \in M_{2,m}(\mathbb{C}) \right\}.$$

Then

$$\mathfrak{p}^{\mathbb{C}} = \left\{ \begin{pmatrix} O & V \\ {}^t W & O \end{pmatrix} \,\middle|\, V, W \in M_{2,m}(\mathbb{C}) \right\}.$$

We define an inner product by $\langle X, Y \rangle = -\mathrm{Tr}(XY)$ for $X, Y \in \mathfrak{p}$. Then the isotropy representation of K is defined by $\mathrm{Ad}_{\mathfrak{p}}(k)X = k_1 X {}^t \overline{k_2}$ for $k = \begin{pmatrix} k_1 & \\ & k_2 \end{pmatrix} \in S(U(2) \times U(m))$ and $X \in \mathfrak{p}$. We take a maximal abelian subspace of $\mathfrak{p}$ as

$$\mathfrak{a} = \left\{ \begin{pmatrix} O & H \\ -{}^t H & O \end{pmatrix} \,\middle|\, H = \begin{pmatrix} a_1 & 0 & 0 \cdots 0 \\ 0 & a_2 & 0 \cdots 0 \end{pmatrix} \in M_{2,m}(\mathbb{R}) \right\}.$$

The group action of $K = S(U(2) \times U(m))$ is naturally induced on Q^{4m-1} in $\mathfrak{p}^{\mathbb{C}}$.

Theorem 3.2. *Let τ be a regular curve in the complex plane $\mathbb{C}$. Define a curve σ in $\mu^{-1}(0) \cap \Phi(\Sigma)$ by $\sigma(s) = \begin{pmatrix} O & H \\ -{}^t H & O \end{pmatrix} \in \mathfrak{a}^{\mathbb{C}}$, where*

$$H = \begin{pmatrix} \cos \tau(s) & 0 & 0 \cdots 0 \\ 0 & \sin \tau(s) & 0 \cdots 0 \end{pmatrix}.$$

Then the K-orbit $L = K \cdot \sigma$ through a curve σ is a cohomogeneity one Lagrangian submanifold under the group action of K in Q^{4m-1}. Moreover, L is a special Lagrangian submanifold of phase θ if and only if τ satisfies

$$\mathrm{Im}\left(e^{\sqrt{-1}\theta} \tau'(s) (\cos 2\tau(s))^2 (\sin 2\tau(s))^{2m-3} \right) = 0. \tag{7}$$

3.3. $(G, K) = (Sp(m+2), Sp(2) \times Sp(m))$

We consider the case of $(G, K) = (Sp(m+2), Sp(2) \times Sp(m))$. The canonical decomposition of $\mathfrak{g}$ is given by $\mathfrak{g} = \mathfrak{k} \oplus \mathfrak{p}$, where

$$\mathfrak{k} = \left\{ \begin{pmatrix} A_{11} & O & B_{11} & O \\ O & A_{22} & O & B_{22} \\ -\overline{B}_{11} & O & \overline{A}_{11} & O \\ O & -\overline{B}_{22} & O & \overline{A}_{22} \end{pmatrix} \,\middle|\, \begin{matrix} A_{11} \in \mathfrak{u}(2), A_{22} \in \mathfrak{u}(m), \\ B_{11} \in M_2(\mathbb{C}), {}^t B_{11} = B_{11}, \\ B_{22} = M_m(\mathbb{C}), {}^t B_{22} = B_{22} \end{matrix} \right\}$$

and

$$\mathfrak{p} = \left\{ \begin{pmatrix} O & X_{12} & O & Y_{12} \\ -{}^t \overline{X}_{12} & O & -{}^t Y_{12} & O \\ O & -\overline{Y}_{12} & O & \overline{X}_{12} \\ -{}^t \overline{Y}_{12} & O & -{}^t X_{12} & O \end{pmatrix} \,\middle|\, X_{12}, Y_{12} \in M_{2,m}(\mathbb{C}) \right\}.$$

Then

$$\mathfrak{p}^{\mathbb{C}} = \left\{ \begin{pmatrix} O & V_{12} & O & V_{14} \\ -{}^t W_{12} & O & {}^t V_{14} & O \\ O & -W_{14} & O & W_{12} \\ -{}^t W_{14} & O & -{}^t V_{12} & O \end{pmatrix} \middle| \; V_{12}, V_{14}, W_{12}, W_{14} \in M_{2,m}(\mathbb{C}) \right\}.$$

We define an inner product by $\langle X, Y \rangle = -\mathrm{Tr}(XY)$ for X, $Y \in \mathfrak{p}$. Then the isotropy representation of K is defined by $\mathrm{Ad}_{\mathfrak{p}}(k)X = k_1 X {}^t \overline{k_2}$ for $k = \begin{pmatrix} k_1 & O \\ O & k_2 \end{pmatrix} \in Sp(2) \times Sp(m)$ and $X \in \mathfrak{p}$. Then

$$\mathfrak{a} = \left\{ \begin{pmatrix} O & H & O & O \\ -{}^t H & O & O & O \\ O & O & O & H \\ O & O & -{}^t H & O \end{pmatrix} \middle| \; H = \begin{pmatrix} a_1 & 0 & 0 \cdots 0 \\ 0 & a_2 & 0 \cdots 0 \end{pmatrix} \right\}$$
$$a_1, a_2 \in \mathbb{R}$$

is a maximal abelian subspace of $\mathfrak{p}$. The group action of $K = Sp(2) \times Sp(m)$ is naturally induced on Q^{8m-1} in $\mathfrak{p}^{\mathbb{C}}$.

Theorem 3.3. *Let τ be a regular curve in the complex plane $\mathbb{C}$. Define a curve σ in $\mu^{-1}(0) \cap \Phi(\Sigma)$ by*

$$\sigma(s) = \begin{pmatrix} O & H & O & O \\ -{}^t H & O & O & O \\ O & O & O & H \\ O & O & -{}^t H & O \end{pmatrix} \in \mathfrak{a}^{\mathbb{C}},$$

where

$$H = \begin{pmatrix} \cos \tau(s) & 0 & 0 \cdots 0 \\ 0 & \sin \tau(s) & 0 \cdots 0 \end{pmatrix}.$$

Then the K-orbit $L = K \cdot \sigma$ through a curve σ is a cohomogeneity one Lagrangian submanifold under the group action of K in Q^{8m-1}. Conversely, such a cohomogeneity one Lagrangian submanifold in Q^{8m-1} is obtained in this way. Moreover, L is a special Lagrangian submanifold of phase θ if and only if τ satisfies

$$\mathrm{Im}\left(e^{\sqrt{-1}\theta} \tau'(s)(\cos 2\tau(s))^4 (\sin 2\tau(s))^{4m-5} \right) - 0. \tag{8}$$

3.4. $(G, K) = (SO(5) \times SO(5), \; SO(5))$

We consider the case of $(G, K) = (SO(5) \times SO(5), \; SO(5))$. The canonical decomposition of $\mathfrak{g}$ is given by $\mathfrak{g} = \mathfrak{k} \oplus \mathfrak{p}$, where

$$\mathfrak{k} = \{(X, X) \mid X \in \mathfrak{o}(5)\} \cong \mathfrak{o}(5) \quad \text{and} \quad \mathfrak{p} = \{(X, -X) \mid X \in \mathfrak{o}(5)\}.$$

Then $\mathfrak{p}^{\mathbb{C}} = \mathfrak{o}(5, \mathbb{C})$. We use the inner product by $\langle X, Y \rangle = -\frac{1}{2} \mathrm{Tr}(XY)$ for $X, Y \in \mathfrak{p}$. Then the isotropy representation of K is defined by $\mathrm{Ad}_{\mathfrak{p}}(k)X = kXk^{-1}$ for $k \in SO(5)$ and $X \in \mathfrak{p}$. Let

$$
\mathfrak{a} = \left\{ (H, -H) \;\middle|\; H = \begin{pmatrix} \begin{array}{cc|cc|c} 0 & a_1 & & & \\ -a_1 & 0 & & & \\ \hline & & 0 & a_2 & \\ & & -a_2 & 0 & \\ \hline & & & & 0 \end{array} \end{pmatrix}, a_1, a_2 \in \mathbb{R} \right\}.
$$

Then $\mathfrak{a}$ is a maximal abelian subalgebra in $\mathfrak{p}$. The group action of $K = SO(5)$ is naturally induced on Q^9 in $\mathfrak{p}^{\mathbb{C}}$.

Theorem 3.4. *Let τ be a regular curve in the complex plane $\mathbb{C}$. Define a curve σ in $\mu^{-1}(0) \cap \Phi(\Sigma)$ by $\sigma(s) = (H, -H) \in \mathfrak{a}^{\mathbb{C}}$, where*

$$
H = \begin{pmatrix} 0 & \cos \tau(s) & & & \\ -\cos \tau(s) & 0 & & & \\ & & 0 & \sin \tau(s) & \\ & & -\sin \tau(s) & 0 & \\ & & & & 0 \end{pmatrix}.
$$

Then the K-orbit $L = K \cdot \sigma$ through a curve σ is a cohomogeneity one Lagrangian submanifold under the group action of K in Q^9. Conversely, such a cohomogeneity one Lagrangian submanifold in Q^9 is obtained in the way. Moreover, L is a special Lagrangian submanifold of phase θ if and only if there exists a constant $c \in \mathbb{R}$ so that τ satisfies

$$
\mathrm{Im} \left(e^{\sqrt{-1}\theta} \tau'(s) (\cos 2\tau(s))^2 (\sin 2\tau(s))^2 \right) = 0. \tag{9}
$$

3.5. $(G, K) = (SO(10), U(5))$

We consider the case of $(G, K) = (SO(10), U(5))$. The canonical decomposition of $\mathfrak{g}$ is given by $\mathfrak{g} = \mathfrak{k} \oplus \mathfrak{p}$, where

$$
\mathfrak{k} = \left\{ \begin{pmatrix} A & B \\ -B & A \end{pmatrix} \;\middle|\; {}^t A = -A, {}^t B = B \right\} \cong \mathfrak{u}(5)
$$

and

$$
\mathfrak{p} = \left\{ \begin{pmatrix} X & Y \\ Y & -X \end{pmatrix} \;\middle|\; X, Y \in \mathfrak{so}(5) \right\}.
$$

Then

$$
\mathfrak{p}^{\mathbb{C}} = \left\{ \begin{pmatrix} V & W \\ W & -V \end{pmatrix} \in \mathfrak{o}(10, \mathbb{C}) \;\middle|\; V, W \in \mathfrak{o}(5, \mathbb{C}) \right\}.
$$

We define an inner product by $\langle X, Y \rangle = -\frac{1}{2}\mathrm{Tr}(XY)$ for $X, Y \in \mathfrak{p}$. Then the isotropy representation of K is defined by $\mathrm{Ad}_\mathfrak{p}(k)X = kX^t\overline{k}$ for $k \in U(5)$ and $X \in \mathfrak{p}$. Then

$$\mathfrak{a} = \left\{ \begin{pmatrix} H & 0 \\ 0 & -H \end{pmatrix} \middle| H = \begin{pmatrix} \begin{array}{cc|c|c} 0 & a_1 & & \\ -a_1 & 0 & & \\ \hline & & \begin{array}{cc} 0 & a_2 \\ -a_2 & 0 \end{array} & \\ \hline & & & 0 \end{array} \end{pmatrix}, a_1, a_2 \in \mathbb{R} \right\}$$

is a maximal abelian subspace of $\mathfrak{p}$. The group action of $K = U(5)$ is naturally induced on Q^{19} in $\mathfrak{p}^{\mathbb{C}}$.

Theorem 3.5. *Let τ be a regular curve in the complex plane $\mathbb{C}$. Define a curve σ in $\mu^{-1}(0) \cap \Phi(\Sigma)$ by $\sigma(s) = \begin{pmatrix} H & 0 \\ 0 & -H \end{pmatrix} \in \mathfrak{a}^{\mathbb{C}}$, where*

$$H = \begin{pmatrix} \begin{array}{cc|cc|c} 0 & \cos\tau(s) & & & \\ -\cos\tau(s) & 0 & & & \\ \hline & & 0 & \sin\tau(s) & \\ & & -\sin\tau(s) & 0 & \\ \hline & & & & 0 \end{array} \end{pmatrix}.$$

Then the K-orbit $L = K \cdot \sigma$ through a curve σ is a cohomogeneity one Lagrangian submanifold under the group action of K in Q^{19}. Conversely, such a cohomogeneity one Lagrangian submanifold in Q^{19} is obtained in this way. Moreover, L is a special Lagrangian submanifold of phase θ if and only if τ satisfies

$$\mathrm{Im}\left(e^{\sqrt{-1}\theta}\tau'(s)(\cos 2\tau(s))^4(\sin 2\tau(s))^5\right) = 0. \tag{10}$$

In the forthcoming paper we will study the remaining cases when G/K are of exceptional type.

Acknowledgments

This paper is dedicated to the 60th birthday of Professor Sadahiro Maeda. The author sincerely would like to thank Professor Sadahiro Maeda for his warm encouragement.

References

1. R. Harvey and H. B. Lawson, *Calibrated geometries*, Acta Math. **148** (1982), 47–157.

2. W. Y. Hsiang and H. B. Lawson, *Minimal submanifolds of low cohomogeneity*, J. Diff. Geom. **5** (1971), 1–38.

3. M. Ionel and M. Min-Oo, *Cohomogeneity one special Lagrangian 3-folds in the deformed conifold and the resolved conifolds*, Illinois J. Math. **52**, No. 3, (2008), 839–865.

4. D.D. Joyce, *Riemannian holonomy groups and calibrated geometry*, Oxford Graduate Texts in Mathematics **12**, Oxford University Press, Oxford, (2007), x+303pp.

5. K. Hashimoto and T. Sakai, *Cohomogeneity one special Lagrangian submanifolds in the cotangent bundle of the sphere*, Tohoku Math. J. **64**, No. 1, (2012), 141–169.

6. S. Helgason, *Differential geometry, Lie groups, and symmetric spaces*, Academic Press, 1978, and American Mathematical Society, 2001.

7. H. Ozeki and Takeuchi, *On some types of isoparametric hypersurfaces in spheres II*, Tohoku Math. J. **28** (1976), 7–55.

8. H. F. Münzner, *Isoparametrische Hyperfläche in Sphären*, Math. Ann. **251** (1980), 57–71.

9. H. F. Münzner, *Isoparametrische Hyperfläche in Sphären. II*, Math. Ann. **256** (1981), 215–232.

10. M. Stenzel, *Ricci-flat metrics on the complexification of a compact rank one symmetric space*, Manuscripta Math. **80**, No. 2, (1993), 151–163.

11. R. Szöke, *Complex structures on tangent bundles of Riemannian manifolds*, Math. Ann. **291** (1991), 409–428.

12. R. Takagi and T. Takahashi, *On the principal curvatures of homogeneous hypersurfaces in a sphere*, Differential geometry (in honor of Kentaro Yano), pp. 469–481. Kinokuniya, Tokyo, 1972.

Received November 25, 2012.
Revised February 20, 2013.

Proceedings of the Workshop on
Differential Geometry of Submanifolds
and its Related Topics
Saga, August 4-6, 2012

SELF-SHRINKERS OF THE MEAN CURVATURE FLOW

Qing-Ming CHENG

*Department of Applied Mathematics, Faculty of Sciences,
Fukuoka University, Fukuoka 814-0180, Japan
E-mail: cheng@fukuoka-u.ac.jp*

Yejuan PENG

*Department of Mathematics, Graduate School of Science
and Engineering, Saga University,
Saga 840-8502, Japan
E-mail: yejuan666@gmail.com*

Dedicated to Professor Sadahiro Maeda for his sixtieth birthday

In this paper, we study complete self-shrinkers of the mean curvature flow in Euclidean space $\mathbb{R}^{n+p}$, and obtain a rigidity theorem by generalizing the maximum principle for $\mathcal{L}$-operator. Here, $\mathcal{L}$-operator is introduced by Colding and Minicozzi in [Generic mean curvature flow I; generic singularities, Ann. Math. **175** (2012) 755-833].

Keywords: mean curvature flow, self-shrinker, $\mathcal{L}$-operator, generalized maximum principle for $\mathcal{L}$-operator, polynomial volume growth.

1. Introduction

1.1. *The mean curvature flow*

Let M^n be an n-dimensional manifold and assume that $X : M^n \to \mathbb{R}^{n+p}$ be an n-dimensional submanifold in the $(n + p)$-dimensional Euclidean space $\mathbb{R}^{n+p}$. We say that $M_0 = X(M^n)$ is moved by its mean curvature if there exists a family $F(\cdot, t)$ of smooth immersions: $F(\cdot, t) : M^n \to \mathbb{R}^{n+p}$ with $F(\cdot, 0) = X(\cdot)$ such that they satisfy

$$\left(\frac{\partial F(p, t)}{\partial t}\right)^N = H(p, t), \tag{1}$$

where $H(p,t)$ denotes the mean curvature vector of submanifold $M_t = F(M^n, t)$ at point $F(p,t)$. The simplest mean curvature flow is given by the one-parameter family of the shrinking spheres $M_t \subset \mathbb{R}^{n+1}$ centered at the origin and with radius $\sqrt{-2n(t-T)}$ for $t \leq T$. This is a smooth flow except at the origin at time $t = T$ when the flow becomes extinct.

For $p = 1$ and $M_0 = X(M^n)$ is an n-dimensional compact convex hypersurface in $\mathbb{R}^{n+1}$, Huisken[10] proved that the mean curvature flow $M_t = F(M^n, t)$ remains smooth and convex until it becomes extinct at a point in the finite time. If we rescale the flow about the point, the resultings converge to the round sphere.

When M_0 is non-convex, the other singularities of the mean curvature flow can occur. In fact, singularities are unavoidable as the flow contracts any closed embedded submanifolds in Euclidean space eventually leading to extinction of the evolving submanifolds. In fact, Grayson[9] constructed a rotationally symmetric dumbbell with a sufficiently long and narrow bar, where the neck pinches off before the two bells become extinct. For the rescalings of the singularity at the neck, the resultings blow up, can not extinctions. Hence, the resultings are not spheres, certainly. In fact, the resultings of the singularity converges to shrinking cylinders. The one-parameter family of the shrinking spheres $M_t \subset \mathbb{R}^{n+1}$ centered at the origin and with radius $\sqrt{-2n(t-T)}$ for $t \leq T$ is self-similar in the sense that M_t is given by

$$M_t = \sqrt{-(t-T)} M_{T-1}.$$

A mean curvature flow M_t is called a self-shrinker if it satisfies

$$M_t = \sqrt{-t} M_{-1}.$$

Self-shrinkers play an important role in the study of the mean curvature flow because they describe all possible blow ups at a given singularity of a mean curvature flow. To explain this, the tangent flow is needed, which generalized the tangent cone construction from minimal surfaces. The idea is by rescaling a mean curvature flow in space and time to obtain a new mean curvature flow, which expands a small neighborhood of the point that we want to focus on. The monotonicity formula of Huisken gives a uniform control over these rescalings. At point $F_0 = F(p_0, T)$, defining

$$\rho(F, t) = (4\pi(T-t))^{-n/2} e^{-\frac{|F - F_0|^2}{4(T-t)}}$$

for the mean curvature flow $F(\cdot, t)$, the monotonicity formula of Huisken

$$\frac{d}{dt} \int_{M_t} \rho(F, t) dv_t = - \int_{M_t} \rho(F, t) \left| H + \frac{(F(p, t) - F_0)^N}{2(T - t)} \right|^2 dv_t \quad (2)$$

holds, where dv_t is the measure on M_t.

Given a singularity (F_0, t_0) of the mean curvature flow. First translate M_t such that to move (F_0, t_0) to $(0, 0)$ and then take a sequence of the parabolic dilations $(F, t) \to (c_j F, c_j^2 t)$ with $c_j \to \infty$ to get the mean curvature flow $M_t^j = c_j (M_{c_j^{-2} t + t_0} - F_0)$. By using the monotonicity formula of Huisken, a subsequence of M_t^j converges weakly to a mean curvature flow T_t, which is called a tangent flow at the point (F_0, t_0). A tangent flow is a self-shrinker according to the monotonicity formula of Huisken.

1.2. *Characterizations of self-shrinkers*

The self shrinker is defined by a mean curvature flow M_t satisfying

$$M_t = \sqrt{-t} M_{-1}.$$

If a submanifold $X : M^n \to \mathbb{R}^{n+p}$ satisfies

$$H = -X^N,$$

then $M_t = \sqrt{-t} X(M^n)$ flows by the mean curvature vector and the mean curvature vector of M_t is given by

$$H_t = \frac{X_t^N}{t}.$$

On the other hand, if M_t is a self-shrinker, then, the mean curvature vector H_t of M_t is given by

$$H_t = \frac{X_t^N}{t}.$$

For simplicity, one calls a submanifold $X : M^n \to \mathbb{R}^{n+p}$ a self-shrinker if

$$H = -X^N, \quad (3)$$

where H denotes the mean curvature vector of the submanifold $X : M^n \to \mathbb{R}^{n+p}$.

Let $X : M^n \to \mathbb{R}^{n+p}$ be a submanifold. We say that

$$X_s : M^n \to \mathbb{R}^{n+p}$$

is a variation of $X : M^n \to \mathbb{R}^{n+p}$ if X_s is a one parameter family of immersions with $X_0 = X$. Define a functional

$$\mathcal{F}_s = (2\pi)^{-n/2} \int_M e^{-\frac{|X_s|^2}{2}} dv.$$

By computing the first variation formula, we obtain that $X : M^n \to \mathbb{R}^{n+p}$ is a critical point of $\mathcal{F}_s$ if and only if $X : M^n \to \mathbb{R}^{n+p}$ is a self-shrinker, that is,

$$H = -X^N.$$

Furthermore, we know that M is a minimal submanifold in $\mathbb{R}^{n+p}$ with respect to the metric $g_{ij} = e^{-\frac{|X|^2}{n}} \delta_{ij}$ if and only if $X : M^n \to \mathbb{R}^{n+p}$ is a self-shrinker, that is,

$$H = -X^N.$$

Thus, self-shrinker can be characterized by one of the followings:

(1) A submanifold $X : M^n \to \mathbb{R}^{n+p}$ is a self-shrinker if $H = -X^N$.
(2) A submanifold $X : M^n \to \mathbb{R}^{n+p}$ is a self-shrinker if $X : M^n \to \mathbb{R}^{n+p}$ is a critical point of the functional $\mathcal{F}_s = (2\pi)^{-n/2} \int_M e^{-\frac{|X_s|^2}{2}} dv$.
(3) A submanifold $X : M^n \to \mathbb{R}^{n+p}$ is a self-shrinker if M^n is a minimal submanifold in $\mathbb{R}^{n+p}$ with respect to the metric $g_{ij} = e^{-\frac{|X|^2}{n}} \delta_{ij}$.

1.3. *Examples of self-shrinkers*

Example 1.1. $\mathbb{R}^n \subset \mathbb{R}^{n+1}$ is a complete self-shrinker.

Example 1.2. $S^n(\sqrt{n}) \subset \mathbb{R}^{n+1}$ is a compact self-shrinker.

Example 1.3. For any positive integers $m_1, \cdots, m_p$ such that $m_1 + \cdots + m_p = n$, submanifold

$$M^n = \mathbb{S}^{m_1}(\sqrt{m_1}) \times \cdots \times \mathbb{S}^{m_p}(\sqrt{m_p}) \subset \mathbb{R}^{n+p}$$

is an n-dimensional compact self-shrinker in $\mathbb{R}^{n+p}$ with

$$S = p, \quad |H|^2 = n,$$

where S denotes the squared norm of the second fundamental form.

Example 1.4. For positive integers $m_1, \cdots, m_p, q \geq 1$, with $m_1 + \cdots + m_p + q = n$, submanifold

$$M^n = \mathbb{S}^{m_1}(\sqrt{m_1}) \times \cdots \times \mathbb{S}^{m_p}(\sqrt{m_p}) \times \mathbb{R}^q \subset \mathbb{R}^{n+p}$$

is an n-dimensional complete noncompact self-shrinker in $\mathbb{R}^{n+p}$ which satisfies

$$S = p, \quad |H|^2 = p.$$

Definition 1.1. We say that a submanifold M^n in $\mathbb{R}^{n+p}$ has polynomial volume growth if there exist constants C and d such that for all $r \geq 1$ such that

$$Vol(B_r(0) \cap M) \leq Cr^d$$

holds, where $B_r(0)$ is the Euclidean ball with radius r and centered at the origin.

Remark 1.1. The above examples have polynomial volume growth and non-negative mean curvature.

2. Complete self-shrinkers in $\mathbb{R}^{n+1}$

2.1. *Compact self-shrinkers in $\mathbb{R}^{n+1}$*

For $n = 1$, Abresch and Langer[2] classified all smooth closed self-shrinker curves in $\mathbb{R}^2$ and showed that the round circle is the only embedded self-shrinkers.

For $n \geq 2$, Huisken[11] studied compact self-shrinkers and gave a complete classification of self-shrinkers with non-negative mean curvature.

Theorem 2.1. *If $X : M^n \to \mathbb{R}^{n+1}(n \geq 2)$ is an n-dimensional compact self-shrinker with non-negative mean curvature H in $\mathbb{R}^{n+1}$, then $X(M^n) = S^n(\sqrt{n})$.*

Remark 2.1. The condition of non-negative mean curvature is essential. In fact, let Δ and ∇ denote the Laplacian and the gradient operator on the self-shrinker, respectively and $\langle \cdot, \cdot \rangle$ denotes the standard inner product of $\mathbb{R}^{n+1}$. Because

$$\Delta H - \langle X, \nabla H \rangle + SH - H = 0,$$

we obtain $H > 0$ from the maximum principle if the mean curvature H is non-negative.

Remark 2.2. Angenent[1] has constructed compact self-shrinker torus $S^1 \times S^{n-1}$ in $\mathbb{R}^{n+1}$.

We consider the following linear operator:

$$\mathcal{L}u = \Delta u - \langle X, \nabla u \rangle.$$

In [5], we have studied eigenvalues of the $\mathcal{L}$-operator. The sharp universal estimates for eigenvalues of the $\mathcal{L}$-operator on compact self-shrinkers are obtained. Since

$$\mathcal{L}u = \Delta u - \langle X, \nabla u \rangle = e^{\frac{|X|^2}{2}} div(e^{-\frac{|X|^2}{2}} \nabla u),$$

we know that, from the Stokes formula, the operator $\mathcal{L}$ is self-adjoint in a weighted L^2-space, that is, for a submanifold $X : M^n \to \mathbb{R}^{n+p}$,

$$\int_M u(\mathcal{L}v) e^{-\frac{|X|^2}{2}} dv = - \int_M \langle \nabla v, \nabla u \rangle e^{-\frac{|X|^2}{2}} dv \qquad (4)$$

holds, where u, v are C^2 functions. Now, we can prove Theorem 2.1.

Proof. From Remark 2.1, we can assume $H > 0$ because of $H \geq 0$.

Huisken considered the function $\dfrac{S}{H^2}$ and computed

$$\mathcal{L}\frac{S}{H^2} = \Delta \frac{S}{H^2} - \langle X, \nabla \frac{S}{H^2} \rangle$$

$$= \frac{2}{H^4} |h_{ij} \nabla_k H - h_{ijk} H|^2 - \frac{2}{H} \langle \nabla H, \nabla \frac{S}{H^2} \rangle,$$

where h_{ij} and h_{ijk} denote components of the second fundamental form and the first covariant derivative of it.

Since the operator $\mathcal{L}$ is self-adjoint in a weighted L^2-space and M is compact, we can get that by (4)

$$\int_M S\left(\mathcal{L}\frac{S}{H^2} + \frac{2}{H}\langle \nabla H, \nabla \frac{S}{H^2}\rangle \right) e^{-\frac{|X|^2}{2}} dv = - \int_M H^2 \langle \nabla \frac{S}{H^2}, \nabla \frac{S}{H^2} \rangle e^{-\frac{|X|^2}{2}} dv.$$

Hence, we have

$$\int_M \frac{2S}{H^4} |h_{ij} \nabla_k H - h_{ijk} H|^2 e^{-\frac{|X|^2}{2}} dv + \int_M H^2 \langle \nabla \frac{S}{H^2}, \nabla \frac{S}{H^2} \rangle e^{-\frac{|X|^2}{2}} dv = 0.$$

Therefore, we obtain

$$h_{ij} \nabla_k H - h_{ijk} H = 0$$

on M and $S = \beta H^2$, where β is some positive constant. We split the tensor $h_{ij} \nabla_k H - h_{ijk} H$ into its symmetric and antisymmetric parts

$$0 = |h_{ij} \nabla_k H - h_{ijk} H|^2$$

$$= |h_{ijk} H - \frac{1}{2}(h_{ij} \nabla_k H + h_{ik} \nabla_j H) - \frac{1}{2}(h_{ij} \nabla_k H - h_{ik} \nabla_j H)|^2$$

$$\geq \frac{1}{4}|(h_{ij} \nabla_k H - h_{ik} \nabla_j H)|^2 \geq 0,$$

then

$$|h_{ij}\nabla_k H - h_{ik}\nabla_j H|^2 \equiv 0 \quad \text{on M.}$$

Now we have only to consider points where the gradient of the mean curvature does not vanish. At a given point of M^n we now rotate $e_1, \cdots, e_n$ such that $e_1 = \nabla H/|\nabla H|$ points in the direction of the gradient of the mean curvature. Then

$$0 = |h_{ij}\nabla_k H - h_{ik}\nabla_j H|^2 = 2|\nabla H|^2 (S - \sum_{i=1}^n h_{1i}^2).$$

Thus, at each point of M^n we have either $|\nabla H|^2 = 0$ or $S = \sum_{i=1}^n h_{1i}^2$. If $|\nabla H|^2 = 0$, then the second fundamental form is parallel. By a theorem of Lawson[13] and $H > 0$, we know that M^n is the round sphere $S^n(\sqrt{n})$. So we can suppose that there is a point in M where $S = \sum_{i=1}^n h_{1i}^2$. Since

$$S = h_{11}^2 + 2\sum_{i=1}^n h_{1i}^2 + \sum_{i,j\neq 1}^n h_{ij}^2$$

this is only possible if $h_{ij} = 0$ unless $i = j = 1$. Then we have $S = H^2$ at this point and therefore everywhere on M. Then, by $\Delta H = \langle X, \nabla H \rangle - SH + H$ and divergence theorem, we can conclude that

$$(1-n)\int_M H dv = 0,$$

which is a contradiction for $n \geq 2$. $\qquad\square$

2.2. *Self-shrinkers with polynomial volume growth*

Huisken[12] studied complete and non-compact self-shrinkers in $\mathbb{R}^{n+1}$. In general, since the Stokes formula can not be applied to complete and non-compact self-shrinkers in $\mathbb{R}^{n+1}$, he assumes that self-shrinkers in $\mathbb{R}^{n+1}$ have polynomial volume growth so that the similar formula can be used. He proved

Theorem 2.2. *Let* $X : M^n \to \mathbb{R}^{n+1}$ *be an n-dimensional complete non-compact self-shrinker in* $\mathbb{R}^{n+1}$ *with* $H \geq 0$ *and polynomial volume growth. If the squared norm S of the second fundamental form is bounded, then M^n is isometric to one of the following:*

(1) $S^m(\sqrt{m}) \times \mathbb{R}^{n-m} \subset \mathbb{R}^{n+1}$,

(2) $\Gamma \times \mathbb{R}^{n-1}$,

where Γ is one of curves of Abresch and Langer.

Remark 2.3. Whether the condition that the squared norm S of the second fundamental form is bounded can be dropped.

Colding and Minicozzi[4] have studied this problem. They have removed the assumption on the second fundamental form and proved the following

Theorem 2.3. *Let $X : M^n \to \mathbb{R}^{n+1}$ be an n-dimensional complete embedded self-shrinker in $\mathbb{R}^{n+1}$ with $H \geq 0$ and polynomial volume growth. Then M^n is isometric to one of the following:*

(1) $S^n(\sqrt{n})$,
(2) $\mathbb{R}^n$,
(3) $S^m(\sqrt{m}) \times \mathbb{R}^{n-m} \subset \mathbb{R}^{n+1}$, $\quad 1 \leq m \leq n-1$.

In the theorems of Huisken and Colding and Minicozzi, they all assumed the condition of polynomial volume growth. How do they use the condition of polynomial volume growth? In fact, the condition of polynomial volume growth is essential in the proof of theorems.

Proposition 2.1. *Let $X : M^n \to \mathbb{R}^{n+p}$ be a complete submanifold. If C^2-functions u, v satisfy*

$$\int_M \left(|u\nabla v| + |\nabla u||\nabla v| + |u\mathcal{L}v| \right) e^{-\frac{|X|^2}{2}} dv < +\infty,$$

then, one has

$$\int_M u(\mathcal{L}v)e^{-\frac{|X|^2}{2}} dv = -\int_M \langle \nabla v, \nabla u \rangle e^{-\frac{|X|^2}{2}} dv. \qquad (5)$$

Proof. Let η be a function with compact support on M. Then

$$\int_M \eta u(\mathcal{L}v)e^{-\frac{|X|^2}{2}} dv = -\int_M u\langle \nabla \eta, \nabla v \rangle e^{-\frac{|X|^2}{2}} dv - \int_M \eta \langle \nabla u, \nabla v \rangle e^{-\frac{|X|^2}{2}} dv.$$

Let $\eta = \eta_j$ be a cut-off function linearly to zero from B_j to B_{j+1}, where $B_j = M \cap B_j(0)$, $B_j(0)$ is the Euclidean ball of radius j centered at the origin. Then

$$\int_M \eta_j u(\mathcal{L}v)e^{-\frac{|X|^2}{2}} dv = -\int_M u\langle \nabla \eta_j, \nabla v \rangle e^{-\frac{|X|^2}{2}} dv - \int_M \eta_j \langle \nabla u, \nabla v \rangle e^{-\frac{|X|^2}{2}} dv.$$

Since $|\eta_j|$ and $|\nabla \eta_j|$ are bounded by one, let $j \to \infty$, dominated convergence theorem gives (5). $\qquad \square$

Proof of Theorem 2.3. From $\Delta H - \langle X, \nabla H \rangle + SH - H = 0$, we know $H \equiv 0$ or $H > 0$ by making use of the maximum principle. If $H \equiv 0$, we know that M^n is isometric to $\mathbb{R}^n$ because M^n is a complete self-shrinker. Hence, one only needs to consider the case of $H > 0$.

The condition of polynomial volume growth yields the following:

Proposition 2.2. *If $X : M^n \to \mathbb{R}^{n+1}$ is an n-dimensional complete self-shrinker with $H > 0$ and polynomial volume growth, then*

$$\int_M (S|\nabla \log H| + |\nabla S||\nabla \log H| + S|\mathcal{L} \log H|) e^{-\frac{|X|^2}{2}} dv < \infty,$$

$$\int_M (S^{\frac{1}{2}}|\nabla S^{\frac{1}{2}}| + |\nabla S^{\frac{1}{2}}|^2 + S^{\frac{1}{2}}|\mathcal{L}S^{\frac{1}{2}}|) e^{-\frac{|X|^2}{2}} dv < \infty.$$

Before we prove Proposition 2.2, we need the following two propositions:

Proposition 2.3. *Let $X : M^n \to \mathbb{R}^{n+1}$ be an n-dimensional complete self-shrinker with $H > 0$. If ϕ is in the weighted $W^{1,2}$, i.e.*

$$\int_M (|\phi|^2 + |\nabla \phi|^2) e^{-\frac{|X|^2}{2}} dv < +\infty,$$

then

$$\int_M \phi^2 (S + \frac{1}{2}|\nabla \log H|^2) e^{-\frac{|X|^2}{2}} dv \leq \int_M (2|\nabla \phi|^2 + \phi^2) e^{-\frac{|X|^2}{2}} dv. \quad (6)$$

Proof. Since $X : M^n \to \mathbb{R}^{n+1}$ is an n-dimensional complete self-shrinker, we have

$$\mathcal{L} \log H = 1 - S - |\nabla \log H|^2. \quad (7)$$

If φ is a smooth function with compact support, then

$$\int_M \langle \nabla \varphi^2, \nabla \log H \rangle e^{-\frac{|X|^2}{2}} dv = -\int_M \varphi^2 \mathcal{L} \log H e^{-\frac{|X|^2}{2}} dv$$

$$= -\int_M \varphi^2 (1 - S - |\nabla \log H|^2) e^{-\frac{|X|^2}{2}} dv.$$

Using the following inequality

$$\langle \nabla \varphi^2, \nabla \log H \rangle \leq 2|\nabla \varphi|^2 + \frac{1}{2}\varphi^2 |\nabla \log H|^2$$

we can obtain that

$$\int_M \varphi^2 (S + \frac{1}{2}|\nabla \log H|^2) e^{-\frac{|X|^2}{2}} dv \leq \int_M (2|\nabla \varphi|^2 + \varphi^2) e^{-\frac{|X|^2}{2}} dv. \quad (8)$$

Let η_j be one on $B_j = M \cap B_j(0)$ and cut off linearly to zero from ∂B_j to ∂B_{j+1}, where $B_j(0)$ is the Euclidean ball of radius j centered at the origin. Applying (8) with $\varphi = \eta_j \phi$, letting $j \to \infty$ and using the monotone convergence theorem gives (6). $\square$

Proposition 2.4. *Let* $X : M^n \to \mathbb{R}^{n+1}$ *be an n-dimensional complete self-shrinker with* $H > 0$. *If* M^n *has polynomial volume growth, then*

$$\int_M (S + S^2 + |\nabla S^{\frac{1}{2}}| + \sum_{i,j,k}(h_{ijk})^2)e^{-\frac{|X|^2}{2}}\,dv < \infty. \tag{9}$$

Proof. Give any compactly supported function ϕ, we have by (7)

$$\int_M \langle \nabla \phi^2, \nabla log H \rangle e^{-\frac{|X|^2}{2}}\,dv = -\int_M \phi^2 \mathcal{L} log H e^{-\frac{|X|^2}{2}}\,dv$$

$$= -\int_M \phi^2(1 - S - |\nabla log H|^2)e^{-\frac{|X|^2}{2}}\,dv.$$

By the following inequality

$$\langle \nabla \phi^2, \nabla log H \rangle \leq |\nabla \phi|^2 + \phi^2 |\nabla log H|^2$$

we can easily get

$$\int_M \phi^2 S e^{-\frac{|X|^2}{2}}\,dv \leq \int_M (\phi^2 + |\nabla \phi|^2)e^{-\frac{|X|^2}{2}}\,dv.$$

Let $\phi = \eta S^{\frac{1}{2}}$, where $\eta \geq 0$ has compact support. Then

$$\int_M \eta^2 S^2 e^{-\frac{|X|^2}{2}}\,dv \leq \int_M (\eta^2 S + |\nabla \eta|^2 S + 2\eta S^{\frac{1}{2}}|\nabla \eta||\nabla S^{\frac{1}{2}}| + \eta^2|\nabla S^{\frac{1}{2}}|^2)e^{-\frac{|X|^2}{2}}\,dv$$

For $\varepsilon > 0$, we have

$$2\eta S^{\frac{1}{2}}|\nabla \eta||\nabla S^{\frac{1}{2}}| \leq \varepsilon \eta^2|\nabla S^{\frac{1}{2}}|^2 + \frac{1}{\varepsilon}S|\nabla \eta|^2.$$

Thus,

$$\int_M \eta^2 S^2 e^{-\frac{|X|^2}{2}}\,dv \leq (1 + \varepsilon)\int_M \eta^2|\nabla S^{\frac{1}{2}}|^2 e^{-\frac{|X|^2}{2}}\,dv$$

$$+ \int_M ((1 + \frac{1}{\varepsilon})|\nabla \eta|^2 + \eta^2)S e^{-\frac{|X|^2}{2}}\,dv. \tag{10}$$

We fix a point p and choose a frame e_i, $i = 1, \cdots, n$ such that $h_{ij} = \lambda_i \delta_{ij}$. By using $\nabla S = 2S^{\frac{1}{2}}\nabla S^{\frac{1}{2}}$, we have

$$|\nabla S^{\frac{1}{2}}|^2 \leq \sum_{i,k} h_{iik}^2 \leq \sum_{i,j,k} h_{ijk}^2. \tag{11}$$

Since

$$
\begin{aligned}
|\nabla S^{\frac{1}{2}}|^2 &\leq \sum_{i,k} h_{iik}^2 = \sum_{i\neq k} h_{iik}^2 + \sum_i h_{iii}^2 \\
&= \sum_{i\neq k} h_{iik}^2 + \sum_i (H_{,i} - \sum_{j\neq i} h_{jji})^2 \\
&\leq \sum_{i\neq k} h_{iik}^2 + n\sum_i H_{,i}^2 + n\sum_{j\neq i} h_{jji}^2 \\
&= n|\nabla H|^2 + (n+1)\sum_{i\neq k} h_{iik}^2 \\
&= n|\nabla H|^2 + \frac{n+1}{2}(\sum_{i\neq k} h_{iki}^2 + \sum_{i\neq k} h_{kii}^2),
\end{aligned}
$$

then, we have

$$
(1 + \frac{2}{n+1})|\nabla S^{\frac{1}{2}}|^2 \leq \sum_{i,j,k} h_{ijk}^2 + \frac{2n}{n+1}|\nabla H|^2.
$$

Thus, we have

$$
\mathcal{L}S = 2S - 2S^2 + 2\sum_{i,j,k} h_{ijk}^2
$$

$$
\geq 2S - 2S^2 + 2(1 + \frac{2}{n+1})|\nabla S^{\frac{1}{2}}|^2 - \frac{4n}{n+1}|\nabla H|^2. \tag{12}
$$

Taking integral at the both sides of (12) with $\frac{1}{2}\eta^2$, one yields

$$
\frac{1}{2}\int_M \eta^2 \mathcal{L}S e^{-\frac{|X|^2}{2}} dv \geq \int_M \eta^2\Big(S - S^2 + (1 + \frac{2}{n+1})|\nabla S^{\frac{1}{2}}|^2
$$
$$
- \frac{2n}{n+1}|\nabla H|^2\Big)e^{-\frac{|X|^2}{2}} dv.
$$

By combining this with

$$
\begin{aligned}
\frac{1}{2}\int_M \eta^2 \mathcal{L}S e^{-\frac{|X|^2}{2}} dv &= -\frac{1}{2}\int_M \langle\nabla\eta^2, \nabla S\rangle e^{-\frac{|X|^2}{2}} dv \\
&= -\int_M \eta\langle\nabla\eta, \nabla S\rangle e^{-\frac{|X|^2}{2}} dv \\
&\leq \int_M (\frac{1}{\varepsilon}S|\nabla\eta|^2 + \varepsilon\eta^2|\nabla S^{\frac{1}{2}}|^2)e^{-\frac{|X|^2}{2}} dv,
\end{aligned}
$$

we have

$$
\int_M (\eta^2 S^2 + \frac{2n}{n+1}\eta^2|\nabla H|^2 + \frac{1}{\varepsilon}S|\nabla\eta|^2)e^{-\frac{|X|^2}{2}} dv
$$
$$
\geq \int_M \eta^2|\nabla S^{\frac{1}{2}}|^2(1 + \frac{2}{n+1} - \varepsilon)e^{-\frac{|X|^2}{2}} dv. \tag{13}
$$

Since $|\eta| \leq 1$ and $|\nabla \eta| \leq 1$, from (10) and (13), we can get

$$\int_M \eta^2 S^2 e^{-\frac{|X|^2}{2}} dv \leq \frac{1+\varepsilon}{1+\frac{2}{n+1}-\varepsilon} \int_M \eta^2 S^2 e^{-\frac{|X|^2}{2}} dv$$
$$+ C_\varepsilon \int_M (|\nabla H|^2 + S) e^{-\frac{|X|^2}{2}} dv.$$

Choosing $\varepsilon > 0$ sufficiently small, such that $\frac{1+\varepsilon}{1+\frac{2}{n+1}-\varepsilon} < 1$, then

$$\int_M \eta^2 S^2 e^{-\frac{|X|^2}{2}} dv \leq C \int_M (|\nabla H|^2 + S) e^{-\frac{|X|^2}{2}} dv$$
$$\leq C \int_M S(1+|X|^2) e^{-\frac{|X|^2}{2}} dv. \tag{14}$$

We should notice that 1 and $|X|$ are in the weighted $W^{1,2}$ space, so we can let $\phi \equiv 1$ and $\phi \equiv |X|$ in Proposition 2.3, by polynomial volume growth, then one can easily get that

$$\int_M S e^{-\frac{|X|^2}{2}} dv < \infty, \quad \int_M S|X|^2 e^{-\frac{|X|^2}{2}} dv < \infty. \tag{15}$$

(14), (15) and dominated convergence theorem give that

$$\int_M S^2 e^{-\frac{|X|^2}{2}} dv < \infty.$$

Together with (13), we have $\int_M |\nabla S^{\frac{1}{2}}|^2 e^{-\frac{|X|^2}{2}} dv < \infty$. By taking integral $\frac{1}{2}\mathcal{L}S = S - S^2 + \sum\limits_{i,j,k} h_{ijk}^2$ with $\eta^2 e^{-\frac{|X|^2}{2}}$, it's not difficult to prove that

$$\int_M \sum_{i,j,k} h_{ijk}^2 e^{-\frac{|X|^2}{2}} dv < \infty. \qquad \square$$

Proof. We begin to prove Proposition 2.2. In fact, Proposition 2.4 implies that $S^{\frac{1}{2}}$ is in the weighted $W^{1,2}$ space, so by Proposition 2.3, we get

$$\int_M S|\nabla log H|^2 e^{-\frac{|X|^2}{2}} dv < \infty,$$

$$\int_M |\nabla S||\nabla log H| e^{-\frac{|X|^2}{2}} dv \leq \int_M (|\nabla S^{\frac{1}{2}}|^2 + S|\nabla log H|^2) e^{-\frac{|X|^2}{2}} dv < \infty.$$

From (7), we have

$$\int_M S|\mathcal{L}log H| e^{-\frac{|X|^2}{2}} dv = \int_M S|1 - S - |\nabla log H|^2| e^{-\frac{|X|^2}{2}} dv < \infty.$$

Since

$$\mathcal{L}S^{\frac{1}{2}} = S^{\frac{1}{2}} - S^{\frac{3}{2}} + S^{-\frac{1}{2}}\sum_{i,j,k} h_{ijk}^2 - S^{-\frac{1}{2}}|\nabla S^{\frac{1}{2}}|^2, \tag{16}$$

we obtain that

$$\int_M S^{\frac{1}{2}}|\mathcal{L}S^{\frac{1}{2}}|e^{-\frac{|X|^2}{2}}\,dv = \int_M (S - S^2 + \sum_{i,j,k} h_{ijk}^2 - |\nabla S^{\frac{1}{2}}|)e^{-\frac{|X|^2}{2}}\,dv.$$

The second part follows from Proposition 2.4. □

Now, we continue to prove Theorem 2.3. By applying Proposition 2.1 to S and $\log H$, and $S^{\frac{1}{2}}$, respectively, we have by Proposition 2.2

$$\int_M (S|\nabla \log H| + |\nabla S||\nabla \log H| + S|\mathcal{L}\log H|)e^{-\frac{|X|^2}{2}}\,dv < \infty,$$

$$\int_M (S^{\frac{1}{2}}|\nabla S^{\frac{1}{2}}| + |\nabla S^{\frac{1}{2}}|^2 + S^{\frac{1}{2}}|\mathcal{L}S^{\frac{1}{2}}|)e^{-\frac{|X|^2}{2}}\,dv < \infty.$$

We can see that they satisfy the conditions of Proposition 2.1, then

$$\int_M \langle \nabla S, \nabla \log H\rangle e^{-\frac{|X|^2}{2}}\,dv = -\int_M S\mathcal{L}\log H e^{-\frac{|X|^2}{2}}\,dv$$

and

$$\int_M \langle \nabla S^{\frac{1}{2}}, \nabla S^{\frac{1}{2}}\rangle e^{-\frac{|X|^2}{2}}\,dv = -\int_M S^{\frac{1}{2}}\mathcal{L}S^{\frac{1}{2}}e^{-\frac{|X|^2}{2}}\,dv.$$

Since

$$\mathcal{L}\log H = 1 - S - |\nabla \log H|^2, \quad \mathcal{L}S^{\frac{1}{2}} \geq S^{\frac{1}{2}} - S^{\frac{3}{2}},$$

we obtain

$$\int_M |S^{\frac{1}{2}}\nabla \log H - \nabla S^{\frac{1}{2}}|^2 e^{-\frac{|X|^2}{2}}\,dv \leq 0,$$

that is,

$$|S^{\frac{1}{2}}\nabla \log H - \nabla S^{\frac{1}{2}}|^2 \equiv 0.$$

Hence,

$$|\nabla S^{\frac{1}{2}}|^2 = \sum_{i,j,k} h_{ijk}^2, \quad S = \beta H^2$$

for some positive constant β.

Claim: $\nabla_k H = 0$ and $h_{ijk} = 0$, or $S = H^2$. In fact, if we fix a point p and choose a frame $e_i, i = 1, \cdots, n$ such that $h_{ij} = \lambda_i \delta_{ij}$, we have proved

(11). From (16) and $LS^{\frac{1}{2}} \geq S^{\frac{1}{2}} - S^{\frac{3}{2}}$, we know that the equality holds if and only if

$$|\nabla S^{\frac{1}{2}}|^2 = \sum_{i,j,k} h_{ijk}^2.$$

Together with (11),

$$|\nabla S^{\frac{1}{2}}|^2 = \sum_{i,k} h_{iik}^2 = \sum_{i,j,k} h_{ijk}^2,$$

i.e., (i) for each k, there exists a constant a_k such that $h_{iik} = a_k \lambda_i$ for every i, (ii) if $i \neq j$, $h_{ijk} = 0$. By Codazzi equation, we know (ii) implies that $h_{ijk} = 0$ unless $i = j = k$.

If $\lambda_i \neq 0$, for $j \neq i, 0 = h_{iij} = a_j \lambda_i$, we have $a_j = 0$. If the rank of (h_{ij}) is at least two at p, then $a_j = 0$ for all $j \in \{1, \cdots, n\}$. Thus, (i), (ii) imply $\nabla_k H = 0$ and $h_{ijk} = 0$. If the rank of (h_{ij}) is one at p, then H is the only nonzero eigenvalue of (h_{ij}), and $S = H^2$. It is not difficult to prove that, the above two cases also hold everywhere on M. Thus, if $\nabla_k H = 0$ and $h_{ijk} = 0$, by a theorem of Lawson[13] and $H > 0$, we know that M^n is the round sphere $S^n(\sqrt{n})$. If $S = H^2$, by $\Delta H = \langle X, \nabla H \rangle - SH + H$ and using twice divergence theorem, we can conclude that

$$(1-n) \int_M H dv = 0,$$

which is a contradiction for $n \geq 2$. It completes the proof of Theorem 2.3.

2.3. Without condition of polynomial volume growth

In the theorems of Huisken and Colding and Minicozzi, they all assumed the condition of polynomial volume growth.

Open problem. Does the theorem of Colding and Minicozzi holds without the assumption of polynomial volume growth?

In fact, without the assumption of polynomial volume growth, we can give a characterization of complete self-shrinkers (see [6]).

Theorem 2.4. Let $X : M^n \to \mathbb{R}^{n+1}$ be an n-dimensional complete self-shrinker with $\inf H^2 > 0$. If S is bounded, then $\inf S \leq 1$.

Corollary 2.1. Let $X : M^n \to \mathbb{R}^{n+1}$ be an n-dimensional complete self-shrinker with $\inf H^2 > 0$. If S is constant, then M^n is isometric to one of the following:

(1) $S^n(\sqrt{n})$,

(2) $S^m(\sqrt{m}) \times \mathbb{R}^{n-m} \subset \mathbb{R}^{n+1}, 1 \leq m \leq n-1$.

Remark 2.4. Angenent[1] has constructed compact self-shrinker torus $S^1 \times S^{n-1}$ in $\mathbb{R}^{n+1}$ with inf $H^2 = 0$.

In order to prove our theorem, first of all, we generalize the generalized maximum principle of Omori-Yau to $\mathcal{L}$-operator on complete self-shrinkers.

Theorem 2.5. *(Generalized maximum principle for $\mathcal{L}$-operator[6]) Let $X : M^n \to \mathbb{R}^{n+p}$ be an n-dimensional complete self-shrinker with Ricci curvature bounded from below. If f is a C^2-function bounded from above, there exists a sequence of points $\{p_k\} \subset M^n$, such that*

$$\lim_{k \to \infty} f(X(p_k)) = \sup f, \quad \lim_{k \to \infty} |\nabla f|(X(p_k)) = 0, \quad \limsup_{k \to \infty} \mathcal{L}f(X(p_k)) \leq 0.$$

Remark 2.5. Since the $\mathcal{L}$-operator is defined by

$$\mathcal{L}f = \Delta f - \langle X, \nabla f \rangle,$$

in order to prove the generalized maximum principle for $\mathcal{L}$-operator, we need to consider the second term

$$-\langle X, \nabla f \rangle.$$

By the same proof of Yau, we can prove

$$-\langle X(p_k), \nabla f(X(p_k)) \rangle \to 0, \quad (k \to \infty).$$

3. Complete self-shrinkers in $\mathbb{R}^{n+p}$

In order to prove their theorems, Huisken and Colding-Minicozzi assumed that complete self-shrinkers in $\mathbb{R}^{n+1}$ satisfy the condition of polynomial volume growth. One wants to ask what means the condition of polynomial volume growth?

For complete self-shrinkers with arbitrary co-dimensions in $\mathbb{R}^{n+p}$, Ding-Xin[8] and X. Cheng-Zhou[7] have completely characterized this condition:

Theorem 3.1. *An n-dimensional complete non-compact self-shrinker M^n in $\mathbb{R}^{n+p}$ has polynomial volume growth if and only if M^n is proper.*

For n-dimensional complete proper self-shrinkers M^n in $\mathbb{R}^{n+p}$, X. Cheng and Zhou[7] has proved that the volume growth satisfies

$$Vol(B_r(0) \cap M^n) \leq Cr^{n-\beta},$$

where $\beta \leq \inf |H|^2$ is constant.

Remark 3.1. Estimate of the volume growth for n-dimensional complete proper self-shrinkers M^n in $\mathbb{R}^{n+p}$ are optimal. In fact, self-shrinker

$$\mathbb{S}^k(\sqrt{k}) \times \mathbb{R}^{n-k}, \quad 0 \le k \le n, \quad |H| = \sqrt{k},$$

satisfies

$$Vol(B_r(0) \cap M) = Cr^{n-k}.$$

For n-dimensional complete self-shrinkers M^n in $\mathbb{R}^{n+p}$, Cao and Li[3] have studied their rigidity, and proved a gap theorem on the squared norm S of the second fundamental form:

Theorem 3.2. *Let* $X : M^n \to \mathbb{R}^{n+p}(p \ge 1)$ *be a complete self-shrinker with polynomial volume growth in* $\mathbb{R}^{n+p}$. *If the squared norm* S *of the second fundamental form satisfies*

$$S \le 1.$$

Then M *is one of the following:*

(1) $S^n(\sqrt{n})$,
(2) $S^m(\sqrt{m}) \times \mathbb{R}^{n-m} \subset \mathbb{R}^{n+1}, 1 \le m \le n-1$,
(3) $\mathbb{R}^n$.

In the theorem of H. Cao and H. Li,[3] they also assume that complete self-shrinkers have polynomial volume growth. Without assumption that complete self-shrinkers have polynomial volume growth, we have proved the following in [6]:

Theorem 3.3. *Let* $X : M^n \to \mathbb{R}^{n+p}$ $(p \ge 1)$ *be an* n-*dimensional complete self-shrinker in* $\mathbb{R}^{n+p}$, *then, one of the following holds:*

(1) $\sup S \ge 1$,
(2) $S \equiv 0$ *and* M^n *is* $\mathbb{R}^n$.

Remark 3.2. Let $X : M^n \to \mathbb{R}^{n+p}$ $(p \ge 1)$ be an n-dimensional complete self-shrinker in $\mathbb{R}^{n+p}$. If $\sup S < 1$, then M^n is $\mathbb{R}^n$.

Remark 3.3. The round sphere $S^n(\sqrt{n})$ and the cylinder $S^k(\sqrt{k}) \times \mathbb{R}^{n-k}, 1 \le k \le n-1$ are complete self-shrinkers in $\mathbb{R}^{n+1}$ with $S = 1$. Thus, our result is sharp.

References

1. S. Angenent, *Shrinking doughnuts*, Nonlinear diffusion equations and their equilibrium states, 3 (Gregrnog, 1989), Progr. Nonlinear Differential Equations Appl., vol. 7, Birkhaüser Boston, Massachusetts, 1992, pp. 21-38.

2. U. Abresch and J. Langer, *The normalized curve shortening flow and homothetic solutions*, J. Diff. Geom., **23**(1986), 175-196.

3. H.-D. Cao and H. Li, *A Gap theorem for self-shrinkers of the mean curvature flow in arbitrary codimension*, Calc. Var. Partial Differential Equations, **46** (2013),879-889, DOI: 10.1007/s00526-012-0508-1.

4. Tobias H. Colding and William P. Minicozzi II, *Generic Mean Curvature Flow I; Generic Singularities*, Ann. of Math., **175** (2012), 755-833.

5. Q.-M. Cheng and Y. Peng, *Estimates for eigenvalues of $\mathcal{L}$ operator on self-shrinkers*, to appear in Comm. Contemp. Math., DOI: 10.1142/S0219199713500119.

6. Q.-M. Cheng and Y. Peng, *Complete self-shrinkers of the mean curvature flow*, arXiv:1202.1053.

7. X. Cheng and D. Zhou, *Volume estimate about shrinkers*, Proc. Amer. Math. Soc. **141** (2013), 687-696.

8. Q. Ding, Y. L. Xin, *Volume growth, eigenvalue and compactness for self-shrinkers*, arXiv:1101.1411, 2011.

9. M. A. Grayson, *A short note on the evolution of a surface by its mean curvature*, Duke Math. J. **58** (1989) 555-558.

10. G. Huisken, *Flow by mean curvature convex surfaces into spheres*, J. Differential Geom., **20** (1984), 237-266.

11. G. Huisken, *Asymptotic Behavior for Singularities of the Mean Curvature Flow*, J. Differential Geom., **31** (1990), 285-299.

12. G. Huisken, *Local and global behaviour of hypersurfaces moving by mean curvature*, Differential geometry: partial differential equations on manifolds (Los Angeles, CA, 1990), Proc. Sympos. Pure Math., 54, Part 1, Amer. Math. Soc., Providence, RI, (1993), 175-191.

13. H. B. Lawson, *Local rigidity theorems for minimal hypersurfaces*, Ann. of Math., **89** (1969), 187-197.

Received February 26, 2013.

Proceedings of the Workshop on
Differential Geometry of Submanifolds
and its Related Topics
Saga, August 4-6, 2012

SPECTRUM OF POLY-LAPLACIAN AND FRACTIONAL LAPLACIAN

Lingzhong ZENG*

*Department of Mathematics, Graduate School
of Science and Engineering, Saga University,
1 Honzyo, Saga 840-8502, Japan
E-mail: lingzhongzeng@yeah.net*

Dedicated to Professor Sadahiro Maeda for his sixtieth birthday

In this paper, we study eigenvalues of poly-Laplacian and fractional Laplacian on bounded domains in Euclidean space. In particular, we focus all attentions on lower bounds for sum of eigenvalues of poly-Laplacian and fractional Laplacian. In addition, for the case of Laplacian, we also discuss lower bounds of eigenvalues of Dirichlet problem on a polytope.

Keywords: eigenvalue, lower bound, Laplacian, poly-Laplacian, fractional Laplacian, polytope.

1. Eigenvalue Problems

The first eigenvalue problem we consider is that of Dirichlet problem of poly-Laplacian with arbitrary order. Let Ω be a bounded domain with piecewise smooth boundary $\partial\Omega$ in an n-dimensional Euclidean space $\mathbb{R}^n$. Let $\lambda_i^{(l)}$ be the i-th eigenvalue of Dirichlet problem of poly-Laplacian with arbitrary order:

$$
\begin{cases}
(-\Delta)^l u = \lambda u, & \text{in } \Omega, \\
u = \dfrac{\partial u}{\partial \nu} = \cdots = \dfrac{\partial^{l-1} u}{\partial \nu^{l-1}} = 0, & \text{on } \partial\Omega,
\end{cases}
\tag{1}
$$

where Δ is the Laplacian in $\mathbb{R}^n$ and ν denotes the outward unit normal vector field of the boundary $\partial\Omega$. It is well known that the spectrum of eigenvalue problem (1) is real and discrete: $0 < \lambda_1^{(l)} \le \lambda_2^{(l)} \le \lambda_3^{(l)} \le \cdots \nearrow$

---

*Current address: Department of Mathematics, Jiangxi Normal University, Nanchang, Jiangxi 330022, P. R. China.

$+\infty$, where each $\lambda_i^{(l)}$ has finite multiplicity. We denote $V(\Omega)$ the volume of Ω and B_n the volume of unit ball in $\mathbb{R}^n$.

Good sources of information on this problem can be found in those books [2, 8, 9, 25]. Certainly, some classical papers of Payne [29], and Protter [31] are quite useful, and we also refer the readers to [13, 18, 20, 26, 30].

Next, we consider the eigenvalue problem of fractional Laplacian restricted to Ω. We denote the fractional Laplacian operators by $(-\Delta)^{\alpha/2}|_\Omega$, where $\alpha \in (0, 2]$. This fractional Laplacian is defined by

$$(-\Delta)^{\alpha/2} u(x) =: \mathbf{P.V.} \int_{\mathbb{R}^n} \frac{u(x) - u(y)}{|x - y|^{n+\alpha}} dy,$$

where **P.V.** denotes the principal value and $u : \mathbb{R}^n \to \mathbb{R}$. We define a characteristic function $\chi_\Omega : t \mapsto \chi_\Omega(t)$ on Ω, such that $\chi_\Omega(t) = 1$ when $x \in \Omega$ and $\chi_\Omega(t) = 0$ when $x \in \mathbb{R}^n \backslash \Omega$, then the special pseudo-differential operator can be represented as a Fourier transform of u [21, 35], namely, $(-\Delta)^{\alpha/2}|_\Omega u := \mathbf{F}^{-1}[|\xi|^\alpha \mathbf{F}[u\chi_\Omega]]$, where $\mathbf{F}[u]$ denotes the Fourier transform of u:

$$\mathbf{F}[u](\xi) = \widehat{u}(\xi) = \frac{1}{(2\pi)^n} \int_{\mathbb{R}^n} e^{-ix\cdot\xi} u(x) dx.$$

It is well known that fractional Laplacian operator $(-\Delta)^{\alpha/2}$ can be considered as a infinitesimal generator of the symmetric α-stable process [4, 5, 6, 7, 41]. Let $\Lambda_j^{(\alpha)}$ and $u_j^{(\alpha)}$ denote the j-th eigenvalue and the corresponding normalized eigenvector of $(-\Delta)^{\alpha/2}|_\Omega$, respectively. Eigenvalues $\Lambda_j^{(\alpha)}$ (including multiplicities) satisfy: $0 < \Lambda_1^{(\alpha)} \le \Lambda_2^{(\alpha)} \le \Lambda_3^{(\alpha)} \le \cdots \nearrow +\infty$.

2. Some Technical Lemmas

In this section, we will give some technical Lemmas which will play an important role in the proof of Theorem 4.1 and Theorem 5.1.

Lemma 2.1. *Suppose that* $\psi : [0, \infty) \to [0, 1]$ *such that*

$$0 \le \psi(s) \le 1 \ and \ \int_0^\infty \psi(s) ds = 1.$$

Then, there exists $\epsilon \ge 0$ *such that*

$$\int_\epsilon^{\epsilon+1} s^d ds = \int_0^\infty s^d \psi(s) ds \ and \ \int_\epsilon^{\epsilon+1} s^{d+2\ell} ds \le \int_0^\infty s^{d+2\ell} \psi(s) ds.$$

Lemma 2.2. *For $s > 0$, $\tau > 0$, $2 \le b \in \mathbb{N}$, $l \in \mathbb{N}^+$, we have the following inequality:*

$$bs^{b+2l} - (b + 2l)\tau^{2l}s^b + 2l\tau^{b+2l} - 2l\tau^{b+2(l-1)}(s - \tau)^2 \ge 0. \qquad (2)$$

Lemma 2.3. *Let $b(\ge 2)$ be a positive real number and $\mu > 0$. If $\psi : [0, +\infty) \to [0, +\infty)$ is a decreasing function such that*

$$-\mu \le \psi'(s) \le 0 \quad and \quad A := \int_0^\infty s^{b-1}\psi(s)ds > 0.$$

Let

$$\Upsilon = \frac{1}{b+\alpha}(bA)^{\frac{b+\alpha}{b}}\psi(0)^{-\frac{\alpha}{b}} + \frac{\alpha}{12b(b+\alpha)\mu^2}(bA)^{\frac{b+\alpha-2}{b}}\psi(0)^{\frac{2b-\alpha+2}{b}}$$

then, we have

$$\int_0^\infty s^{b+\alpha-1}\psi(s)ds \ge \Upsilon + \frac{\alpha(b+\alpha-2)^2}{288b^2(b+\alpha)^2\mu^4}(bA)^{\frac{b+\alpha-4}{b}}\psi(0)^{\frac{4b-\alpha+4}{b}}, \qquad (3)$$

when $b \ge 4$; we have

$$\int_0^\infty s^{b+\alpha-1}\psi(s)ds \ge \Upsilon + \frac{\alpha(b+\alpha-2)^2}{384b^2(b+\alpha)^2\mu^4}(bA)^{\frac{b+\alpha-4}{b}}\psi(0)^{\frac{4b-\alpha+4}{b}}, \qquad (4)$$

when $2 \le b < 4$. In particular, the inequality (2) holds when $\alpha = 2$ and $b \ge 2$.

Proof. If we consider the following function

$$\varrho(t) = \frac{\psi\left(\frac{\psi(0)}{\mu}t\right)}{\psi(0)},$$

then it is not difficult to see that $\varrho(0) = 1$ and $-1 \le \varrho'(t) \le 0$. Without loss of generality, we can assume $\psi(0) = 1$ and $\mu = 1$. For the sake of convenience, we define $E_\alpha := \int_0^\infty s^{b+\alpha-1}\psi(s)ds$ and put $h(s) := -\psi'(s)$. Then, for any $s \ge 0$, we have $0 \le h(s) \le 1$ and $\int_0^\infty h(s)ds = \psi(0) = 1$. By making use of integration by parts, we get

$$\int_0^\infty s^b h(s)ds = b\int_0^\infty s^{b-1}\psi(s)ds = bA \text{ and } \int_0^\infty s^{b+\alpha}h(s)ds \le (b+\alpha)E_\alpha,$$

since $\psi(s) > 0$. By Lemma 2.1, we infer that there exists an $\epsilon > 0$ such that

$$\int_\epsilon^{\epsilon+1} s^b ds = \int_0^\infty s^b h(s)ds = bA, \qquad (5)$$

and

$$\int_\epsilon^{\epsilon+1} s^{b+\alpha}ds \le \int_0^\infty s^{b+\alpha}h(s)ds \le (b+\alpha)E_\alpha. \qquad (6)$$

Let $\Theta(s) = bs^{b+\alpha} - (b+\alpha)\tau^\alpha s^b + \alpha\tau^{b+\alpha} - \alpha\tau^{b+\alpha-2}(s-\tau)^2$, then, by Lemma 2.2, we have $\Theta(s) \geq 0$. Integrating the function $\Theta(s)$ from ϵ to $\epsilon+1$, we deduce from (5) and (6) that, for any $\tau > 0$,

$$E_\alpha = \int_0^\infty s^{b+\alpha-1}\psi(s)ds \geq \frac{f(\tau)}{b(b+\alpha)},$$

where

$$f(\tau) := (b+\alpha)\tau^\alpha bA - \alpha\tau^{b+\alpha} + \frac{\alpha}{12}\tau^{b+\alpha-2}. \tag{7}$$

Taking

$$\tau = (bA)^{\frac{1}{b}}\left(1 + \frac{b+\alpha-2}{12(b+\alpha)}(bA)^{-\frac{2}{b}}\right)^{\frac{1}{b}},$$

and substituting it into (7), we obtain

$$\begin{aligned} f(\tau) = (bA)^{\frac{b+\alpha}{b}}&\left(b - \frac{\alpha(b+\alpha-2)}{12(b+\alpha)}(bA)^{-\frac{2}{b}}\right)\left(1 + \frac{b+\alpha-2}{12(b+\alpha)}(bA)^{-\frac{2}{b}}\right)^{\frac{\alpha}{b}} \\ &+ \frac{\alpha}{12}(bA)^{\frac{b+\alpha-2}{b}}\left(1 + \frac{b+\alpha-2}{12(b+\alpha)}(bA)^{-\frac{2}{b}}\right)^{\frac{b+\alpha-2}{b}}. \end{aligned} \tag{8}$$

By using Taylor formula, we can estimate the lower bound of $f(\tau)$ given by (8) and obtain that

$$f(\tau) \geq b(bA)^{\frac{b+\alpha}{b}} + \frac{\alpha}{12}(bA)^{\frac{b+\alpha-2}{b}} + \frac{\alpha(b+\alpha-2)^2}{288b(b+\alpha)}(bA)^{\frac{b+\alpha-4}{b}},$$

when $b \geq 4$;

$$f(\tau) \geq b(bA)^{\frac{b+\alpha}{b}} + \frac{\alpha}{12}(bA)^{\frac{b+\alpha-2}{b}} + \frac{\alpha(b+\alpha-2)^2}{384b(b+\alpha)}(bA)^{\frac{b+\alpha-4}{b}},$$

when $2 \leq b < 4$. In particular, when $\alpha = 2$ and $b \geq 2$, one can yield

$$f(\tau) \geq b(bA)^{\frac{b+\alpha}{b}} + \frac{\alpha}{12}(bA)^{\frac{b+\alpha-2}{b}} + \frac{\alpha(b+\alpha-2)^2}{288b(b+\alpha)}(bA)^{\frac{b+\alpha-4}{b}}.$$

This completes the proof of the Lemma 2.3. $\qquad\square$

Similarly, by using the same method as the proof of Lemma 2.3, we can obtain the following lemma:

Lemma 2.4. *Let $b \geq 2$ be a positive real number and $\mu > 0$. If $\psi : [0, \;$ $\infty) \to [0, +\infty)$ is a decreasing function such that*

$$-\mu \leq \psi'(s) \leq 0 \quad and \quad A := \int_0^\infty s^{b-1}\psi(s)ds > 0,$$

then, for any positive integer $l \geq 2$, we have

$$\int_0^\infty s^{b+2l-1}\psi(s)ds \geq \frac{1}{b+2l}(bA)^{\frac{b+2l}{b}}\psi(0)^{-\frac{2l}{b}}$$

$$+ \frac{l}{6b(b+2l)\mu^2}(bA)^{\frac{b+2(l-1)}{b}}\psi(0)^{\frac{2b-2l+2}{b}}$$

$$+ \frac{l(b+2(l-1))^2}{144b^2(b+2l)^2\mu^4}(bA)^{\frac{b+2l-4}{b}}\psi(0)^{\frac{4b-2l+4}{b}}.$$

3. Symmetric Rearrangements and Related Properties

In this section, we will introduce some definitions of symmetric rearrangements and related properties which will be used in the next sections.

Let $\Omega \subset \mathbb{R}^n$ be a bounded domain. Its *symmetric rearrangement* Ω^* is an open ball with the same volume as Ω (see [2, 25]),

$$\Omega^* = \left\{ x \in \mathbb{R}^n \;\middle|\; |x| < \left(\frac{V(\Omega)}{B_n}\right)^{\frac{1}{n}} \right\}.$$

By using a symmetric rearrangement of Ω, one can obtain

$$I(\Omega) = \int_\Omega |x|^2 dx \geq \int_{\Omega^*} |x|^2 dx = \frac{n}{n+2}V(\Omega)\left(\frac{V(\Omega)}{\omega_n}\right)^{\frac{2}{n}}. \qquad (9)$$

Let h be a nonnegative bounded continuous function on Ω. We consider its distribution function $\mu_h(t)$ defined by $\mu_h(t) = Vol(\{x \in \Omega | h(x) > t\})$, where the distribution function can be viewed as a function from $[0, \infty)$ to $[0, Vol(\Omega)]$. For any $x \in \Omega^*$, the *symmetric decreasing rearrangement* h^* of h is defined by $h^*(x) = \inf\{t \geq 0 | \mu_h(t) < B_n|x|^n\}$. For the symmetric rearrangements of set and function, we have the following properties [2, 11]:

Proposition 3.1. *Let h^* be the symmetric decreasing rearrangement of h, and $g(|x|) = h^*(x)$, then*

$$\int_{\mathbb{R}^n} h(x)dx = \int_{\mathbb{R}^n} h^*(x)dx = nB_n \int_0^\infty s^{n-1}g(s)ds = nB_n A,$$

where $A = \int_0^\infty s^{n-1}g(s)ds$.

Proposition 3.2. *Let h^* be the symmetric decreasing rearrangement of h, and $g(|x|) = h^*(x)$, then, for any $\beta > 0$,*

$$\int_{\mathbb{R}^n} |x|^\beta h(x)dx \geq \int_{\mathbb{R}^n} |x|^\beta h^*(x)dx = nB_n \int_0^\infty s^{n+\beta-1}g(s)ds.$$

Using coarea formula and isoperimetric inequality, we have the following properties:

Proposition 3.3. *Let h^* be the symmetric decreasing rearrangement of h, and $g(|x|) = h^*(x)$. If $\delta := \sup |\nabla h|$, then, for almost every s, $-\delta \leq g'(s) \leq 0$.*

The other useful information on the symmetric rearrangements can be found in [8, 9, 32].

4. Lower Bounds of Eigenvalues of Poly-Laplacian

In this section, we will consider the lower bounds of eigenvalues of poly-Laplacian. When $l = 1$, the eigenvalue problem (1) is said to be a *fixed membrane problem*. For this case, one has the following Weyl's asymptotic formula

$$\lambda_k^{(1)} \sim \frac{4\pi^2}{(B_n V(\Omega))^{\frac{2}{n}}} k^{\frac{2}{n}}, \quad k \to +\infty. \tag{10}$$

From the above asymptotic formula, one can derive

$$\frac{1}{k}\sum_{i=1}^k \lambda_i^{(1)} \sim \frac{n}{n+2} \frac{4\pi^2}{(B_n V(\Omega))^{\frac{2}{n}}} k^{\frac{2}{n}}, \quad k \to +\infty. \tag{11}$$

Pólya [34] proved that, if Ω is a tiling domain in $\mathbb{R}^n$, then

$$\lambda_k^{(1)} \geq \frac{4\pi^2}{(B_n V(\Omega))^{\frac{2}{n}}} k^{\frac{2}{n}}, \quad \text{for } k = 1, 2, \cdots. \tag{12}$$

Furthermore, he proposed a conjecture as follows:

Conjecture of Pólya. *If Ω is a bounded domain in $\mathbb{R}^n$, then the k-th eigenvalue λ_k of the fixed membrane problem satisfies*

$$\lambda_k^{(1)} \geq \frac{4\pi^2}{(B_n V(\Omega))^{\frac{2}{n}}} k^{\frac{2}{n}}, \quad \text{for } k = 1, 2, \cdots. \tag{13}$$

On the conjecture of Pólya, Berezin [3] and Lieb [24] gave a partial solution. In particular, Li and Yau [23] proved that

$$\frac{1}{k}\sum_{i=1}^{k}\lambda_i^{(1)} \geq \frac{n}{n+2}\frac{4\pi^2}{(B_nV(\Omega))^{\frac{2}{n}}}k^{\frac{2}{n}}, \quad \text{for } k=1,2,\cdots. \tag{14}$$

The formula (10) shows that the result of Li and Yau is sharp in the sense of average. From this formula (14), one can infer

$$\lambda_k^{(1)} \geq \frac{n}{n+2}\frac{4\pi^2}{(B_nV(\Omega))^{\frac{2}{n}}}k^{\frac{2}{n}}, \quad \text{for } k=1,2,\cdots, \tag{15}$$

which gives a partial solution for the conjecture of Pólya with a factor $n/(n+2)$. Recently, Melas [27] improved the estimate (14) to the following:

$$\frac{1}{k}\sum_{i=1}^{k}\lambda_i^{(1)} \geq \frac{n}{n+2}\frac{4\pi^2}{(B_nV(\Omega))^{\frac{2}{n}}}k^{\frac{2}{n}} + \frac{1}{24(n+2)}\frac{V(\Omega)}{I(\Omega)}, \quad \text{for } k=1,2,\cdots, \tag{16}$$

where $I(\Omega) = \min_{a\in\mathbb{R}^n}\int_\Omega |x-a|^2 dx$ is called *the moment of inertia* of Ω. Furthermore, Wei, Sun and author improved the estimate (16). See Theorem 1.1 in [37].

When $l=2$, the eigenvalue problem (1) is said to be a *clamped plate problem*. For the eigenvalues of the clamped plate problem, Agmon [1] and Pleijel [33] obtained

$$\lambda_k^{(2)} \sim \frac{16\pi^4}{(B_nV(\Omega))^{\frac{4}{n}}}k^{\frac{4}{n}}, \quad k\to+\infty. \tag{17}$$

Furthermore, Levine and Protter [22] proved that the eigenvalues of the clamped plate problem satisfy the following inequality:

$$\frac{1}{k}\sum_{i=1}^{k}\lambda_i^{(2)} \geq \frac{n}{n+4}\frac{16\pi^4}{(B_nV(\Omega))^{\frac{4}{n}}}k^{\frac{4}{n}}. \tag{18}$$

The formula (17) shows that the coefficient of $k^{\frac{4}{n}}$ is the best possible constant. By adding to its right hand side two terms of lower order in k, Cheng and Wei [14] gave the following estimate which is an improvement of (18):

$$\frac{1}{k}\sum_{i=1}^{k}\lambda_i^{(2)} \geq \frac{n}{n+4}\frac{16\pi^4}{(B_nV(\Omega))^{\frac{4}{n}}}k^{\frac{4}{n}}$$
$$+ \frac{n+2}{12n(n+4)}\frac{4\pi^2}{(B_nV(\Omega))^{\frac{2}{n}}}\frac{n}{n+2}\frac{V(\Omega)}{I(\Omega)}k^{\frac{2}{n}} \tag{19}$$
$$+ \frac{(n+2)^2}{1152n(n+4)^2}\left(\frac{V(\Omega)}{I(\Omega)}\right)^2.$$

When l is arbitrary, Levine and Protter [22] proved the following

$$\frac{1}{k}\sum_{i=1}^{k}\lambda_i^{(l)} \geq \frac{n}{n+2l}\frac{\pi^{2l}}{(B_nV(\Omega))^{\frac{2l}{n}}}k^{\frac{2l}{n}}, \quad \text{for} \quad k = 1, 2, \cdots. \tag{20}$$

By adding l terms of lower order of $k^{\frac{2l}{n}}$ to its right hand side, Cheng, Qi and Wei [11] obtained more sharper result than (20):

$$\frac{1}{k}\sum_{i=1}^{k}\lambda_i^{(l)} \geq \frac{n}{n+2l}\frac{(2\pi)^{2l}}{(B_nV(\Omega))^{\frac{2l}{n}}}k^{\frac{2l}{n}} + \frac{n}{(n+2l)}$$

$$\times \sum_{p=1}^{l}\frac{l+1-p}{(24)^pn\cdots(n+2p-2)}\frac{(2\pi)^{2(l-p)}}{(B_nV(\Omega))^{\frac{2(l-p)}{n}}}\left(\frac{V(\Omega)}{I(\Omega)}\right)^p k^{\frac{2(l-p)}{n}}. \tag{21}$$

Recently, Wei and author [38] investigated eigenvalues of the Dirichlet problem (1) of Laplacian with arbitrary order and gave a inequality which is sharper than (21), i.e., they obtained the following

Theorem 4.1. *Let Ω be a bounded domain in an n-dimensional Euclidean space $\mathbb{R}^n$. Assume that $l \geq 2$ and λ_i is the i-th eigenvalue of the eigenvalue problem (1.1). Then the eigenvalues satisfy*

$$\frac{1}{k}\sum_{j=1}^{k}\lambda_j^{(l)} \geq \frac{n}{n+2l}\frac{(2\pi)^{2l}}{(B_nV(\Omega))^{\frac{2l}{n}}}k^{\frac{2l}{n}}$$

$$+ \frac{l}{24(n+2l)}\frac{(2\pi)^{2(l-1)}}{(B_nV(\Omega))^{\frac{2(l-1)}{n}}}\frac{V(\Omega)}{I(\Omega)}k^{\frac{2(l-1)}{n}} \tag{22}$$

$$+ \frac{l(n+2(l-1))^2}{2304n(n+2l)^2}\frac{(2\pi)^{2(l-2)}}{(B_nV(\Omega))^{\frac{2(l-2)}{n}}}\left(\frac{V(\Omega)}{I(\Omega)}\right)^2 k^{\frac{2(l-2)}{n}}.$$

In order to give the proof of Theorem 4.1, we need some adequate preparation. Let $u_j^{(l)}$ be an orthonormal eigenfunction corresponding to the eigenvalue $\lambda_j^{(l)}$. Define $w^{(l)}(\xi) := \sum_{j=1}^{k}|\widehat{\varphi_j^{(l)}}(\xi)|^2$, where $\varphi_j^{(l)}(x) = u_j^{(l)}(x)$ when $x \in \Omega$ and $\varphi_j^{(l)}(x) = 0$ when $x \in \mathbb{R}^n \backslash \Omega$. And then, we have

$$\sum_{j=1}^{k}\lambda_j^{(l)} = \int_\Omega w^{(l)}(\xi)|\xi|^{2l}d\xi. \tag{23}$$

To be convenient, we denote the symmetric decreasing rearrangement of $w^{(l)}$ by $w^{(l)*}$ and define $\phi^{(l)}(|\xi|) := w^{(l)*}(\xi)$. By making use of Proposition

3.2, we have the following inequality:

$$\sum_{j=1}^{k} \lambda_j^{(l)} \geq nB_n \int_0^{\infty} s^{n+2l-1} \phi(s) ds. \tag{24}$$

By Parseval's identity, one can obtain $k = \int_{\mathbb{R}^n} w^{(l)}(\xi) d\xi$. Therefore, by using Proposition 3.1, we can show that $k = nB_n A$, where $A = \int_0^{\infty} s^{n-1} \phi^{(l)}(s) ds$. By using Proposition 3.3, we obtain

$$0 < \phi^{(l)}(0) \leq \sup w^{(l)*}(\xi) = \sup w^{(l)}(\xi) \leq (2\pi)^{-n} V(\Omega). \tag{25}$$

Hence, by Cauchy-Schwarz inequality, we get

$$|\nabla w^{(l)}(\xi)| \leq 2(2\pi)^{-n} \sqrt{I(\Omega) V(\Omega)}.$$

Next, we will give the outline of the proof of Theorem 4.1.

Proof. Take $b = n$, $\psi^{(l)}(s) = \phi^{(l)}(s)$, $A = \frac{k}{nB_n}$, and $\mu = \sigma = 2(2\pi)^{-n} \sqrt{V(\Omega) I(\Omega)}$. By using Lemma 2.4, (24) and inequality (25), we have

$$\sum_{j=1}^{k} \lambda_j^{(l)} \geq \frac{nB_n \left(\frac{k}{B_n}\right)^{\frac{n+2l}{n}}}{n+2l} \phi^{(l)}(0)^{-\frac{2l}{n}} + \frac{lB_n \left(\frac{k}{B_n}\right)^{\frac{n+2(l-1)}{n}}}{6(n+2l)\mu^2} \phi^{(l)}(0)^{\frac{2n-2l+2}{n}}$$
$$+ \frac{l(n+2(l-1))^2 B_n \left(\frac{k}{B_n}\right)^{\frac{n+2l-4}{n}}}{144n(n+2l)^2 \mu^4} \phi^{(l)}(0)^{\frac{4n-2l+4}{n}}. \tag{26}$$

Now letting $t = \phi^{(l)}(0)$ and defining a function $\xi(t)$ as follows:

$$\xi(t) = \frac{nB_n}{n+2l}\left(\frac{k}{B_n}\right)^{\frac{n+2l}{n}} t^{-\frac{2l}{n}} + \frac{lB_n}{6(n+2l)\mu^2}\left(\frac{k}{B_n}\right)^{\frac{n+2(l-1)}{n}} t^{\frac{2n-2l+2}{n}}$$
$$+ \frac{l(n+2(l-1))^2 B_n}{144n(n+2l)^2\mu^4}\left(\frac{k}{B_n}\right)^{\frac{n+2l-4}{n}} t^{\frac{4n-2l+4}{n}}$$

then we can prove that $\xi(t)$ is a decreasing function on $(0, (2\pi)^{-n} V(\Omega)]$. Hence, we can claim that $\xi(t) \geq \xi((2\pi)^{-n} V(\Omega))$. We replace $\phi^{(l)}(0)$ by $(2\pi)^{-n} V(\Omega)$ in (26), which yields (22). $\qquad\square$

Remark 4.1. When $l = 1$, a lower bound has been given by Wei-Sun-Zeng in [37], and the result is more sharper than Males's estimate. When $l = 2$, Theorem 10 reduces to the inequality (19).

5. Lower Bounds of Eigenvalues of Fractional Laplacian

In this section, we discuss the lower bounds of eigenvalues of fractional Laplacian. First of all, we assume $\alpha = 1$. For the case of $\alpha = 1$, E. Harrell and S. Y. Yolcu gave an analogue of the Berezin-Li-Yau type inequality for the eigenvalues of the Klein-Gordon operators $\mathbf{H}_{0,\Omega} := \sqrt{-\Delta}$ restricted to Ω (see [15]):

$$\frac{1}{k}\sum_{j=1}^{k}\Lambda_j^{(\alpha)} \geq \frac{n}{n+1}\left(\frac{2\pi}{(B_n V(\Omega))^{\frac{1}{n}}}\right)k^{\frac{1}{n}}. \tag{27}$$

Very recently, S. Y. Yolcu [40] improved the estimate (27) to the following:

$$\frac{1}{k}\sum_{j=1}^{k}\Lambda_j^{(\alpha)} \geq \frac{n}{n+1}\left(\frac{2\pi}{(B_n V(\Omega))^{\frac{1}{n}}}\right)k^{\frac{1}{n}} + \widetilde{M}_n \frac{V(\Omega)^{1+\frac{1}{n}}}{I(\Omega)}k^{-1/n}, \tag{28}$$

where the constant $\widetilde{M}_n$ depends only on the dimension n. Next, we consider the general case. For any $\alpha \in (0,2]$, S. Y. Yolcu and T. Yolcu [41] generalized (28) as follows:

$$\frac{1}{k}\sum_{j}^{k}\Lambda_j^{(\alpha)} \geq \frac{n}{n+\alpha}\left(\frac{2\pi}{(\omega_n V(\Omega))^{\frac{1}{n}}}\right)^{\alpha}k^{\frac{\alpha}{n}}$$
$$+ \frac{\alpha}{48(n+\alpha)}\frac{(2\pi)^{\alpha-2}}{(B_n V(\Omega))^{\frac{\alpha-2}{n}}}\frac{V(\Omega)}{I(\Omega)}k^{\frac{\alpha-2}{n}}. \tag{29}$$

By using Lemma 2.3, Wei, Sun and author [37] gave an analogue of the Berezin-Li-Yau type inequality of the eigenvalues of $(-\Delta)^{\alpha/2}|_\Omega$ [37]:

Theorem 5.1. *Let Ω be a bounded domain in an n-dimensional Euclidean space $\mathbb{R}^n$. Assume that $\Lambda_i^{(\alpha)}, i = 1, 2, \cdots,$ is the i-th eigenvalue of the fractional Laplacian $(-\Delta)^{\alpha/2}|_\Omega$. Then, the sum of its eigenvalues satisfies*

$$\frac{1}{k}\sum_{j=1}^{k}\Lambda_j^{(\alpha)} \geq \frac{n}{n+\alpha}\frac{(2\pi)^{\alpha}}{(B_n V(\Omega))^{\frac{\alpha}{n}}}k^{\frac{\alpha}{n}}$$
$$+ \frac{\alpha}{48(n+\alpha)}\frac{(2\pi)^{\alpha-2}}{(B_n V(\Omega))^{\frac{\alpha-2}{n}}}\frac{V(\Omega)}{I(\Omega)}k^{\frac{\alpha-2}{n}}$$
$$+ \frac{\alpha(n+\alpha-2)^2}{\mathcal{C}(n)n(n+\alpha)^2}\frac{(2\pi)^{\alpha-4}}{(B_n V(\Omega))^{\frac{\alpha-4}{n}}}\left(\frac{V(\Omega)}{I(\Omega)}\right)^2 k^{\frac{\alpha-4}{n}}, \tag{30}$$

where $\mathcal{C}(n) = 4608$ *when* $n \geq 4$, *and* $\mathcal{C}(n) = 6144$, *when* $n = 2,3$. *In particular, the sum of its eigenvalues satisfies*

$$\frac{1}{k}\sum_{j=1}^{k}\Lambda_j^{(2)} \geq \frac{nk^{\frac{2}{n}}}{n+2}B_n^{-\frac{2}{n}}(2\pi)^2 V(\Omega)^{-\frac{2}{n}} + \frac{1}{24(n+2)}\frac{V(\Omega)}{I(\Omega)}$$

$$+ \frac{nk^{-\frac{2}{n}}}{2304(n+2)^2}B_n^{\frac{2}{n}}(2\pi)^{-2}\left(\frac{V(\Omega)}{I(\Omega)}\right)^2 V(\Omega)^{\frac{2}{n}}, \tag{31}$$

when $\alpha = 2$.

Remark 5.1. Observing Theorem 5.1, it is not difficult to see that the coefficients (with respect to $k^{\frac{\alpha-2}{n}}$) of the second terms in (30) are equal to that of (29). In other word, we can claim that the inequalities (30) are sharper than (29) since the coefficients (with respect to $k^{\frac{\alpha-4}{n}}$) of the third terms in (29) are positive.

Remark 5.2. By using Theorem 4.1, we can obatin an analogue of the Berezin-Li-Yau type inequality for the eigenvalues of the Klein-Gordon operators.

Remark 5.3. By using the same method as the proof of Theorem 4.1 or Theorem 5.1, we can obtain a similar estimates for eigenvalues of the problem (1) of Laplacian, which agrees with the case of $l = 1$ in (22).

6. Improvement of the Second Term

In this section, we focus on the improvement of the second term. Recently, Ilyin [16] obtained the following asymptotic lower bound for eigenvalues of problem (1):

$$\frac{1}{k}\sum_{j=1}^{k}\lambda_j^{(1)} \geq \frac{nk^{\frac{2}{n}}}{n+2}B_n^{-\frac{2}{n}}(2\pi)^2 V(\Omega)^{-\frac{2}{n}} + \frac{n}{48}\frac{V(\Omega)}{I(\Omega)}\left(1-\varepsilon_n(k)\right), \tag{32}$$

where $0 \leq \varepsilon_n(k) \leq O(k^{-\frac{2}{n}})$ is a infinitesimal of $k^{-\frac{2}{n}}$. Moreover, he derived some explicit inequalities for the particular cases of $n = 2,3,4$:

$$\frac{1}{k}\sum_{j=1}^{k}\lambda_j^{(1)} \geq \frac{nk^{\frac{2}{n}}}{n+2}B_n^{-\frac{2}{n}}(2\pi)^2 V(\Omega)^{-\frac{2}{n}} + \frac{n}{48}\beta_n\frac{V(\Omega)}{I(\Omega)}, \tag{33}$$

where $\beta_2 = \frac{119}{120}$, $\beta_3 = 0.986$ and $\beta_3 = 0.983$, see [16]. However, when l is arbitrary, Cheng-Sun-Wei-Zeng [12] generalized the inequality (32) and obtained the following

Theorem 6.1. *Let Ω be a bounded domain in an n-dimensional Euclidean space $\mathbb{R}^n$. Assume that $\lambda_i, i = 1, 2, \cdots$, is the i-th eigenvalue of the eigenvalue problem (1.1). Then the sum of its eigenvalues satisfies*

$$
\frac{1}{k} \sum_{j=1}^{k} \lambda_j^{(l)} \geq \frac{nk^{\frac{2l}{n}}}{n+2l} B_n^{-\frac{2}{n}} (2\pi)^{2l} V(\Omega)^{-\frac{2l}{n}}
$$

$$
+ \frac{nl}{48} \frac{(2\pi)^{2l-2}}{(B_n V(\Omega))^{\frac{2l-2}{n}}} \frac{V(\Omega)}{I(\Omega)} k^{\frac{2l-2}{n}} \left(1 - \varepsilon_n(k)\right).
$$

(34)

Remark 6.1. Taking $l = 1$ in (34), we obtain (32). Moreover, for large k, (34) is sharper than (21).

7. Lower Bounds with a Correction Term

It is important to compare the lower bound (14) with the asymptotical behaviour of the sum on the left-hand side, which reads as follows:

$$
\sum_{j=1}^{k} \lambda_j^{(1)} \sim \frac{nC_n}{n+2} V(\Omega)^{-\frac{2}{n}} k^{\frac{n+2}{n}} + \widetilde{C}_n \frac{|\partial\Omega|}{V(\Omega)^{1+\frac{1}{n}}} k^{1+\frac{1}{n}} \quad \text{as} \quad k \to \infty \quad (35)
$$

where $\widetilde{C}_n$ is a constant only depending on n and $|\partial\Omega|$ denotes the area of $\partial\Omega$. We remark that the first term in (35) is given by Weyl [39], and the second term in (35) was established, under some conditions on Ω, in [17, 28]. Here, the asymptotic formula (35) also implies that the constant in (14) cannot be improved. By observing the second asymptotical term, one might improve (14) by adding an additional positive term of a lower order in k to the right-hand side. Indeed, Kovařík, Vugalter and Weidl [19] noticed that the coefficient of the second term in (35) is explicitly associated with the boundary of Ω. Therefore, they have made an important breakthrough towards this landmark goal in the case of dimension 2. They have added a positive term in the right hand side of (14), which is similar to the second term of in (35) in the asymptotic sense. In other word, they proved the following theorem:

Theorem 7.1. (Lower bound for polygons) *Let Ω be a polygon with m sides. Let l_j be the length of the j-th side of Ω. Then for any $k \in \mathbb{N}$ and* *any* $\alpha \in [0,1]$ *we have*

$$\frac{1}{k}\sum_{j=1}^{k}\lambda_j^{(1)} \geq \frac{2\pi}{V(\Omega)}k^2 + 4\alpha c_3 k^{\frac{3}{2}-\varepsilon(k)}V(\Omega)^{-\frac{3}{2}}\sum_{j=1}^{m}l_j\Theta\left(k - \frac{9V(\Omega)}{2\pi d_j^2}\right)$$
$$+ (1-\alpha)\frac{V(\Omega)}{32I(\Omega)}k, \tag{36}$$

where

$$\varepsilon(k) = \frac{2}{\sqrt{\log_2(2\pi k/c_1)}}, \quad c_1 = \sqrt{\frac{3\pi}{14}}10^{-11}, \quad c_3 = \frac{2^{-3}}{9\sqrt{236}}(2\pi)^{\frac{5}{4}}c_1^{\frac{1}{4}}.$$

Furthermore, motivated by the outstanding work of Kovařík, Vugalter and Weidl [19], Cheng and Qi [10] investigated the n-dimensional case for arbitrary dimension n and obtained the second term of (35) in the asymptotic sense. For an n-dimensional polytope Ω in $\mathbb{R}^n$, we denote by $p_j, j = 1, 2, \cdots, m$, the j-face of Ω. Assume that A_j is the area of the j-th face p_j of Ω. For each $j = 1, 2, \cdots, m$, we choose several non-overlapping $(n-1)$-dimensional convex subdomains s_{r_j} in the interior of p_j such that the area of $\bigcup_{r_j} s_{r_j}$ is greater than or equal to one third of A_j and the distance d_j between $\bigcup_{r_j} s_{r_j}$ and $\partial\Omega \backslash p_j$ is greater than 0. Define the function $\Theta : \mathbb{R} \to \mathbb{R}$ by $\Theta(t) = 0$ if $t \leq 0$ and $\Theta(t) = 1$ if $t > 0$. Then, they obtained the following remarkable result:

Theorem 7.2. *Let Ω be an n-dimensional polytope in $\mathbb{R}^n$ with m faces. Denote by A_j the area of the j-th face of Ω. Then, for any positive integer k, we have*

$$\frac{1}{k}\sum_{j=1}^{k}\lambda_j^{(1)} \geq \frac{n}{n+2}\frac{4\pi^2}{(B_n V(\Omega))^{\frac{2}{n}}}k^{\frac{2}{n}}$$
$$+ \frac{8\pi^2 c_2 c_1^{\frac{1}{2}}}{(n+2)B_n^{\frac{2}{n}}}\frac{k^{\frac{2}{n}}\lambda_k^{-\frac{1}{2}}}{V(\Omega)^{1+\frac{2}{n}}}\left(\frac{V(\Omega)\lambda_k}{c_1}\right)^{-n\varepsilon(k)}\sum_{j=1}^{m}A_j\Theta(\lambda_k - \lambda_0), \tag{37}$$

where

$$\varepsilon(k) = \left[\sqrt{\frac{\log_2\left((V(\Omega)/c_1)^{n-1}\lambda_k^{\frac{n}{2}}\right)}{n+12}}\right]^{-1}, \quad c_1 = \sqrt{\frac{3}{B_n}\left(\frac{4n\pi^2}{n+2}\right)^{\frac{n}{2}}},$$

$$c_2 = \frac{c_1^{-\frac{1}{2}}}{g^2 \cdot 2^n}, \quad g = 1 - 6x^4 + 8x^6 - 3x^8, \ 0 \leq x \leq 1,$$

$$\lambda_0 = \max\left\{ \frac{4n}{\min\{d_j^2\}},\ \left(\frac{c_1}{V(\Omega)}\right)^{\frac{2}{n}},\ 2^{\frac{2(n+12)}{n}}\left(\frac{c_1}{V(\Omega)}\right)^{\frac{2(n-1)}{n}}, \right.$$
$$\left. \left(\frac{12}{\min\{A_j\}}\right)^{\frac{2}{n-1}} \right\}.$$

Remark 7.1. Notice that $\varepsilon(k) \to 0$ and $\lambda_k \sim \frac{4\pi^2}{(B_n V(\Omega))^{\frac{2}{n}}} k^{\frac{2}{n}}$ as $k \to +\infty$. It shows that the second term on the right hand side of the inequality in Theorem 7.2 is very similar to the second term in the asymptotic (35) when k is large enough.

References

1. S. Agmon, On kernels, *eigenvalues and eigenfunctions of operators related to elliptic problems*, Comm. Pure Appl. Math. 18 (1965), 627-663.
2. C. Bandle, *Isoperimetric inequalities and applications*, Pitman Monographs and Studies in Mathematics, vol. 7, Pitman, Boston, (1980).
3. F. A. Berezin, *Covariant and contravariant symbols of operators*, Izv. Akad. Nauk SSSR Ser. Mat. 36 (1972), 1134-1167.
4. R. Blumenthal and R.Getoor, *The asymptotic distribution of the eigenvalues for a class of Markov operators*, Pacific J. Math, 9 (1959) 399-408.
5. R. Bañuelos and T. Kulczycki, *The Cauchy process and the Steklov problem*, J. Funct. Anal., 211(2) (2004) 355-423.
6. R. Bañuelos and T. Kulczycki, *Eigenvalue gaps for the Cauchy process and a Poincaré inequality*, J. Funct. Anal., 234 (2006) 199-225.
7. R. Bañuelos, T. Kulczycki and Bartłomiej Siudeja, *On the trace of symmetric stable processes on Lipschitz domains*, J. Funct. Anal., 257(10) (2009) 3329-3352.
8. I. Chavel, *Eigenvalues in Riemannian Geometry*, Academic Press, New York, (1984).
9. I. Chavel, *Riemannian geometry: A Modern Introduction*, Second Edition, Cambridge University Press, 2006.
10. Q.-M., Cheng and X. Qi, *Lower bound estimates for eigenvalues of the Laplacian*, http://arxiv.org/abs/1208.5226, (2012).
11. Q.-M. Cheng, X. Qi and G. Wei, *A lower bound for eigenvalues of the poly-Laplacian with arbitrary order*, to appear in Pacific J. Math.
12. Q.-M. Cheng, H.-J. Sun, G. Wei and L. Zeng , *Estimates for lower bounds of eigenvalues of the poly-Laplacian and quadratic polynomial operator of the Laplacian*, to appear in Proc. Royal Soc. Edinburgh, Section: A Mathematics.
13. Q.-M. Cheng and G. Wei, *A lower bound for eigenvalues of a clamped plate problem*, Calc. Var. Partial Differential Equations 42 (2011), 579-590.
14. Q.-M. Cheng and G. Wei, *Upper and lower bounds of the clamped plate problem*, to appear in J. Diff. Equa..
15. E. M. Harrell II and S. Yildirim Yolcu, *Eigenvalue inequalities for Klein-Gordon Operators*, J. Funct. Anal., 256(12) (2009) 3977-3995.

16. A. A. Ilyin, *Lower bounds for the spectrum of the Laplacian and Stokes operators*, Discrete Cont. Dyn. S. 28 (2010), 131-146.
17. V.Ivrii, *The second term of the spectral asymptotics for a Laplace-Beltrami operator on manifolds with boundary*. (Russian) Funk. Anal. i Pril. 14(2) (1980), 25-34.
18. J. Jost, X. Li-Jost, Q. Wang, and C. Xia, *Universal bounds for eigenvalues of the polyharmonic operators*, Trans. Amer. Math. Soc., 363(2011), 1821-1854.
19. H. Kovařík, S. Vugalter and T. Weidl, *Two dimensional Berezin-Li-Yau inequalities with a correction term*, Comm. Math. Phys., 287(3) (2009), 959-981.
20. A. Laptev, *Dirichlet and Neumann eigenvalue problems on domains in Euclidean spaces*, J. Funct. Anal. 151 (1997), 531-545.
21. N. S. Landkof, *Foundations of modern potential theory*, New York: Springer-Verlag (1972).
22. H. A. Levine and M. H. Protter, *Unrestricted lower bounds for eigenvalues for classes of elliptic equations and systems of equations with applications to problems in elasticity*, Math. Methods Appl. Sci. 7 (1985), no. 2, 210-222.
23. P. Li and S. T. Yau, *On the Schrödinger equations and the eigenvalue problem*, Comm. Math. Phys. 88 (1983), 309-318.
24. E. Lieb, *The number of bound states of one-body Schrödinger operators and the Weyl problem*, Proc. Sym. Pure Math. 36 (1980), 241-252.
25. E. Lieb and M. Loss, *Analysis* (second version), Graduate studies in mathematics, Volume 14, American Mathematical Society, (2001).
26. A. Laptev and T. Weidl, *Recent results on Lieb-Thirring inequalities*, Journées " Équations aux Dérivées Partielles"(La Chapelle sur Erdre, 2000), Exp. No. XX, 14 pp., Univ. Nantes, Nantes, 2000.
27. A. D. Melas, *A lower bound for sums of eigenvalues of the Laplacian*, Proc. Amer. Math. Soc. 131 (2003), 631-636.
28. R.B.Melrose, *Weyls conjecture for manifolds with concave boundary*. In: Geometry of the Laplace operator (Proc. Sympos. PureMath., Univ. Hawaii, Honolulu, Hawaii, 1979), Proc. Sympos. PureMath., XXXVI, Providence, RI: Amer. Math. Soc., 257C274 (1980).
29. Payne, L. E., *Inequalities for eigenvalues of membranes and plates*, J. Rational Mech. Anal. 4 (1955), 517-529.
30. Payne, L. E., G. Pólya, and H. F. Weinberger, *On the ratio of consecutive eigenvalues*, J. Math. and Phys. 35, 289-298, (1956).
31. Protter, M. H., *Can one hear the shape of a drumg revisited*, SIAM Review 29, (1987), 185-197.
32. G. Pólya and G. Szegö, *Isoperimetric inequalities in mathematical physics*, Annals of mathe- matics studies, number 27, Princeton university press, Princeton, New Jersey, (1951).
33. A. Pleijel, *On the eigenvalues and eigenfunctions of elastic plates*, Comm. Pure Appl. Math., 3 (1950), 1-10.
34. G. Pólya, *On the eigenvalues of vibrating membranes*, Proc. London Math. Soc., 11 (1961), 419-433.
35. E. Valdinoci, *From the long jump random walk to the fractional Laplacian*, ArXiv:0901.3261.

36. T. Weidl, *Improved Berezin-Li-Yau inequalities with a remainder term,* Spectral Theory of Differential Operators, Amer. Math. Soc. Transl., 225(2) (2008), 253-263.
37. G.Wei, H.-J. Sun and L. Zeng, *Lower bounds for Laplacian and fractional Laplacian eigenvalues,* http://arxiv.org/abs/1112.4571v2.
38. G. Wei and L. Zeng, *Estimates for Eigenvalues of Poly-harmonic Operators,* http://arxiv.org/abs/1111.3115.
39. H. Weyl, *Dasasymptotische Verteilungsgesetz der Eigenwerte linearer partieller Differentialgleichungen.* Math. Ann. 71,(1912), 441-479.
40. S. Yildirim Yolcu, *An improvement to a Brezin-Li-Yau type inequality,* Proc. Amer. Math. Soc, 138(11) (2010) 4059-4066.
41. S.Y. Yolcu and T. Yolcu, *Estimates for the sums of eigenvalues of the fractional Laplacian on a bounded domain,* to appear in Communications in Contemporary Mathematics (2013).

Received February 26, 2013.

Proceedings of the Workshop on
Differential Geometry of Submanifolds
and its Related Topics
Saga, August 4-6, 2012

FLAT CENTROAFFINE SURFACES WITH NON-SEMISIMPLE TCHEBYCHEV OPERATOR

Atsushi FUJIOKA*

*Department of Mathematics, Kansai University,
3-3-35, Yamate-cho, Suita, Osaka 564-8680 Japan
E-mail: afujioka@kansai-u.ac.jp*

Dedicated to Professor Sadahiro Maeda for his sixtieth birthday

In this paper, we study centroaffine surfaces in the real affine 3-space with flat centroaffine metric whose Tchebychev operator is not semisimple.

Keywords: centroaffine surface, flat metric, Tchebychev operator.

1. Introduction

Centroaffine differential geometry is a kind of affine differential geometry, in which we study properties of submanifolds in the affine space invariant affine transformations fixing the origin. There are two wide classes of submanifolds which are interesting from the viewpoint of submanifold theory itself as well as integrable systems. One is centroaffine minimal surfaces defined in general for hypersurfaces by Wang [13] as extremals for the area integral of the centroaffine metric. Indeed, Schief [11] showed that Tzitzéica transformation for proper affine spheres can be generalized and discretized for centroaffine minimal surfaces. Another one is flat centroaffine hypersurfaces, *i.e.*, centroaffine hypersurfaces whose centroaffine scalar curvature vanishes. Indeed, Ferapontov [2] showed that the integrability conditions for flat centroaffine hypersurfaces are equivalent to equations of associativity in topological field theories.

---

*This research was financially supported by Grant-in-Aid Scientific Research No. 22540070, Japan Society for the Promotion of Science, and the Kansai University Grant-in-Aid for progress of research in graduate course, 2012.

In this paper, we study centroaffine surfaces with flat centroaffine metric whose Tchebychev operator is not semisimple. In particluar, we shall determine such surfaces with constant Pick invariant in Theorem 4.1. In general case, we shall determine geometric invariants for such surfaces in Theorem 4.2.

2. The Gauss equations for centroaffine surfaces

A centroaffine surface is a surface in the real affine 3-space $\mathbf{R}^3$ such that the position vector is transversal to the tangent plane at each point. Any centroaffine surface is given locally by a smooth immersion f from a 2-dimensional domain to $\mathbf{R}^3$. In the following, we assume that f is nondegenerate, *i.e.*, the centroaffine metric h is nondegenerate. As we shall see in the beginning of Sec. 4, we have only to consider the case that h is indefinite for our purposes. Then as can be seen in Ref. [11], we can take asymptotic line coordinates (u, v) and the Gauss equations for f are given by

$$
\begin{cases}
f_{uu} = \left(\dfrac{\varphi_u}{\varphi} + \rho_u \right) f_u + \dfrac{a}{\varphi} f_v, \\[2mm]
f_{uv} = -\varphi f + \rho_v f_u + \rho_u f_v, \\[2mm]
f_{vv} = \left(\dfrac{\varphi_v}{\varphi} + \rho_v \right) f_v + \dfrac{b}{\varphi} f_u
\end{cases}
\tag{1}
$$

with the integrability conditions:

$$
(\log |\varphi|)_{uv} = -\varphi - \frac{ab}{\varphi^2} + \rho_u \rho_v, \quad a_v + \rho_u \varphi_u = \rho_{uu} \varphi, \quad b_u + \rho_v \varphi_v = \rho_{vv} \varphi, \tag{2}
$$

where $\varphi = h(\partial_u, \partial_v)$,

$$
a = \varphi \det \begin{pmatrix} f \\ f_u \\ f_{uu} \end{pmatrix} \Big/ \det \begin{pmatrix} f \\ f_u \\ f_v \end{pmatrix}, \quad b = \varphi \det \begin{pmatrix} f \\ f_v \\ f_{vv} \end{pmatrix} \Big/ \det \begin{pmatrix} f \\ f_v \\ f_u \end{pmatrix},
$$

and $\pm e^\rho$ is the equiaffine support function from the origin, which is an equiaffine invariant.

It is obvious to see that the cubic differentials $a\,du^3$ and $b\,dv^3$ are centroaffine invariants. Centroaffine transformations preserve the property that ρ is a constant, which was discovered by Tzitzéica [12]. Moreover, ρ is a constant if and only if f is a proper affine sphere centered at the origin. See Ref. [10] for basic facts about affine hyperspheres.

3. Some invariants for centroaffine surfaces

For a centroaffine surface f as in Sec. 2, the centroaffine curvature κ is given by

$$\kappa = -\frac{(\log|\varphi|)_{uv}}{\varphi}.$$

If f is flat, *i.e.*, $\kappa = 0$, changing the coordinates u and v, if necessary, we may assume that $\varphi = 1$. Then Eq. (2) can be expressed as a single partial differential equation of third order for a function g:

$$g_{uuu}g_{vvv} - g_{uuv}g_{uvv} + 1 = 0, \tag{3}$$

where

$$\rho = g_{uv}, \quad a = g_{uuu}, \quad b = g_{vvv}.$$

Equation (3) is known as one of equations of associativity in topological field theories [1, 2].

Let ∇ be the connection induced by the immersion f and $\tilde{\nabla}$ the Levi-Civita connection of the centroaffine metric h. It is easy to see that the Christoffel symbols $\tilde{\Gamma}^k_{ij}$ $(i, j, k = u, v)$ for $\tilde{\nabla}$ with respect to (u, v) vanish except

$$\tilde{\Gamma}^u_{uu} = \frac{\varphi_u}{\varphi}, \quad \tilde{\Gamma}^v_{vv} = \frac{\varphi_v}{\varphi}. \tag{4}$$

If we denote the difference tensor $\nabla - \tilde{\nabla}$ by C, from Eqs. (1) and (4), we have

$$\begin{cases} C(\partial_u, \partial_u) = \rho_u \partial_u + \dfrac{a}{\varphi}\partial_v, \\ C(\partial_u, \partial_v) = \rho_v \partial_u + \rho_u \partial_v, \\ C(\partial_v, \partial_v) = \dfrac{b}{\varphi}\partial_u + \rho_v \partial_v. \end{cases} \tag{5}$$

Then the Tchebychev vector field T is given by

$$T = \frac{1}{2}\mathrm{tr}_h C = \frac{\rho_v}{\varphi}\partial_u + \frac{\rho_u}{\varphi}\partial_v = \mathrm{grad}_h \rho. \tag{6}$$

From the second and third equations of Eq. (2), Eq. (4) and Eq. (6), the Tchebychev operator $\tilde{\nabla}T$ is computed as

$$\tilde{\nabla}T(\partial_u) = \frac{\rho_{uv}}{\varphi}\partial_u + \frac{a_v}{\varphi^2}\partial_v, \quad \tilde{\nabla}T(\partial_v) = \frac{b_u}{\varphi^2}\partial_u + \frac{\rho_{uv}}{\varphi}\partial_v. \tag{7}$$

Centroaffine surfaces with vanishing Tchebychev operator were classified as follows.

Proposition 3.1 (Liu and Wang [5]). *Let f be a centroaffine surface with vanishing Tchebychev operator and put $f = (X, Y, Z)$. Then up to centroaffine congruence, f is one of the following:*

(i) *A piece of a quadric.*

(ii) *A proper affine sphere centered at the origin.*

(iii) $X^\alpha Y^\beta Z^\gamma = 1$, *where* $\alpha, \beta, \gamma \in \mathbf{R}$ *such that* $\alpha\beta\gamma(\alpha + \beta + \gamma) \neq 0$.

(iv) $\left\{ \exp\left(-\alpha \tan^{-1} \dfrac{X}{Y} \right) \right\} (X^2 + Y^2)^\beta Z^\gamma = 1$, *where* $\alpha, \beta, \gamma \in \mathbf{R}$ *such that* $\gamma(2\beta + \gamma)(\alpha^2 + \beta^2) \neq 0$.

(v) $Z = -X(\alpha \log X + \beta \log Y)$, *where* $\alpha, \beta, \gamma \in \mathbf{R}$ *such that* $\beta(\alpha + \beta) \neq 0$.

(vi) $Z = \pm X \log X + \dfrac{Y^2}{X}$.

(vii) $f(u, v) = (e^u, A_1(u)e^v, A_2(u)e^v)$, *where* A_1 *and* A_2 *are linearly independent solutions to the linear ordinary differential equation* $A'' - A' - a(u)A = 0$ *for any function* $a = a(u)$.

Remark 3.1. Proper affine spheres centered at the origin with flat centroaffine metric were given by Magid and Ryan [8], and all the surfaces given by (iii)–(vii) in Proposition 3.1 are known to be flat. If the centroaffine metric for such a surface f is indefinite, using the same notations as in Sec. 2 and changing the coordinates u and v, if necessary, we may assume that $\varphi = 1$, and obtain one of the following:

(a) Both a and b are non-zero constants, and $\rho = c_1 u + c_2 v + c_3$, where $c_1, c_2, c_3 \in \mathbf{R}$ such that $ab - c_1 c_1 + 1 = 0$.

(b) Changing the coordinates u and v, if necessary, we have $a = a(u)$, $b = 0$, and $\rho = c_1 u + c_2 v + c_3$, where $c_1, c_2, c_3 \in \mathbf{R}$ such that $c_1 c_1 = 1$.

In the case of (a), f is given by (iii)–(vi) and the solution to Eq. (3) is

$$g = \frac{a}{6}u^3 + \frac{b}{6}v^3 + \frac{c_1}{2}u^2 v + \frac{c_2}{2}uv^2 + (\text{any quadratic polynomial of } u \text{ and } v).$$

As a special case, if $c_1 = c_2 = 0$, we have the solution corresponding to a piece of a hyperbolic paraboloid or a proper affine sphere given by $(X^2 + Y^2)Z = 1$.

In the case of (b), f is given by (vii) and the solution to Eq. (3) is

$$g = \frac{c_1}{2}u^2 v + \frac{c_2}{2}uv^2 + c_3 uv$$

$$+ (\text{any function of } u) + (\text{any quadratic polynomial of } v).$$

A centroaffine surface is called to be centroaffine minimal if it extremizes the area integral of the centroaffine metric, which is known to be equivalent to the condition that the trace of the Tchebychev operator vanishes [13]. In particular, centroaffine surfaces with vanishing Tchebychev operator is centroaffine minimal. For a centroaffine surface f as above, from Eq. (7), the condition is written as $\rho_{uv} = 0$.

Moreover, the Pick invariant is given by

$$J = \frac{1}{2}\|C\|^2 = \frac{3\rho_u\rho_v}{\varphi} + \frac{ab}{\varphi^3} \tag{8}$$

by use of Eq. (5).

4. Non-semisimple flat centroaffine surfaces

As in Ref. [3], we define as follows.

Definition 4.1. A centroaffine surface is called to be semisimple if and only if the Tchebychev operator is semisimple.

In the following, we study non-semisimple flat centroaffine surfaces. It is easy to see that any non-semisimple centroaffine surface should be indefinite. Hence we may assume that the non-semisimple centroaffine surface f is given by as in Sec. 2.

Theorem 4.1. *Let f be a non-semisimple flat centroaffine surface with constant Pick invariant. Then f is given by*

$$f = \left(\sum_{n=0}^{\infty} A_{n,1}(v)u^n, \sum_{n=0}^{\infty} A_{n,2}(v)u^n, \sum_{n=0}^{\infty} A_{n,3}(v)u^n \right), \tag{9}$$

where the coordinates (u,v) are defined around $(0,v_0)$ such that $v_0 \neq 0$, and $A_{0,1}, A_{0,2}$ and $A_{0,3}$ are linearly independent solutions to the linear ordinary differential equation:

$$vA''' + A'' - A = 0, \tag{10}$$

and

$$A_{n+1,i} = -\frac{v}{n+1}A''_{n,i} \quad (i = 1,2,3). \tag{11}$$

Proof. Since f is non-semisimple and flat, changing the coordinates u and v, if necessary, we may assume that $\varphi = 1$, $a_v \neq 0$, $b = b(v)$. Then Eq. (2) becomes

$$0 = -1 - ab + \rho_u\rho_v, \quad a_v = \rho_{uu}, \quad 0 = \rho_{vv}. \tag{12}$$

From the third equation of Eq. (12), we have

$$\rho = \alpha(u)v + \beta(u). \tag{13}$$

Then from Eq. (8), the first equation of Eq. (12) and Eq. (13), we have

$$J = 4\alpha\alpha'v + 4\alpha\beta' - 1. \tag{14}$$

Since J is a constant, we have $\alpha\alpha' = 0$ and $\alpha\beta'$ is a constant. In particular, α is a constant.

If $\alpha \neq 0$, then β' is a constant. Then from the second equation of Eq. (12) and Eq. (13), we have $a_v = 0$, which is a contradiction. Hence $\alpha = 0$, which implies that f is a centroaffine minimal surface with $J = -1$. Therefore, by results of Refs. [3, 4], f is given by Eq. (9). $\qquad\square$

Remark 4.1. In Ref. [4], the function φ is normalized as $\varphi = -1$, which can be given by changing the sign of u or v in the proof of Theorem 4.1. Hence the sign of the right-hand side of Eq. (11) is opposite to that in Ref. [4].

Remark 4.2. For the surface f given by Eq. (9), we have

$$a = v, \quad b = -\frac{1}{v}, \quad \rho = \frac{1}{2}u^2 + c$$

for $c \in \mathbf{R}$, and the solution to Eq. (3) is

$$g = \frac{1}{6}u^3v - \frac{1}{2}v^2 \log v + (\text{any quadratic polynomial of } u \text{ and } v).$$

Remark 4.3. In Ref. [7], Meijer introduced very general functions, called the Meijer G-functions nowadays, including most of the known special functions. For $m, n, p, q \in \mathbf{Z}$ with $0 \leq m \leq q$, $0 \leq n \leq p$, the Meijer G-function $G_{p,q}^{m,n}$ is characterized by a solution to the ordinary differential equation:

$$\left\{ (-1)^{p-m-n} z \prod_{j=1}^{p}\left(z\frac{d}{dz} - a_j + 1\right) - \prod_{j=1}^{q}\left(z\frac{d}{dz} - b_j\right) \right\} G(z) = 0,$$

and can be expressed by complex integration:

$$G_{p,q}^{m,n}\left(z \,\middle|\, \begin{matrix} a_1, \ldots, a_p \\ b_1, \ldots, b_q \end{matrix}\right) = \frac{1}{2\pi i} \int_C \frac{\displaystyle\prod_{j=1}^{m} \Gamma(b_j + s) \prod_{j=1}^{n} \Gamma(1 - a_j - s)}{\displaystyle\prod_{j=m+1}^{q} \Gamma(1 - b_j - s) \prod_{j=n+1}^{p} \Gamma(a_j + s)} z^{-s} ds.$$

See Refs. [6, 9] for more detail about the conditions for parameters $a_1, \ldots, a_p, b_1, \ldots, b_q$ and the path C.

By use of the Meijer G-functions, the solution to Eq. (10) can be written as

$$A = c_1 G_{0,3}^{2,0}\left(\frac{v^2}{8}\,\bigg|\,{-\atop \frac{1}{2},\frac{1}{2},0}\right) + i c_2 G_{0,3}^{1,0}\left(-\frac{v^2}{8}\,\bigg|\,{-\atop \frac{1}{2},\frac{1}{2},0}\right) + c_3 G_{0,3}^{1,0}\left(-\frac{v^2}{8}\,\bigg|\,{-\atop 0,\frac{1}{2},\frac{1}{2}}\right)$$

for $c_1, c_2, c_3 \in \mathbf{R}$. The second and third terms of the right-hand side can be written by using the generalized hypergeometric function $_0F_2$.

Theorem 4.2. *Let f be a non-semisimple flat centroaffine surface. Then changing the coordinates u and v, if necessary, we have*

$$\varphi = 1, \quad \rho = \alpha(u)v + \beta(u), \quad a = \frac{1}{2}\alpha''v^2 + \beta''v + C_4(\alpha\beta' - 1)$$

with $(\alpha'', \beta'') \neq (0,0)$ and $C_4 \in \mathbf{R}$, and one of the following:

(i) *For some $C_3 \in \mathbf{R}$ with $(C_3, C_4) \in \mathbf{R}^2 \setminus \{(0,0)\}$, we have*

$$b = \frac{1}{C_3 v + C_4}$$

and

$$\frac{1}{2}\alpha'' - C_3\alpha\alpha' = 0, \quad \beta'' - C_4\alpha\alpha' = C_3(\alpha\beta' - 1). \tag{15}$$

(ii) *For some $C_1, C_2, C_3 \in \mathbf{R}$ with $C_1, C_2 \neq 0$, we have*

$$b = \frac{C_1 v + C_2}{(C_1 C_3 + C_2^2)v^2 + (C_1 C_4 + C_2 C_3)v + C_2 C_4} \tag{16}$$

and

$$\begin{cases} \alpha\alpha' = \dfrac{C_1}{C_2}(\alpha\beta' - 1), \\[2mm] \dfrac{1}{2}\alpha'' = \dfrac{C_1 C_3 + C_2^2}{C_2}(\alpha\beta' - 1), \\[2mm] \beta'' = \dfrac{C_1 C_4 + C_2 C_3}{C_2}(\alpha\beta' - 1). \end{cases} \tag{17}$$

Proof. As in the proof of Theorem 4.1, we may assume that $\varphi = 1$, $a_v \neq 0$, $b = b(v)$, and obtain Eqs. (12) and (13). Then from the second equation of Eq. (12) and Eq. (13), we have

$$a = \frac{1}{2}\alpha''v^2 + \beta''v + \gamma(u). \tag{18}$$

Since $a_v \neq 0$, we have $(\alpha'', \beta'') \neq (0,0)$. From the first equation of Eq. (12), Eq. (13) and Eq. (18), we have

$$0 = -1 - \left(\frac{1}{2}\alpha''v^2 + \beta''v + \gamma\right)b + (\alpha'v + \beta')\alpha. \tag{19}$$

If $b = 0$, from Eqs. (14) and (19), we have $J = 3$, which is a contradiction since the Pick invariant of the surface given by Theorem 4.1 is equal to -1.

Since $b \neq 0$, Eq. (19) is equivalent to

$$\frac{1}{b} = \frac{\frac{1}{2}\alpha'' v^2 + \beta'' v + \gamma}{\alpha\alpha' v + \alpha\beta' - 1}. \tag{20}$$

If $\alpha\beta' = 1$, we have $\beta' = \dfrac{1}{\alpha}$. On the other hand, since $b = b(v)$, the right-hand side of Eq. (20) shows that the function $\dfrac{\beta''}{\alpha\alpha'}$ should be a constant. Then

$$\frac{\beta''}{\alpha\alpha'} = -\frac{\alpha'}{\alpha^2}\frac{1}{\alpha\alpha'} = -\frac{1}{\alpha^3},$$

which leads to a contradiction since α becomes a constant.

Now we put

$$P = \frac{1}{2}\frac{\alpha''}{\alpha\beta' - 1}, \quad Q = \frac{\beta''}{\alpha\beta' - 1}, \quad R = \frac{\gamma}{\alpha\beta' - 1}, \quad S = \frac{\alpha\alpha'}{\alpha\beta' - 1}.$$

Then the right-hand side of Eq. (20) is equal to

$$\frac{Pv^2 + Qv + R}{Sv + 1} = (Pv^2 + Qv + R)(1 - Sv + S^2 v^2 - S^3 v^3 + \cdots)$$

$$= R + (Q - RS)v + (P - QS + RS^2)v^2$$

$$- (P - QS + RS^2)Sv^3 + \cdots.$$

Hence there exist $C_1, \ldots, C_4 \in \mathbf{R}$ such that

$$(P - QS + RS^2)S = C_1, \quad P - QS + RS^2 = C_2, \quad Q - RS = C_3, \quad R = C_4.$$

In particular, we have

$$\gamma = C_4(\alpha\beta' - 1).$$

If $C_2 = 0$, we have

$$C_1 = 0, \quad P - C_3 S = 0, \quad Q - C_4 S = C_3,$$

which leads to the case (i).

If $C_2 \neq 0$, we have

$$C_2 S = C_1, \quad P - C_3 S = C_2, \quad Q - C_4 S = C_3,$$

which leads to Eqs. (16) and (17). Moreover, if $C_1 = 0$, from the first equation of Eq. (17), α is a constant, which is not compatible with the second equation of Eq. (17) since $\alpha\beta' \neq 1$. Hence we have the case (ii). □

Remark 4.4. If $C_3 = 0$ in the case of (i) in Theorem 4.2, Eq. (15) can be explicitly solved.

If $C_3 \neq 0$ in the case of (i) in Theorem 4.2, changing the coordinate v, if necessary, we may assume that $C_3 = 1$ and $C_4 = 0$.

Remark 4.5. Theorem 4.2 is a generalization of a result obtained in Ref. [3], where we classified non-semisimple flat centroaffine minimal surfaces.

If the surface f in Theorem 4.2 is centroaffine minimal, we have

$$\rho_{uv} = \alpha' = 0,$$

i.e., α is a constant.

If the case (ii) in Theorem 4.2 holds, since $\alpha\beta' \neq 1$, from the first equation of Eq. (17), we have $C_1 = 0$, which is a contradiction. Hence we have the case (i) in Theorem 4.2.

From the second equation of Eq. (15), we have

$$\beta'' = C_3(\alpha\beta' - 1).$$

If $\alpha = 0$, we have the surface given by Eq. (9), and if $\alpha \neq 0$, we have another case as in Ref. [3].

Acknowledgements

The author would like to thank Professor Hitoshi Furuhata for his kind advice.

References

1. B. Dubrovin, *Geometry of 2D topological field theories. Integrable systems and quantum groups (Montecatini Terme, 1993)*, 120–348, Lecture Notes in Math., **1620**, Springer, Berlin, 1996.
2. E. V. Ferapontov, *Hypersurfaces with flat centroaffine metric and equations of associativity*, Geom. Dedicata **103** (2004), 33–49.
3. A. Fujioka, *Centroaffine minimal surfaces with non-semisimple centroaffine Tchebychev operator*, Results Math. **56** (2009), no. 1–4, 177–195.
4. A. Fujioka, *Centroaffine minimal surfaces whose centroaffine curvature and Pick function are constants*, J. Math. Anal. Appl. **365** (2010), no. 2, 694–700.
5. H. L. Liu and C. P. Wang, *The centroaffine Tchebychev operator. Festschrift dedicated to Katsumi Nomizu on his 70th birthday (Leuven, 1994; Brussels, 1994)*, Results Math. **27** (1995), no. 1–2, 77–92.
6. A. M. Mathai, A handbook of generalized special functions for statistical and physical sciences. Oxford Science Publications. The Clarendon Press, Oxford University Press, New York, 1993.

7. C. S. Meijer, *Über Whittakersche bzw. Besselsche Funktionen und deren Produkte*, Nieuw Arch. Wiskunde **18** (1936), no.4, 10–39.

8. M. A. Magid and P. J. Ryan, *Flat affine spheres in* $\mathbf{R}^3$, Geom. Dedicata **33** (1990), no. 3, 277–288.

9. A. M. Mathai and R. K. Saxena, Generalized hypergeometric functions with applications in statistics and physical sciences. Lecture Notes in Mathematics, Vol. 348. Springer-Verlag, Berlin-New York, 1973.

10. K. Nomizu and T. Sasaki, Affine differential geometry. Geometry of affine immersions. Cambridge Tracts in Mathematics, **111**. Cambridge University Press, Cambridge, 1994.

11. W. K. Schief, *Hyperbolic surfaces in centro-affine geometry. Integrability and discretization. Integrability and chaos in discrete systems (Brussels, 1997)*, Chaos Solitons Fractals **11** (2000), no. 1–3, 97–106.

12. G. Tzitzéica, *Sur une nouvelle classe de surfaces*, Rend. Circ. Mat. Palermo **25** (1908) 180–187; **28** (1909) 210–216.

13. C. P. Wang, *Centroaffine minimal hypersurfaces in* $\mathbf{R}^{n+1}$, Geom. Dedicata **51** (1994), no. 1, 63–74.

Received March 8, 2013.

Proceedings of the Workshop on
Differential Geometry of Submanifolds
and its Related Topics
Saga, August 4-6, 2012

THE TOTAL ABSOLUTE CURVATURE OF OPEN CURVES IN E^N

Kazuyuki ENOMOTO

Faculty of Industrial Science and Technology,
Tokyo University of Science,
Oshamambe, Hokkaido, 049–3514 Japan
E-mail: enomoto_kazuyuki@rs.tus.ac.jp

Jin-ichi ITOH

Faculty of Education,
Kumamoto University,
Kumamoto, Kumamoto, 860–8555 Japan
E-mail: j-itoh@gpo.kumamoto-u.ac.jp

Dedicated to Professor Sadahiro Maeda for his sixtieth birthday

In our previous work [3], we studied the total absolute curvature of open curves in E^3 and determined the infimum of the total absolute curvature in the set of curves with fixed endpoints, end-directions and length. We also showed that, if the total absolute curvature of a sequence of curves in this set tends to the infimum, the limit curve must lie in a plane. Moreover, it was shown that the limit curve is either a subarc of a closed plane convex curve or a piecewise linear curve with at most three edges. In this paper, we generalize this result to open curves in E^N. We show that the limit curve must lie in a 3-dimensional affine subspace, and hence its geometry is described by our theorem in [3].

Keywords: total absolute curvature, open curve.

1. Introduction

Let Σ be a C^2 curve in the N-dimensional Euclidean space E^3. The total absolute curvature $\tau(\Sigma)$ of Σ is the total integral of the curvature of Σ. If Σ is only piecewise C^2, we add the exterior angles at non-smooth points to define $\tau(\Sigma)$. In particular, the total absolute curvature of a piecewise linear curve is just the sum of the exterior angles.

The study of the total absolute curvature of curves has a long history since Fenchel proved that the total absolute curvature of any closed curve in E^N is not less than 2π and the infimum is attained by a plane convex curve ([4]). Fenchel's theorem has been extended in various directions ([1], [5] etc.), but it seems that most results are concerned with closed curves and not much has been done for open curves. In this paper, we study the total absolute curvature of open curves in E^N.

When we are given points p and q in E^N, unit tangent vectors X at p and Y at q and a positive constant L $(> |pq|)$, we denote by $\mathcal{C}(p, X, q, Y, L)$ the set of all piecewise C^2 curves in E^N whose endpoints, end-directions and length are p, q, X, Y and L. We study the shape of a curve which minimizes the total absolute curvature in $\mathcal{C}(p, X, q, Y, L)$. Fenchel's thorem says that, if the curve is closed (i.e. $p = q$, $X = Y$), then $\inf\{\tau(\Sigma) : \Sigma \in \mathcal{C}(p, X, q, Y, L)\} = 2\pi$, which does not depend on L. For the open case, in contrast, $\inf\{\tau(\Sigma) : \Sigma \in \mathcal{C}(p, X, q, Y, L)\}$ usually depends on the length L (see [3] and Remark 5.8 in [3]). We also note that the problem becomes almost trivial if we do not fix the length (see Remark 5.9 in [3]).

Another difference between the closed case and the open case is that for the open case $\inf\{\tau(\Sigma) : \Sigma \in \mathcal{C}(p, X, q, Y, L)\}$ is not necessarily attained by an element of $\mathcal{C}(p, X, q, Y, L)$. When the total absolute curvature of a sequence of curves in $\mathcal{C}(p, X, q, Y, L)$ tends to $\inf\{\tau(\Sigma) : \Sigma \in \mathcal{C}(p, X, q, Y, L)\}$, the limit curve Σ_0 may have end-directions different from X or Y, and then Σ_0 becomes only an element of $\mathcal{C}(p, q, L)$, the set of all piecewise C^2 curves whose endpoints and length are p, q and L. This phenomenon has already been seen in the planar case [2].

In the main thorem of our previous paper [3], we consider the case when the curve lies in E^3 and show that if $\{\Sigma_k : k = 1, 2, 3, \ldots\}$ is a sequence of curves in $\mathcal{C}(p, X, q, Y, L)$ such that $\tau(\Sigma_k)$ tends to $\inf\{\tau(\Sigma) : \Sigma \in \mathcal{C}(p, X, q, Y, L)\}$ as $k \to \infty$, then the limit curve must lie in a plane. Moreover, it is shown that the limit curve is either a subarc of a closed plane convex curve or a piecewise linear curve with at most three edges.

In this paper, we consider the case when the curve lies in E^N and show that if $\{\Sigma_k : k = 1, 2, 3, \ldots\}$ is a sequence of curves in $\mathcal{C}(p, X, q, Y, L)$ such that $\tau(\Sigma_k)$ tends to $\inf\{\tau(\Sigma) : \Sigma \in \mathcal{C}(p, X, q, Y, L)\}$ as $k \to \infty$, then the limit curve must lie in a 3-dimensional affine subspace. This enables us to use our results in [3] to determine the shape of the limit curve.

As in [3], we introduce the notion of the "extended total absolute curvature" $\tilde{\tau}$ and look for the curve in $\mathcal{C}(p, q, L)$ which minimizes the extended total absolute curvature, since the limit curve may have different end-directions. We first study the simple case when the curve is only a piecewise linear curve with two edges, and show that the curve minimizing the extended total absolute curvature among all piecewise linear curves with two edges in $\mathcal{C}(p, q, L)$ and X and Y must lie in a 3-dimensional affine subspace (Lemma 3.4).

We use this result to show that the curve minimizing the extended total absolute curvature among all piecewise linear curves with n edges in $\mathcal{C}(p, q, L)$ and X and Y must lie in a 3-dimensional affine subspace (Lemma 4.2). Finally, through approximation of a piecewise smooth curve by a piecewise linear curve, we show that the curve minimizing the extended total absolute curvature among all piecewise smooth curves in $\mathcal{C}(p, q, L)$ and X and Y must lie in a 3-dimensional affine subspace (Theorem 5.1). This enables us to use the results in [3] to determine the shape of the curve minimizing the total absolute curvature in $\mathcal{C}(p, X, q, Y, L)$ (Theorem 5.2).

2. Preliminaries

Let Σ be a piecewise C^2 curve in the N-dimensional Euclidean space E^N. Let ℓ be the length of Σ and $x(s)$ $(0 \leq s \leq \ell)$ be a parameterization of Σ by its arclength. Let $0 = s_0 < s_1 < \cdots < s_n = \ell$ be the subdivision of $[0, \ell]$ such that $\{x(s) : s_{i-1} < s < s_i\}$ is C^2 for each $i = 1, \ldots, n$. If Σ is C^2 at $x(s)$, the curvature $k(s)$ of Σ is defined by $k(s) = |d^2x/ds^2|$. Let $T = T(s) = dx/ds$ be the unit tangent vector of Σ. For each i, let θ_i $(0 \leq \theta_i \leq \pi)$ be the angle at $x(s_i)$ between $\lim_{s \to s_i - 0} T(s)$ and $\lim_{s \to s_i + 0} T(s)$.

Definition 2.1. The *total absolute curvature* $\tau(\Sigma)$ of Σ is defined by

$$\tau(\Sigma) = \sum_{i=1}^{n} \int_{s_{i-1}}^{s_i} k(s)\, ds + \sum_{i=1}^{n-1} \theta_i.$$

We now define several classes of curves. Here p and q are points in E^N, X and Y are unit tangent vectors of E^N at p and q, respectively, and L is a positive constant greater than $|pq|$. Let

$$\mathcal{C}(p, q) = \{\, \Sigma : x(s) \,|\, x(0) = p,\ x(\ell) = q \,\}$$
$$\mathcal{C}(p, q, L) = \{\, \Sigma \in \mathcal{C}(p, q) \,|\, \ell = L \,\}$$
$$\mathcal{C}(p, X, q, Y) = \{\, \Sigma \in \mathcal{C}(p, q) \,|\, T(0) = X,\ T(\ell) = Y \,\}$$
$$\mathcal{C}(p, X, q, Y, L) = \mathcal{C}(p, q, L) \cap \mathcal{C}(p, X, q, Y).$$

Let n be a positive integer and let $\mathcal{P}_n$ be the set of all piecewise linear curves with n edges. For all $m < n$, we regard $\mathcal{P}_m$ as a subset of $\mathcal{P}_n$ by allowing angles between two edges to be zero. Let $\mathcal{P}_n(p, q, L) = \mathcal{P}_n \cap \mathcal{C}(p, q, L)$.

When $\{\Sigma_k : k = 1, 2, \ldots\}$ is a sequence of curves in $\mathcal{C}(p, X, q, Y)$, the limit curve $\lim_{k \to \infty} \Sigma_k$ may not be an element of $\mathcal{C}(p, X, q, Y)$, since the limit curve may not be tangent to X or Y (Fig. 2.1). If $\Sigma \in \mathcal{C}(p, q)$ is the limit curve of $\{\Sigma_k \in \mathcal{C}(p, X, q, Y) : k = 1, 2, \ldots\}$, then we have

$$\lim_{k \to \infty} \tau(\Sigma_k) = \angle(X, T(0)) + \tau(\Sigma) + \angle(T(\ell), Y).$$

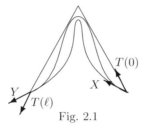

Fig. 2.1

This leads us to the following notion of the extended total absolute curvature.

Definition 2.2. For $\Sigma \in \mathcal{C}(p, q)$, the *extended total absolute curvature* $\tilde{\tau}(\Sigma)$ is defined by

$$\tilde{\tau}(\Sigma) = \angle(X, T(0)) + \tau(\Sigma) + \angle(T(\ell), Y),$$

where $\angle(\ ,\ )$ denotes the angle between two vectors with value in $[0, \pi]$.

Note that the definition of $\tilde{\tau}$ depends on the choice of X and Y, and $\tilde{\tau}(\Sigma)$ may be regarded as the total absolute curvature of $\Sigma \in \mathcal{C}(p, q)$ "as a curve in $\mathcal{C}(p, X, q, Y)$". If Σ happens to lie in $\mathcal{C}(p, X, q, Y)$, then we have $\tilde{\tau}(\Sigma) = \tau(\Sigma)$.

Definition 2.3. An element Σ_0 of $\mathcal{C}(p, q, L)$ is called an *extremal curve in* $\mathcal{C}(p, X, q, Y, L)$ if

$$\tilde{\tau}(\Sigma_0) \leq \tau(\Sigma)$$

holds for any $\Sigma \in \mathcal{C}(p, X, q, Y, L)$.

For $\Sigma \in \mathcal{C}(p, q, L)$, let T_Σ be the piecewise C^1 curve in S^2 which consists of $\cup_{i=1}^{n} \{T(s) : s_{i-1} < s < s_i\}$, the geodesic arcs between $\lim_{s \to s_i - 0} T(s)$ and $\lim_{s \to s_i + 0} T(s)$ $(i = 1, \ldots, n)$, the geodesic arc between X and $T(0)$ and the geodesic arc between Y and $T(L)$. Then the length of T_Σ is equal to $\tilde{\tau}(\Sigma)$.

Definition 2.4. A subarc of a closed plane convex curve (the boundary of a convex domain in a plane) is called a *plane convex arc*. $\{p, X, q, Y, L\}$ is said to *satisfy the convexity condition* if there exists a plane convex arc Σ in $\mathcal{C}(p, X, q, Y, L)$.

If $\{p, X, q, Y, L\}$ satisfies the convexity condition, then every plane convex arc Σ in $\mathcal{C}(p, X, q, Y, L)$ becomes an extremal curve in $\mathcal{C}(p, X, q, Y, L)$. If a sequence of plane convex arcs in $\mathcal{C}(p, X, q, Y, L)$ converges to a plane convex arc Σ_0 in $\mathcal{C}(p, q, L)$, then Σ_0 is an extremal curve in $\mathcal{C}(p, X, q, Y, L)$. For a plane convex arc Σ_0 the curve T_{Σ_0} becomes a subarc of a great circle.

Remark 2.1. If $p = q$, the convexity condition is automatically satisfied by $\{p, X, q, Y, L\}$ for any X, Y and L.

3. Piecewise linear curves with two edges

In this section, we study the shape of a piecewise linear curve with two edges which attains $\inf\{\tilde{\tau}(P) : P \in \mathcal{P}_2(p, q, L)\}$. Since $\mathcal{P}_2(p, q, L)$ is compact and $\tilde{\tau}$ is continuous in $\mathcal{P}_2(p, q, L)$, such a piecewise linear curve always exists.

Any element P of $\mathcal{P}_2(p, q, L)$ is expressed as

$$P = pp_1 \cup p_1 q$$

with $|pp_1| + |p_1 q| = L$. Then we have

$$\tilde{\tau}(P) = \angle(X, \overrightarrow{pp_1}) + \angle(\overrightarrow{pp_1}, \overrightarrow{p_1 q}) + \angle(\overrightarrow{p_1 q}, Y).$$

If we set

$$\xi = \frac{\overrightarrow{pp_1}}{|pp_1|},$$

then $\xi \in S^{N-1}$ is uniquely determined by P. Conversely, for any $\xi \in S^{N-1}$ there exists a unique point p_1 which satisfies

$$\frac{\overrightarrow{pp_1}}{|pp_1|} = \xi, \quad |pp_1| + |p_1 q| = L.$$

By this, $\mathcal{P}_2(p, q, L)$ is identified with S^{N-1}. Thus $\tilde{\tau}(P)$ on $\mathcal{P}_2(p, q, L)$ may be regarded as a function $\tilde{\tau}(\xi)$ on S^{N-1}. Let

$$f(\xi) = \frac{\overrightarrow{p_1 q}}{|p_1 q|}.$$

Then f defines a bijective map on the unit sphere S^{N-1}.

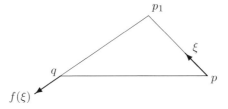

Fig. 3.1

If d denotes the distance in S^{N-1}, we have

$$\angle(X, \overrightarrow{pp_1}) = d(X, \xi), \quad \angle(\overrightarrow{pp_1}, \overrightarrow{p_1 q}) = d(\xi, f(\xi)), \quad \angle(\overrightarrow{p_1 q}, Y) = d(f(\xi), Y).$$

Hence $\tilde{\tau}$ on $\mathcal{P}_2(p, q, L)$, as a function in S^{N-1}, is written as

$$\tilde{\tau}(\xi) = d(X, \xi) + d(\xi, f(\xi)) + d(f(\xi), Y). \tag{1}$$

When $p \neq q$, we define a unit vector Z by

$$Z = \frac{\overrightarrow{pq}}{|pq|}.$$

Since ξ, $f(\xi)$ and Z lie in a plane as vectors in E^N, they lie on a great circle as points in S^{N-1}.

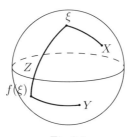

Fig 3.2

$\{p, X, q, Y, L\}$ satisfies the convexity condition if and only if we have either $p = q$ or $p \neq q$ and $\{X, f^{-1}(Y), Z, f(X), Y, -Z\}$ lies on a great circle in S^2 in this order (Figure 3.3).

In the following argument, we assume $p \neq q$. Let $d_X = d(X, \cdot)$, $d_Y = d(Y, \cdot)$ and $d_Z = d(Z, \cdot)$. Then (1) is rewritten as

$$\tilde{\tau}(\xi) = d_X(\xi) + d_Z(\xi) + d_Z(f(\xi)) + d_Y(f(\xi)). \tag{2}$$

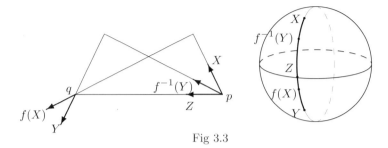

Fig 3.3

At any ξ with $\xi \neq \pm X$, d_X is differentiable and its gradient vector ∇d_X is defined. ∇d_X is identical with the unit tangent vector of the oriented geodesic from X to ξ. $\tilde{\tau}(\xi)$ is differentiable for all ξ with $\xi \neq \pm X$ and $\xi \neq f^{-1}(\pm Y)$. Note that $d_Z(\xi) + d_Z(f(\xi)) = d(\xi, f(\xi))$ is differentiable for all ξ.

Let T be a tangent vector of S^{N-1} at ξ. Let f_* be the differential of f. To describe a property of $f_* T$, we take a positively oriented orthonormal frame $\{E_1, E_2, \ldots, E_{N-1}\}$ of the tangent bundle of S^{N-1} defined in $S^{N-1} \backslash \{Z, -Z\}$ with $E_1 = \nabla d_Z$. In the following lemma, λ is a function defined by

$$\lambda = \frac{|pp_1|}{|p_1 q|}.$$

We set $D = |pq|$. Then we have

$$|pp_1| = \frac{L^2 - D^2}{2(L - D \cos d(Z, \xi))},$$

and λ, as a function in ξ, is given by

$$\lambda(\xi) = \frac{|pp_1|}{L - |pp_1|}$$
$$= \frac{L^2 - D^2}{L^2 + D^2 - 2LD \cos d(Z, \xi)}. \tag{3}$$

Lemma 3.1. *For any tangent vector T of S^{N-1} at ξ, we have*

$$f_* T = \lambda(\xi)(-\langle T, E_1(\xi) \rangle E_1(f(\xi)) + \sum_{k=2}^{N-1} \langle T, E_k(\xi) \rangle E_k(f(\xi))).$$

Proof. Let $\xi(t)$ be the point on the oriented geodesic $Z\xi$ such that $d(Z, \xi(t)) = t$. Then $\xi'(t) = E_1(\xi(t))$. Let $p_1(t)$ be the point in E^N which corresponds to $\xi(t)$. Since $p_1(t)$ lies in one plane for all t, $f(\xi(t))$ moves along the geodesic through Z and ξ, which implies that $f_*(E_1(\xi(t)))$ is parallel to $E_1(f(\xi(t)))$. $d(Z, f(\xi(t))) = \angle(\overrightarrow{pq}, \overrightarrow{p_1(t)q})$ is related to $t = d(Z, \xi(t)) = \angle(\overrightarrow{pq}, \overrightarrow{pp_1(t)})$ as

$$d(Z, f(\xi(t))) = \arccos\left(\frac{2LD - (L^2 + D^2)\cos t}{L^2 + D^2 - 2LD\cos t}\right).$$

Thus we have

$$
\begin{aligned}
f_*(E_1(\xi(t))) &= \frac{d}{dt} d(Z, f(\xi(t)))\, E_1(f(\xi(t))) \\
&= -\frac{L^2 - D^2}{L^2 + D^2 - 2LD\cos t}\, E_1(f(\xi(t))) \\
&= -\lambda(\xi(t))\, E_1(f(\xi(t))).
\end{aligned}
\tag{4}
$$

There exists a small circle C_k centered at Z of radius $d(Z, \xi)$ which is an integral curve of E_k ($k = 2, \ldots, N-1$). When ξ moves along C_k, $f(\xi)$ moves along a small circle centered at Z of radius $d(Z, f(\xi))$ which lies in the same 2-dimensional great sphere as C_k does. Then we see that

$$
\begin{aligned}
f_*(E_k(\xi)) &= \frac{\sin d(Z, f(\xi))}{\sin d(Z, \xi)}\, E_k(f(\xi)) \\
&= \frac{|pp_1|}{|p_1q|}\, E_k(f(\xi)) = \lambda(\xi)\, E_k(f(\xi)).
\end{aligned}
\tag{5}
$$

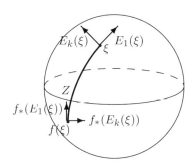

Fig 3.4

It follows from (4) and (5) that

$$f_* T = \langle T, E_1(\xi) \rangle f_*(E_1(\xi)) + \sum_{k=1}^{N-1} \langle T, E_k(\xi) \rangle f_*(E_k(\xi))$$

$$= \lambda(\xi)(-\langle T, E_1(\xi) \rangle E_1(f(\xi)) + \sum_{k=1}^{N-1} \langle T, E_k(\xi) \rangle E_k(f(\xi))). \qquad \square$$

Lemma 3.2. *Suppose $\nabla \tilde{\tau} = 0$ at ξ_0. Then either (1) or (2) holds:*

(1) $\{p, X, q, Y, L\}$ satisfies the convexity condition and ξ_0 is a point on the geodesic segment $X f^{-1}(Y)$
(2) $d(Z, \xi_0) = \arccos(D/L)$.

Proof. Let T be a tangent vector of S^2 at ξ. By (2), we have

$$T\tilde{\tau} = \langle \nabla d_X(\xi), T \rangle + \langle \nabla d_Z(\xi), T \rangle + \langle \nabla d_Z(f(\xi)), f_* T \rangle$$
$$+ \langle \nabla d_Y(f(\xi)), f_* T \rangle.$$

If we set $T = E_1(\xi) = \nabla d_Z(\xi)$, using Lemma 3.1, we have

$$E_1 \tilde{\tau} = \langle \nabla d_X(\xi), E_1(\xi) \rangle + 1 + \lambda(\xi)(-1 - \langle \nabla d_Y(f(\xi)), E_1(f(\xi)) \rangle). \qquad (6)$$

If we set $T = E_k(\xi)$ $(k = 1, \ldots, N-1)$, we have

$$E_k \tilde{\tau} = \langle \nabla d_X(\xi), E_k(\xi) \rangle + \lambda(\xi) \langle \nabla d_Y(f(\xi)), E_k(f(\xi)) \rangle. \qquad (7)$$

If $\nabla \tilde{\tau} = 0$ at ξ_0, then (6) gives

$$\langle \nabla d_X(\xi_0), E_1(\xi_0) \rangle + 1 = \lambda(\xi_0)(1 + \langle \nabla d_Y(f(\xi_0)), E_1(f(\xi_0)) \rangle) \qquad (8)$$

and (7) gives

$$\langle \nabla d_X(\xi_0), E_k(\xi_0) \rangle = -\lambda(\xi_0) \langle \nabla d_Y(f(\xi_0)), E_k(f(\xi_0)) \rangle. \qquad (9)$$

Combining (8) and (9) with

$$\langle \nabla d_X(\xi_0), E_1(\xi_0) \rangle^2 + \sum_{k=1}^{N-1} \langle \nabla d_X(\xi_0), E_k(\xi_0) \rangle^2 = |\nabla d_X(\xi_0)|^2 = 1 \qquad (10)$$

and

$$\langle \nabla d_Y(f(\xi_0)), E_1(f(\xi_0)) \rangle^2 + \sum_{k=1}^{N-1} \langle \nabla d_Y(f(\xi_0)), E_k(f(\xi_0)) \rangle^2$$
$$= |\nabla d_Y(f(\xi_0))|^2 = 1, \qquad (11)$$

we obtain

$$((\langle \nabla d_X(\xi_0), E_1(\xi_0)\rangle + 1)(1 - \lambda(\xi_0)) = 0. \tag{12}$$

This implies that either

$$\nabla d_X(\xi_0) = -\nabla d_Z(\xi_0) \tag{13}$$

or

$$\lambda(\xi_0) = 1 \tag{14}$$

holds.

If (13) holds, then (8) gives

$$\nabla d_Y(f(\xi_0)) = -\nabla d_Z(f(\xi_0)). \tag{15}$$

(13) and (15) imply that we have $\{X, \xi_0, Z, f(\xi_0), Y\}$ on a great circle in this order. Since $d(Z, X) \geq d(Z, \xi_0)$, we have $d(Z, f(X)) \leq d(Z, f(\xi_0))$. Hence if such ξ_0 exists, we have $d(Z, f(X)) \leq d(Z, Y)$, or equivalently, $d(Z, X) \geq d(Z, f^{-1}(Y))$. Thus $\{p, X, q, Y, L\}$ satisfies the convexity condition.

If (14) holds, we have $d(Z, \xi_0) = \arccos(D/L)$. $\qquad\square$

Remark 3.1.

(1) If $\{p, X, q, Y, L\}$ satisfies the convexity condition, we have $\nabla \tilde{\tau} = 0$ at any point on the geodesic segment $Xf^{-1}(Y)$.
(2) In the case (2) in Lemma 3.2, ξ_0 lies on the circle of radius $\arccos(D/L)$ centered at Z. Then we have $|pp_1| = |p_1q|$ by (3) and (14).

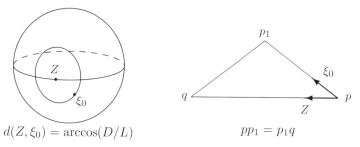

$$d(Z, \xi_0) = \arccos(D/L) \qquad\qquad pp_1 = p_1q$$

Fig 3.5

Lemma 3.3. *Suppose $\nabla\tilde{\tau} = 0$ at ξ_0. Then $\overrightarrow{pp_1}$, $\overrightarrow{p_1q}$, X and Y lie in a 3-dimensional linear subspace.*

Proof. If $\{p, X, q, Y, L\}$ satisfies the convexity condition, then ξ_0 is a point on the geodesic segment $Xf^{-1}(Y)$, which implies that both $\overrightarrow{pp_1}$ and $\overrightarrow{p_1q}$ lie in the 2-dimensional linear subspace spanned by X and Y.

Otherwise, we have $d(Z, \xi_0) = \arccos(D/L)$. Since ξ_0 is a critical point of $\tilde{\tau}$, it is a critical point of $\tilde{\tau}$ restricted on the small sphere S with center at Z and radius $\arccos(L/D)$. We will denote this function by $\tilde{\tau}'$. Let Y' be the unit vector which is symmetric to Y with respect to Z. For any $\xi \in S$ we have $f(\xi) \in S$ and $d(f(\xi), Y) = d(\xi, Y')$. Hence we get

$$\tilde{\tau}'(\xi) = d(X, \xi) + d(\xi, f(\xi)) + d(f(\xi), Y)$$
$$= d(X, \xi) + 2\arccos(D/L) + d(\xi, Y')$$

If $\xi_0 = X$ or $\xi_0 = Y$, then $\overrightarrow{pp_1}$, $\overrightarrow{p_1 q}$, X and Y lie in a 3-dimensional linear subspace. If $\xi_0 \neq X$ and $\xi_0 \neq Y$, then ∇d_X and ∇d_Y are defined at ξ_0. Let V be any vector tangent to S at ξ_0. Then we have

$$V\tilde{\tau}' = 0, \tag{16}$$

which gives

$$\langle \nabla d_X(\xi_0), V \rangle + \langle V, \nabla d_{Y'}(\xi_0) \rangle = \langle \nabla d_X(\xi_0) + \nabla d_{Y'}(\xi_0), V \rangle = 0. \tag{17}$$

Hence the vector $\nabla d_X(\xi_0) + \nabla d_{Y'}(\xi_0)$ must be parallel to $\nabla d_Z(\xi_0)$, which is a normal vector of S at ξ_0, and we obtain

$$\nabla d_X(\xi_0) + \nabla d_{Y'}(\xi_0) = a\nabla d_Z(\xi_0) \tag{18}$$

for some a.

Since $\nabla d_X(\xi_0)$ is tangent to the great circle passing through X and ξ_0, it is contained in the linear subspace spanned by X and ξ_0, and hence it can be expressed as

$$\nabla d_X(\xi_0) = b_1 X + b_2 \xi_0. \tag{19}$$

Since $\xi_0 \neq X$, we have $b_1 \neq 0$. By similar reasons, we have

$$\nabla d_{Y'}(\xi_0) = c_1 Y' + c_2 \xi_0 \tag{20}$$

with $c_1 \neq 0$ and

$$\nabla d_Z(\xi_0) = d_1 Z + d_2 \xi_0. \tag{21}$$

Combining (18), (19), (20) and (21), we have

$$b_1 X + c_1 Y' - ad_1 Z + (b_2 + c_2 - ad_2)\xi_0 = 0. \tag{22}$$

Since $b_1 \neq 0$, (22) implies that X, Y', Z and ξ_0 lie in a 3-dimensional linear subspace. Since Y lies in the 2-dimensional linear subspace spanned by Y' and Z, X, Y, Z and ξ_0 lie in the same 3-dimensional linear subspace. This immediately proves the lemma. $\square$

Lemma 3.4. *Let $P = pp_1 \cup p_1q$ be an extremal curve in $\mathcal{P}_2(p, X, q, Y, L)$. Then P, X and Y lie in a 3-dimensional affine subspace.*

Proof. If $P = pp_1 \cup p_1q$ be an extremal curve in $\mathcal{P}_2(p, X, q, Y, L)$, then one of the following holds.

(1) $\nabla\tilde{\tau} = 0$,
(2) $\overrightarrow{pp_1}/|pp_1| = \pm X$,
(3) $\overrightarrow{p_1q}/|p_1q| = \pm Y$.

In the case (1), by Lemma 3.3, P, X and Y lie in a 3-dimensional affine subspace. In (2) and (3), it is obvious that P, X and Y lie in a 3-dimensional affine subspace. □

We need the following lemma to treat the case when P has 3 edges.

Lemma 3.5. *Let $P = pp_1 \cup p_1q$ be an extremal curve in $\mathcal{P}_2(p, X, q, Y, L)$. Let Π be the 2-dimensional affine subspace containing P. Then X lies in Π if and only if Y lies in Π.*

Proof. If X lies in Π, then X, Z and $\xi_0 = \overrightarrow{pp_1}/|pp_1|$ lie on a great circle. Then we have $\nabla d_X(\xi_0) = \pm d_Z(\xi_0)$. By (22), we see $\nabla d_Y(\xi_0) = \pm d_Z(\xi_0)$. This means that Y also lies in Π.

By a similar argument, the converse is proved. □

4. Piecewise linear curves with n edges

In this section, we study the shape of a piecewise linear curve which attains $\inf\{\tilde{\tau}(P) : P \in \mathcal{P}_n(p, q, L)\}$. Since $\mathcal{P}_n(p, q, L)$ is compact and $\tilde{\tau}$ is continuous, such a piecewise linear curve always exists.

We start with the case when $n = 3$ and prove the following lemma.

Lemma 4.1. *Let $P = pp_1 \cup p_1p_2 \cup p_2q$ be an extremal curve in $\mathcal{C}(p, X, q, Y, L)$. Then*

(1) *P, X and Y lie in a 3-dimensional affine subspace.*
(2) *Moreover, if P lies in a 2-dimensional affine subspace Π, both X and Y must lie in Π.*

Proof. Let $P_0 = pp_1 \cup p_1p_2 \cup p_2q$ be an extremal curve in $\mathcal{P}_3(p, X, q, Y, L)$. Let $P_0^1 = pp_1 \cup p_1p_2$ and $P_0^2 = p_1p_2 \cup p_2q$. Let $X_1 = \overrightarrow{pp_1}/|pp_1|$, $X_2 = \overrightarrow{p_1p_2}/|p_1p_2|$, $X_3 = \overrightarrow{p_2q}/|p_2q|$, $L_1 = |pp_1| + |p_1p_2|$ and $L_2 = |p_1p_2| + |p_2q|$. P_0^1 must be an extremal curve in $\mathcal{P}_2(p, X, p_2, X_3, L_1)$, since if not, the replacement of P_0^1 by an extremal curve in $\mathcal{P}_2(p, X, p_2, X_3, L_1)$ would produce an element of $\mathcal{P}_3(p, q, L)$ whose $\tilde{\tau}$ is smaller than $\tilde{\tau}(P_0)$, which contradicts the minimality of $\tilde{\tau}(P_0)$. By Lemma 3.4, both

(1) X, $\overrightarrow{pp_1}$, $\overrightarrow{p_1p_2}$ and $\overrightarrow{p_2q}$ lie in a 3-dimensional affine space

and

(2) $\overrightarrow{pp_1}$, $\overrightarrow{p_1p_2}$ and $\overrightarrow{p_2q}$ and Y lie in a 3-dimensional affine space

hold. If $\overrightarrow{pp_1}$, $\overrightarrow{p_1p_2}$ and $\overrightarrow{p_2p_3}$ are linearly independent and span a 3-dimensional affine subspace, then both X and Y must lie in this subspace, which proves the lemma. If $\overrightarrow{pp_1}$, $\overrightarrow{p_1p_2}$ and $\overrightarrow{p_2p_3}$ are linearly dependent and contained in a 2-dimensional affine subspace Π, then, by Lemma 3.5, both X and Y must lie in Π. $\qquad\square$

Now we study the shape of a piecewise linear curve which attains $\inf\{\tilde{\tau}(P) : P \in \mathcal{P}_n(p, q, L)\}$.

Lemma 4.2. *Suppose that $P_0 \in \mathcal{P}_n(p, q, L)$ is an extremal curve in $\mathcal{P}_n(p, X, q, Y, L)$. Then P_0, X and Y must lie in a 3-dimensional affine space.*

Proof. We write P_0 as $P_0 = p_0p_1 \cup p_1p_2 \cup \cdots \cup p_{n-2}p_{n-1} \cup p_{n-1}p_n$ with $p_0 = p$, $p_n = q$.

Here we fix some notations. Let

$$P_0^k = p_{k-1}p_k \cup p_kp_{k+1} \cup p_{k+1}p_{k+2} \quad (k = 1, \ldots, n-2)$$
$$L_k = |p_{k-1}p_k| + |p_kp_{k+1}| + |p_{k+1}p_{k+2}| \quad (k = 1, \ldots, n-2)$$
$$X_k = \overrightarrow{p_{k-1}p_k}/|p_{k-1}p_k| \quad (k = 1, \ldots, n), \quad X_0 = X, \quad X_{n+1} = Y.$$

Since P_0 is an extremal curve in $\mathcal{P}_n(p, X, q, Y, L)$, for every k with $1 \le k \le n-2$, P_0^k must be an extremal curve in $\mathcal{P}_3(p_{k-1}, X_{k-1}, p_{k+2}, X_{k+3}, L_k)$.

Suppose that P_0^k lies in a 2-dimensional affine subspace Π for some k. Then, by Lemma 4.1(2), both X_{k-1} and X_{k+3} must lie in Π, which implies that both P_0^{k-1} and P_0^{k+1} lie in Π. Repeating this argument, we conclude that the whole curve P_0 together with X and Y lie in Π, which proves the lemma in this case.

In the remaining case, any of P_0^k does not lie in a 2-dimensional affine subspace. Let S_k be the 3-dimensional affine subspace containing P_0^k. Then, by Lemma 4.1(1), both X_{k-1} and X_{k+3} must lie in S_k, which implies that both P_0^{k-1} and P_0^{k+1} lie in S_k. Thus we have $P_0^{k-1} \subset S_{k-1} \cup S_k$ and $P_0^{k+1} \subset S_k \cup S_{k+1}$. Since P_0^{k-1} or P_0^{k+1} does not contain in a 2-dimensional affine subspace, we find $S_{k-1} = S_k = S_{k+1}$. Repeating this argument, we conclude that the whole curve P_0 together with X and Y lie in S_k, which proves the lemma in this case. $\qquad\square$

5. Smooth curves

Now we state our main result of the present paper.

Theorem 5.1. *Let Σ_0 be an extremal curve in $\mathcal{C}(p, X, q, Y, L)$. Then Σ_0, X and Y must lie in a 3-dimensional affine space.*

Proof. If $x(s)$ $(0 \leq s \leq L)$ is a parameterization of Σ_0 by arclength, we can construct a piecewise linear curve with n edges by connecting $x((i-1)L/n)$ and $x(iL/n)$ for $i = 1, \cdots, n$. By taking a point p_i in a neiborhood of $x(iL/n)$ for $i = 1, \cdots, n-1$, we can construct another piecewise linear curve P_n whose length is L. If Σ_0, X and Y do not lie in a 3-dimensional affine subspace, we can construct P_n so that P_n, X and Y do not lie in a 3-dimensional affine subspace for each n. Then P_n cannot be an extremal curve in $\mathcal{P}_n(p, X, q, Y, L)$ by Lemma 4.2 and we must have

$$\tilde{\tau}(P_n) > \inf\{\tilde{\tau}(P) : P \in \mathcal{P}_n(p, q, L)\} + \delta \tag{23}$$

for some $\delta > 0$. For any positive constant $\varepsilon < \delta$, if we take n sufficiently large, we know

$$\tilde{\tau}(P_n) < \tilde{\tau}(\Sigma_0) + \varepsilon. \tag{24}$$

It follows from (23) and (24) that

$$\begin{aligned}
\inf\{\tilde{\tau}(P) : P \in \mathcal{P}_n(p, q, L)\} &< \tilde{\tau}(P_n) - \delta \\
&< \tilde{\tau}(\Sigma_0) + \varepsilon - \delta \\
&< \tilde{\tau}(\Sigma_0) \\
&= \inf\{\tilde{\tau}(\Sigma) : \Sigma \in \mathcal{C}(p, q, L)\}.
\end{aligned}$$

This is a contradiction, since we must have

$$\inf\{\tilde{\tau}(P) : P \in \mathcal{P}_n(p, q, L)\} \geq \inf\{\tilde{\tau}(\Sigma) : \Sigma \in \mathcal{C}(p, q, L)\}.$$

Thus Σ_0, X and Y must lie in a 3-dimensional affine subspace. $\square$

Now we are ready to use the results in [3] to show the following theorem.

Theorem 5.2. *Let Σ_0 be an extremal curve in $\mathcal{C}(p, X, q, Y, L)$. Then Σ_0 is actually an element of $\mathcal{P}_2(p, q, L)$ except for the following two cases:*

(1) *$\{p, X, q, Y, L\}$ satisfies the convexity condition and Σ_0 is a plane convex arc.*

(2) *$p \neq q$, $X = Y$, $\angle(X, \overrightarrow{pq}) \geq \arccos(D/L)$ and Σ_0 is a Z-curve tangent to X at p and to Y at q.*

Corollary 5.1. *For any* $\{p, X, q, Y, L\}$, *there exists an element of* $\mathcal{P}_2(p, q, L)$ *which is an extremal curve in* $\mathcal{C}(p, X, q, Y, L)$.

Corollary 5.2. *If* X, Y *and* $\overrightarrow{pq}$ *do not lie in a plane, then every extremal curve in* $\mathcal{C}(p, X, q, Y, L)$ *is an element of* $\mathcal{P}_2(p, q, L)$.

Remark 5.1. One may apply Theorem 5.2 for closed curves by setting $p = q$ and $X = Y$. As a result, one can derive the classical theorems by Fenchel [4] ($N = 3$) and Borsuk [1] from Theorem 5.2.

References

1. K. Borsuk, *Sur la courbure totale des courbes*, Ann. Soc. Mth. Polon., **20** (1947), 251–256.
2. K. Enomoto, *The total absolute curvature of open plane curves of fixed length*, Yokohama Math. J. **48** (2000), 83–96.
3. K. Enomoto, J. Itoh and R. Sinclair, *The total absolute curvature of open curves in* E^3, Illinois J. Math. **52** (2008), 47–76
4. W. Fenchel, *Über Krümmung und Windung geschlossener Raumkurven*, Math. Ann. **101** (1929), 238–252.
5. J. Milnor, *On the total curvature of knots*, Ann. Math. **52** (1953), 248–257.

Received March 29, 2013.

Proceedings of the Workshop on
Differential Geometry of Submanifolds
and its Related Topics
Saga, August 4-6, 2012

ANTIPODAL SETS OF COMPACT SYMMETRIC SPACES AND THE INTERSECTION OF TOTALLY GEODESIC SUBMANIFOLDS

Makiko Sumi TANAKA

*Department of Mathematics, Faculty of Science and Technology,
Tokyo University of Science,
Noda, Chiba 278-8510, Japan
E-mail: tanaka_makiko@ma.noda.tus.ac.jp*

Dedicated to Professor Sadahiro Maeda for his sixtieth birthday

A real form in a Hermitian symmetric space M of compact type is the fixed point set of an involutive anti-holomorphic isometry of M, which is connected and a totally geodesic Lagrangian submanifold. We prove that the intersection of two real forms is an antipodal set, in which the geodesic symmetry at each point is the identity. Using this we investigate the intersection of two real forms in irreducible M as well as non-irreducible M and determine the intersection numbers of them. This is a survey article on the joint research with Hiroyuki Tasaki.

Keywords: Riemannian symmetric space, polar, meridian, antipodal set, 2-number, real form, Hermitian symmetric space.

1. Introduction

Tasaki [21] investigated the intersection of two real forms, which are totally geodesic Lagrangian submanifolds, in the complex hyperquadric and gave the explicit description of the intersection. He found that the intersection is an antipodal set if they intersect transversely. An antipodal set, which was introduced by Chen and Nagano [3], is a finite subset in which the geodesic symmetry at each point is the identity. We extended Tasaki's result in [17] and we proved that the intersection of two real forms in a Hermitian symmetric space of compact type is an antipodal set. We also proved that if two real forms are congruent moreover, the intersection is a great antipodal set.

2. Chen-Nagano theory

The fundamental principle of Chen-Nagano theory is that a pair of a polar and the corresponding meridian determines a compact Riemannian symmetric space. One of advantages of the theory is that it is useful for inductive arguments on polars or meridians.

2.1. Polars and meridians

Let M be a Riemannian symmetric space and let x be a point in M. The geodesic symmetry at x is denoted by s_x, i.e., s_x is an isometry of M which satisfies (i) s_x is involutive and (ii) x is an isolated fixed point of s_x. For a point o in M, let $F(s_o, M)$ denote the fixed point set of s_o:

$$F(s_o, M) = \{x \in M \mid s_o(x) = x\}$$

and we express it as a disjoint union of the connected components

$$F(s_o, M) = \bigcup_{j=0}^{r} M_j^+,$$

where we set $M_0^+ = \{o\}$. We call M_j^+ a *polar* of M with respect to o. $M_0^+ = \{o\}$ is called a *trivial* polar. When a polar consists of one point, we call it a *pole*. Each polar is a totally geodesic submanifold since it is a connected component of the fixed point set of an isometry of M. In general, if N is a totally geodesic submanifold in M, the geodesic symmetry s_x of M at a point x in N preserves N and the restriction of s_x to N gives the geodesic symmetry of N with respect to the induced metric, hence N is a Riemannian symmetric space.

Example 2.1. Let M be the n-dimensional sphere $S^n (\subset \mathbb{R}^{n+1})$. The geodesic symmetry at $x \in S^n$ is induced by the reflection with respect to $\mathbb{R}x$ in $\mathbb{R}^{n+1}$, that is, $s_x(y) = y$ for $y \in S^n \cap \mathbb{R}x$ and $s_x(y) = -y$ for $y \in S^n \cap (\mathbb{R}x)^{\perp}$, where $(\mathbb{R}x)^{\perp}$ denotes the orthogonal complement to $\mathbb{R}x$ in $\mathbb{R}^{n+1}$. Hence for a point $o \in S^n$ we have $F(s_o, S^n) = \{o, -o\}$ and the polars are $M_o^+ = \{o\}$ and $M_1^+ = \{-o\}$ which are poles.

Example 2.2. Let $\mathbb{R}H^n$ be the n-dimensional real hyperbolic space. We have $F(s_o, \mathbb{R}H^n) = \{o\}$ for a point $o \in \mathbb{R}H^n$. In general, if M is a Riemannian symmetric space of noncompact type, we have $F(s_o, M) = \{o\}$ for a point $o \in M$.

Example 2.3. Let M be the n-dimensional real projective space $\mathbb{R}P^n$. The geodesic symmetry at $x \in \mathbb{R}P^n$ is induced by the reflection with respect to x in $\mathbb{R}^{n+1}$. Let $e_1, \ldots, e_{n+1}$ be an orthogonal basis of $\mathbb{R}^{n+1}$ and let $o = \langle e_1 \rangle \in \mathbb{R}P^n$ where $\langle e_1 \rangle$ denotes 1-dimensional linear subspace spanned by e_1 in $\mathbb{R}^{n+1}$. Then we have $F(s_o, \mathbb{R}P^n) = \{o\} \cup \{\text{1-dimensional linear subspaces in } \langle e_2, \ldots, e_{n+1} \rangle \}$ and the polars are $M_0^+ = \{o\}$ and $M_1^+ \cong \mathbb{R}P^{n-1}$.

Example 2.4. Let M be the Grassmann manifold $G_k(\mathbb{K}^n)$ ($\mathbb{K} = \mathbb{R}, \mathbb{C}, \mathbb{H}$) consisting of k-dimensional $\mathbb{K}$-subspaces in $\mathbb{K}^n$ where $k \leq n$. $G_k(\mathbb{K}^n)$ is a point if $k = n$. The geodesic symmetry at $x \in G_k(\mathbb{K}^n)$ is induced by the reflection ρ_x with respect to x in $\mathbb{K}^n$, that is, ρ_x is a linear transformation of $\mathbb{K}^n$ which is the identity on x and -1 times the identity on $x^\perp$. Let $e_1, \ldots, e_n$ be an orthogonal basis of $\mathbb{K}^n$ and let $o = \langle e_1, \ldots, e_k \rangle \in G_k(\mathbb{K}^n)$. Then we have

$$F(s_o, G_k(\mathbb{K}^n)) = \bigcup_{j=0}^{k} G_{k-j}(o) \times G_j(o^\perp),$$

hence $M_0^+ = \{o\}$, $M_j^+ \cong G_{k-j}(\mathbb{K}^k) \times G_j(\mathbb{K}^{n-j})$ for $1 \leq j \leq k - 1$ and $M_k^+ \cong G_k(\mathbb{K}^{n-j})$.

Example 2.5. Let M be a compact Lie group equipped with bi-invariant Riemannian metric. The geodesic symmetry s_x at $x \in M$ is given by $s_x(y) = xy^{-1}x$ ($y \in M$). Let e denote the unit element of M. Then we have

$$F(s_e, M) = \{x \in M \mid x^2 = e\}.$$

If $M = U(n)$, the unitary group, we have

$$F(s_e, U(n)) = \bigcup_{j=0}^{n} \mathrm{Ad}(U(n)) \mathrm{diag}(\underbrace{-1, \ldots, -1}_{j}, \underbrace{1, \ldots, 1}_{n-j}),$$

hence $M_0^+ = \{o\}$ and $M_j^+ \cong G_j(\mathbb{C}^n)$ ($1 \leq j \leq n$). Here

$$\mathrm{diag}(\underbrace{-1, \ldots, -1}_{j}, \underbrace{1, \ldots, 1}_{n-j}) = \begin{bmatrix} -E_j & \\ & E_{n-j} \end{bmatrix},$$

where E_k denotes the unit matrix with degree k. $M_n^+ = \{\mathrm{diag}(-1, \ldots, -1)\}$ is a pole.

From now on we assume that every Riemannian symmetric space is compact and connected.

Since $x \in M$ is an isolated fixed point of s_x, the differential $(ds_x)_x$: $T_x M \to T_x M$ of s_x at x is -1 times the identity. If $s_x(y) = y$, $(ds_x)_y$: $T_y M \to T_y M$ is involutive and the eigenvalues are ± 1. Let p be a point in a polar M_j^+ with respect to o. Let $T_p M = V_+ \oplus V_-$ be the diagonal sum decomposition where $V_\pm$ is the eigenspace of $(ds_o)_p$ with eigenvalue ± 1 respectively. Then $V_+ = T_p M_j^+$. Since $(ds_p)_p$ is -1 times the identity of $T_p M$, $V_\pm$ is the eigenspace of $d(s_p \circ s_o)_p$ corresponding to the eigenvalue ± 1 respectively. Then there exists the totally geodesic submanifold $M_j^-(p)$ whose tangent space at p coincides with V_-. We call $M_j^-(p)$ the *meridian* to M_j^+ at p. $M_j^-(p)$ is the connected component of $F(s_p \circ s_o, M)$ containing p.

Let G denote the identity component of the isometry group of M and let K denote the isotropy subgroup at o.

Proposition 2.1 (Chen-Nagano [2]). *Every polar M_j^+ of M with respect to o is a K-orbit.*

If $p, q \in M_j^+$, then $M_j^-(p)$ and $M_j^-(q)$ are K-congruent. Since $F(s_o, M)$ is G-congruent to $F(s_{o'}, M)$ for $o, o' \in M$, each polar with respect to o is G-congruent to the corresponding polar with respect to o'. Therefore it is sufficient to consider a polar M_j^+ with respect to some chosen point $o \in M$ and the meridian $M_j^-(p)$ to M_j^+ at some chosen point $p \in M_j^+$, which is simply denoted by M_j^-.

Remark 2.1. Each connected component of the fixed point set of an involutive isometry of a Riemannian manifold is called a *reflective submanifold*. Polars and meridians are reflective submanifolds.

We can know the polars and meridians in each irreducible compact Riemannian symmetric space in [11].

Theorem 2.1 (Nagano [12]). *Let M and N be compact irreducible Riemannian symemtric spaces. M and N are isometric if and only if there exist a pair (M_j^+, M_j^-) of a non-trivial polar M_j^+ of M and the meridian M_j^- and a pair (N_k^+, N_k^-) of a non-trivial polar N_k^+ of N and the meridian N_k^- such that M_j^+ is isometric to N_k^+ and M_j^- is isometric to N_k^-.*

This theorem indicates that the structure of a compact irreducible Riemannian symmetric space M is determined by a pair (M_j^+, M_j^-) theoretically. The following is one of evidential facts.

Proposition 2.2 (Nagano [12]). *Let M be an irreducible Riemannian symmetric space. Then, M is a Hermitian symmetric space of compact* *type if and only if M has a pair (M_j^+, M_j^-) of a non-trivial polar M_j^+ and the corresponding meridian M_j^- which are Hermitian symmetric spaces of compact type.*

Here a Riemnnian symmetric space M is of *compact type* if the isometry group of M is a compact semisimple Lie group. And M is a *Hermitian symmetric space* if M is a Hermitian manifold which satisfies that for each point x in M there exists an involutive holomorphic isometry $s_x : M \to M$ such that x is an isolated fixed point of s_x.

2.2. Antipodal sets and 2-numbers

Let M be a compact Riemannian symmetric space. A subet S in M is called an *antipodal set* if $s_x(y) = y$ for any $x, y \in S$. An antipodal set is discrete and finite. We define the *2-number* $\#_2 M$ of M by

$$\#_2 M := \sup\{\#S \mid S \text{ is an antipodal set in } M\}.$$

It is known that $\#_2 M$ is finite ([3], [19]). If an antipodal set S satisfies $\#S = \#_2 M$, then S is called a *great* antipodal set.

Example 2.6. $S = \{o, -o\}$ for a point $o \in S^n$ is an antipodal set in S^n. Since there is no antipodal set consisting of more than two points, $\#_2 S^n = 2$ and S is a great antipodal set.

Example 2.7. $\#_2 \mathbb{R}P^n = n + 1$ and $\{\langle e_1 \rangle, \ldots, \langle e_{n+1} \rangle\}$ is a great antipodal set in $\mathbb{R}P^n$ for an orthogonal basis $e_1, \ldots, e_{n+1}$ of $\mathbb{R}^{n+1}$.

Example 2.8. $\#_2 G_k(\mathbb{K}^n) = \binom{n}{k}$ for $\mathbb{K} = \mathbb{R}, \mathbb{C}, \mathbb{H}$ and

$$\{\langle e_{i_1}, \ldots, e_{i_k} \rangle \mid 1 \le i_1 < \cdots < i_k \le n\}$$

is a great antipodal set for an orthogonal basis $e_1, \ldots, e_n$ of $\mathbb{K}^n$.

Example 2.9. Let G be a compact Lie group. If an antipodal set S contains the unit element e, each element in S is involutive and any two elements in S commute. Thus if S is a great antipodal set, S is a commutative subgroup of G and so S is isomorphic to $(\mathbb{Z}_2)^r$ for some natural number r.

Remark 2.2. Borel and Serre [1] defined p-rank for a compact Lie group G for a prime number p. The *p-rank* of G is by definition the maximal possible rank of a subgroup isomorphic to $(\mathbb{Z}_p)^r$. If $p = 2$, we have $\#_2 G = 2^{r_2(G)}$ where $r_2(G)$ denotes the 2-rank of G.

Let S be a great antipodal set of M. Then $S \subset F(s_o, M)$ for a point $o \in S$. Let $F(s_o, M) = \bigcup_{j=0}^{r} M_j^+$ where M_j^+ is a polar with respect to o, then $S \cap M_j^+$ is an antipodal set in M_j^+. Hence we have $\#(S \cap M_j^+) \leq \#_2 M_j^+$. This implies the inequality

$$\#_2 M \leq \sum_{j=0}^{r} \#_2 M_j^+.$$

If a compact Riemannian symmetric space M is realized as an orbit under the action of a linear isotropy representation of a Riemannian symmetric space of compact type, M is called a *symmetric R-space*.

Theorem 2.2 (Takeuchi [6]). *If M is a symmetric R-space, the equality*

$$\#_2 M = dim\, H_*(M, \mathbb{Z}_2)$$

holds, where $H_*(M, \mathbb{Z}_2)$ is a $\mathbb{Z}_2$-homology group of M.

Theorem 2.3 (Takeuchi [6]). *If M is a symmetric R-space, the equality*

$$\#_2 M = \sum_{j=0}^{r} \#_2 M_j^+$$

holds.

Since it is known that a Hermitian symmetric space of compact type is realized as an adjoint orbit of compact semisimple Lie group, every Hermitian symmetric space of compact type is a symmetric R-space.

Theorem 2.4 (Chen-Nagano [3]). *For a compact Riemannian symmetric space M we have $\#_2 M \geq \chi(M)$. In particular, if M is a Hermitian symmetric space of compact type, then we have $\#_2 M = \chi(M)$. Here $\chi(M)$ denotes the Euler number of M.*

Theorem 2.5 (Tanaka-Tasaki [19]). *Let M be a symmetric R-space. Then*

(A) *Every antipodal set is contained in a great antipodal set.*
(B) *Any two great antipodal sets are congruent,*

where two subsets are congruent if they are transformed to each other by an element of the identity component of the isometry group.

It is known that there is an antipodal set in $SU(4)/\mathbb{Z}_4$ which is maximal but not great hence $SU(4)/\mathbb{Z}_4$ does not satify the property (A) ([19]). It is known that there are great antipodal sets which are not congruent in the oriented real Grassmann manifold $\tilde{G}_3(\mathbb{R}^{15})$ (and also in $\tilde{G}_3(\mathbb{R}^{16})$) in [22], which is the case where the property (B) is not satisfied.

3. Real forms in Hermitian symmetric spaces of compact type

3.1. *Definitions and examples*

Let M be a Hermitian symmetric space of compact type and let τ be an involutive anti-holomorphic isometry of M. Then $L := F(\tau, M)$ is called a *real form* in M. It is known that a real form is connected. A real form is a totally geodesic Lagrangian submanifold.

Example 3.1. Let M be the complex projective space $\mathbb{C}P^n$ and τ denotes the complex conjugation. Then $F(\tau, M) = \mathbb{R}P^n$ is a real form. Generally, the real Grassmann manifold $G_k(\mathbb{R}^n)$ is a real form in the complex Grassmann manifold $G_k(\mathbb{C}^n)$.

Example 3.2. The unitary group $U(n)$ is also a real form in $G_n(\mathbb{C}^{2n})$. The quaternion Grassmann manifold $G_l(\mathbb{H}^n)$ is also a real form in $G_{2l}(\mathbb{C}^{2n})$.

The classification of real forms in irreducible Hermitian symmetric spaces of compact type was given by Leung ([8]) and Takeuchi ([15]).

Let M be a Hermitian symmetric space and let τ be an anti-holomorphic isometry of M. Then the map

$$M \times M \ni (x, y) \mapsto (\tau^{-1}(y), \tau(x)) \in M \times M$$

is an involutive anti-holomorphic isometry of $M \times M$. So the map defines the real form

$$D_\tau(M) := \{(x, \tau(x)) \mid x \in M\}$$

and we call it a *diagonal real form* determined by τ.

Let $I(M)$ and $A(M)$ denote the isometry group and the holomorphic isometry group of a Hermitian symmetric space M of compact type respectively. Let $I_0(M)$ and $A_0(M)$ denote the identity components of $I(M)$ and $A(M)$ respectively. Then we have $I_0(M) = A_0(M)$ ([5] Chap. VIII, Lemma 4.3).

Proposition 3.1 (Tanaka-Tasaki [20]). *Let M be an irreducible Hermitian symmetric space of compact type. Then each element in $I(M) - A(M)$ is an anti-holomorphic isometry and the correspondence $\tau \in I(M) - A(M) \mapsto D_\tau(M)$ gives one-to-one correspondence between the connected components of $I(M) - A(M)$ and the congruence classes of diagonal real forms.*

Proposition 3.2 (Tanaka-Tasaki [20]). *Let M be a Hermitian symmetric space of compact type and let $M = M_1 \times \cdots \times M_k$ be the decomposition into a product of irreducible factors. (Each M_j is an irreducible Hermitian symmetric space of compact type.) Then a real form in M is a product of real forms in irreducible factors and diagonal real forms in $M_i \times M_j$'s where M_i and M_j are holomorphically isometric.*

3.2. *Symmetric R-spaces and real forms*

It is known that real forms in a Hermitian symmetric space of compact type are characterized as symmetric R-spaces.

Theorem 3.1 (Takeuchi [15]). *Every symmetric R-space is realized as a real form in a Hermitian symmetric space of compact type. Conversely, every real form in a Hermitian symmetric space of compact type is a symmetric R-space.*

We have another characterization of symmetric R-spaces.

Theorem 3.2 (Takeuchi [15], Loos [9]). *A compact Riemannian symmetric space M is a symmetric R-space if and only if the unit lattice of a maximal torus of M is generated by orthogonal vectors with the same length.*

Here the unit lattice Γ of a maximal torus A is given by

$$\Gamma = \{X \in T_o A \mid \operatorname{Exp}_o X = o\}$$

for $o \in A$. We make use of this theorem in the proof of Theorem 4.2.

4. The intersection of real forms in a Hermitian symmetric space of compact type

Theorem 4.1 (Tanaka-Tasaki [20]). *Let M be a Hermitian symmetric space of compact type and L_1, L_2 be real forms in M. $M = M_1 \times \cdots \times M_m$ is a decomposition of M into a product of irreducible factors. Then L_i ($i = 1, 2$) is decomposed as*

$$L_i = L_{i,1} \times L_{i,2} \times \cdots \times L_{i,n}$$

where $1 \le n \le m$. For each a ($1 \le a \le n$) the pair of $L_{1,a}$ and $L_{2,a}$ is one of the following.

(1) *Two real forms in M_i for some i ($1 \le i \le m$).*

(2) *After renumbering irreducible factors of M if necessary,*

$$N_1 \times D_{\tau_2}(M_2) \times D_{\tau_4}(M_4) \times \cdots \times D_{\tau_{2s}}(M_{2s})$$

and

$$D_{\tau_1}(M_1) \times D_{\tau_3}(M_3) \times \cdots \times D_{\tau_{2s-1}}(M_{2s-1}) \times N_{2s+1}$$

where $\tau_i : M_i \to M_{i+1}$ ($1 \le i \le 2s$) is an anti-holomorphic isometric map which determines $D_{\tau_i}(M_i)$ and $N_1 \subset M_1$ and $N_{2s+1} \subset M_{2s+1}$ are real forms.

(3) *After renumbering irreducible factors of M if necessary,*

$$N_1 \times D_{\tau_2}(M_2) \times D_{\tau_4}(M_4) \times \cdots \times D_{\tau_{2s-2}}(M_{2s-2}) \times N_{2s}$$

and

$$D_{\tau_1}(M_1) \times D_{\tau_3}(M_3) \times \cdots \times D_{\tau_{2s-3}}(M_{2s-3}) \times D_{\tau_{2s-1}}(M_{2s-1})$$

where $\tau_i : M_i \to M_{i+1}$ ($1 \le i \le 2s-1$) is an anti-holomorphic isometric map which determines $D_{\tau_i}(M_i)$ and $N_1 \subset M_1$ and $N_{2s} \subset M_{2s}$ are real forms.

(4) *After renumbering irreducible factors of M if necessary,*

$$D_{\tau_1}(M_1) \times D_{\tau_3}(M_3) \times \cdots \times D_{\tau_{2s-1}}(M_{2s-1})$$

and

$$D_{\tau_2}(M_2) \times D_{\tau_4}(M_4) \times \cdots \times D_{\tau_{2s}}(M_{2s})$$

where $\tau_i : M_i \to M_{i+1}$ ($1 \le i \le 2s - 1$) and $\tau_{2s} : M_{2s} \to M_1$ are anti-holomorphic isometric maps which determine $D_{\tau_i}(M_i)$ ($1 \le i \le 2s$).

Let ☐ denote an irreducible Hermitian symmetric space of compact type and ○ denote a real form in it. Let ☐☐ denote a product of two irreducible Hermitian symmetric spaces of compact type, ○○ denote a product of real forms in each irreducible factors and ○+○ denote a diagonal real form. We use similar notations for real forms in a product of more than two irreducible Hermitian symmetric spaces of compact type. Then we can express (1) - (4) in Theorem 4.1 with these diagrams as follows.

(1)
○
○

(2)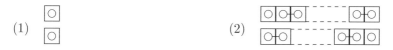

(3)
○ ○+○ ⋯ ○+○ ○
○+○ ⋯ ○+○

(4)

Remark 4.1. A pair of two diagonal real forms is a special case ($s = 1$) of the case (4).

Theorem 4.2 (Tanaka-Tasaki [17], [18]). *Let M be a Hermitian symmetric space of compact type. If two real forms L_1 and L_2 in M intersect transversely, the intersection $L_1 \cap L_2$ is an antipodal set in L_1 and L_2.*

Two real forms in a Hermitian symmetric space of compact type have a non-empty intersection by the following proposition.

Proposition 4.1 (Cheng [4], Tasaki [21]). *Let M be a compact Kähler manifold whose holomorphic sectional curvatures are positive. If L_1 and L_2 are totally geodesic compact Lagrangian submanifolds in M, then $L_1 \cap L_2 \neq \emptyset$.*

Here we give an outline of the proof of Theorem 4.2. Since $L_1 \cap L_2 \neq \emptyset$ by Proposition 4.1, we can take a point $o \in L_1 \cap L_2$. We may consider the case where $\#(L_1 \cap L_2) \geq 2$. What we will show is that for any $p \in L_1 \cap L_2 - \{o\}$ there exists a closed geodesic γ in L_i ($i = 1, 2$) containing o and p on which o and p are antipodal. We take a maximal torus A_i of L_i containing o and p and we also take a maximal torus A_i' of M containing A_i ($i = 1, 2$). Then we can consider the restricted root system of M with respect to A_i' and we can choose a fundamental cell of A_i'. By using the result by Takeuchi [14] we finally conclude that the face of the fundamental cell which contains p is 0-dimensional, that is, p is a vertex of the fundamental cell in the case where M is irreducible. Since M is a symmetric R-space, o and p are antipodal in M by Theorem 3.2. Thus o and p are also antipodal in L_1 and in L_2.

In the argument above we use the fact that the restricted root system of M is of type C or type BC if M is irreducible. If M is not irreducible, to investigate the intersection of L_1 and L_2 is reduced to investigate the intersection of $L_{1,a}$ and $L_{2,a}$ ($1 \leq a \leq n$) of the cases (1) to (4) in Theorem 4.1. The case (1) is an irreducible case. For the case (2), the intersection is

$$\{(x, \tau_1(x), \tau_2\tau_1(x), \ldots, \tau_{2s} \cdots \tau_1(x)) \mid x \in N_1 \cap (\tau_{2s} \cdots \tau_1)^{-1}(N_{2s+1})\}.$$

Since $(\tau_{2s} \cdots \tau_1)^{-1}(N_{2s+1})$ is a real form in M_1, to investigate the intersection is reduced to investigate the intersection of two real forms N_1 and $(\tau_{2s} \cdots \tau_1)^{-1}(N_{2s+1})$ in M_1. For the case (3), the intersection is

$$\{(x, \tau_1(x), \tau_2\tau_1(x), \ldots, \tau_{2s-1} \cdots \tau_1(x)) \mid x \in N_1 \cap (\tau_{2s-1} \cdots \tau_1)^{-1}(N_{2s})\}.$$

Since $(\tau_{2s-1} \cdots \tau_1)^{-1}(N_{2s})$ is a real form in M_1, to investigate the intersection is reduced to investigate the intersection of two real forms N_1 and $(\tau_{2s-1} \cdots \tau_1)^{-1}(N_{2s})$ in M_1. For the case (4), the intersection is

$$\{(x, \tau_1(x), \tau_2 \tau_1(x), \dots, \tau_{2s-1} \cdots \tau_1(x))$$
$$\mid (x, \tau_{2s}^{-1}) \in D_{\tau_{2s-1} \cdots \tau_1}(M_1) \cap D_{\tau_{2s}^{-1}}(M_1)\}.$$

Since $D_{\tau_{2s-1} \cdots \tau_1}(M_1)$ and $D_{\tau_{2s}^{-1}}(M_1)$ are diagonal real forms in $M_1 \times M_{2s} \cong M_1 \times M_1$, to investigate the intersection is reduced to investigate the intersection of these two diagonal real forms. In this case a similar argument as in the irreducible case goes and we complete the outline of the proof.

When two real forms are congruent, we have the following.

Theorem 4.3 (Tanaka-Tasaki [17]). *Let M be a Hermitian symmetric space of compact type and L_1, L_2 be real forms in M which are congruent. We assume that the intersection $L_1 \cap L_2$ is discrete. Then $L_1 \cap L_2$ is a great antipodal set of both in L_1 and in L_2, that is, $\#(L_1 \cap L_2) = \#_2 L_1 = \#_2 L_2$.*

As for the intersection numbers of two real forms in an irreducible Hermitian symmetric space of compact type we have the following.

Theorem 4.4 (Tanaka-Tasaki [17]). *Let M be an irreducible Hermitian symmetric space of compact type and let L_1 and L_2 be two real forms of M whose intersection is discrete. We assume that $\#_2 L_1 \leq \#_2 L_2$.*

(1) *If $M = G_{2m}(\mathbb{C}^{4m})$ $(m \geq 2)$, L_1 is congruent to $G_m(\mathbb{H}^{2m})$ and L_2 is congruent to $U(2m)$, then*

$$\#(L_1 \cap L_2) = 2^m < \binom{2m}{m} = \#_2 L_1 < 2^{2m} = \#_2 L_2.$$

(2) *Otherwise, $L_1 \cap L_2$ is a great antipodal set of L_1, that is,*

$$\#(L_1 \cap L_2) = \#_2 L_1 \ (\leq \#_2 L_2).$$

When a Hermitian symmetric space of compact type is not irreducible, it is enough to consider the intersection of two diagonal real forms as we mentioned before. As for the intersection numbers of two diagonal real forms we have the following.

Theorem 4.5 (Tanaka-Tasaki [20]). *Let M be an irreducible Hermitian symmetric space of compact type and $D_{\tau_1}, D_{\tau_2^{-1}} \subset M \times M$ be the diagonal real forms defined by anti-holomorphic isometries τ_1, τ_2 of M. We assume that the intersection of $D_{\tau_1}(M)$ and $D_{\tau_2^{-1}}(M)$ is discrete. Then we have the following.*

(1) *If* $M = Q_{2m}(\mathbb{C})$ $(m \geq 2)$ *and* $\tau_2 \tau_1$ *does not belong to* $A_0(M)$,

$$\#(D_{\tau_1}(M) \cap D_{\tau_2^{-1}}(M)) = 2m < 2m + 2 = \#_2 M.$$

(2) *If* $M = G_m(\mathbb{C}^{2m})$ $(m \geq 2)$ *and* $\tau_2 \tau_1$ *does not belong to* $A_0(M)$,

$$\#(D_{\tau_1}(M) \cap D_{\tau_2^{-1}}(M)) = 2^m < \binom{2m}{m} = \#_2 M.$$

(3) *Otherwise,* $D_{\tau_1}(M) \cap D_{\tau_2^{-1}}(M)$ *is a great antipodal set both in* $D_{\tau_1}(M)$ *and in* $D_{\tau_2^{-1}}(M)$, *that is,*

$$\#(D_{\tau_1}(M) \cap D_{\tau_2^{-1}}(M)) = \#_2 M.$$

Here $Q_{2m}(\mathbb{C})$ denotes the complex hyperquadric of complex dimension $2m$. The following lemma brings three cases in Theorem 4.5.

Lemma 4.1 (Murakami [10], Takeuchi [13]). *Let* M *be an irreducible Hermitian symmetric space of compact type. Then* $I(M)/I_0(M)$ *and* $A(M)/A_0(M)$ *are as follows.*

(A) *If* $M = Q_{2m}(\mathbb{C})$ $(m \geq 2)$ *or* $M = G_m(\mathbb{C}^{2m})$ $(m \geq 2)$, *then*

$$I(M)/I_0(M) \cong \mathbb{Z}_2 \times \mathbb{Z}_2 \quad and \quad A(M)/A_0(M) \cong \mathbb{Z}_2.$$

(B) *Otherwise,*

$$I(M)/I_0(M) \cong \mathbb{Z}_2 \quad and \quad A(M) = A_0(M).$$

If M is a Hermitian symmetric space of compact type, every polar is a Hermitian symmetric space of compact type by Proposition 2.2.

Lemma 4.2. *Let* M *be a Hermitian symmetric space of compact type, and denote by*

$$F(s_o, M) = \bigcup_{j=0}^{r} M_j^+,$$

where M_j^+'*s are the polars of* M *with respect to the origin* o.

(1) *If* L *is a real form of* M *through* o, *then the polars of* L *with respect to* o *are described by*

$$F(s_o, L) = \bigcup_{j=0}^{r} L \cap M_j^+,$$

and the following equality holds.

$$\#_2 L = \sum_{j=0}^{r} \#_2(L \cap M_j^+).$$

(2) *If L_1, L_2 are real forms of M through o whose intersection is discrete, then we have*

$$L_1 \cap L_2 = \bigcup_{j=0}^{r} \left\{ (L_1 \cap M_j^+) \cap (L_2 \cap M_j^+) \right\},$$

$$\#(L_1 \cap L_2) = \sum_{j=0}^{r} \# \left\{ (L_1 \cap M_j^+) \cap (L_2 \cap M_j^+) \right\}.$$

We prove Theorem 4.3, Theorem 4.4 and Theorem 4.5 by inductive argument on polars by using Lemma 4.2.

Iriyeh, Sakai and Tasaki applied Theorem 4.2 to Floer homology theory and obtained the following result.

Theorem 4.6 (Iriyeh-Sakai-Tasaki [6]). *Let M be a monotone Hermitian symmetric space of compact type. Let L_1 and L_2 be real forms in M which intersect transversely and have minimal Maslov number ≥ 3. Then we have*

$$HF(L_1, L_2 : \mathbb{Z}_2) \cong \bigoplus_{p \in L_1 \cap L_2} \mathbb{Z}_2[p],$$

where $HF(L_1, L_2 : \mathbb{Z}_2)$ is a Floer homology group with coefficients $\mathbb{Z}_2$.

Moreover, when a Hermitian symmetric space is irreducible, they obtaine the following result by Theorem 4.4.

Theorem 4.7 (Iriyeh-Sakai-Tasaki [6]). *Let M be an irreducible Hermitian symmetric space of compact type and Let L_1 and L_2 be real forms in M which intersect transversely.*

(1) *If $M = G_{2m}(\mathbb{C}^{4m})$ $(m \geq 2)$, L_1 is congruent to $G_m(\mathbb{H}^{2m})$ and L_2 is congruent to $U(2m)$, then we have*

$$HF(L_1, L_2 : \mathbb{Z}_2) \cong (\mathbb{Z}_2)^{2^m}.$$

(2) *Otherwise, we have*

$$HF(L_1, L_2 : \mathbb{Z}_2) \cong (\mathbb{Z}_2)^{\min\{\#_2 L_1, \#_2 L_2\}}.$$

Remark 4.2. Recently Iriyeh, Sakai and Tasaki [7] investigated the relationship between the intersection of real flag manifolds in the complex flag manifold and generalized antipodal sets.

References

1. A. Borel and J. P. Serre, *Sur certains sousgroupes des groupes de Lie compacts*, Comm. Math. Helv. **27** (1953), 128–139.
2. B.-Y. Chen and T. Nagano, *Totally geodesic submanifolds of symmetric spaces, II*, Duke Math. J. **45** (1978), 405–425.
3. B.-Y. Chen and T. Nagano, *A Riemannian geometric invariant and its applications to a problem of Borel and Serre*, Trans. Amer. Math. Soc. **308** (1988), 273–297.
4. Xu Cheng, *The totally geodesic coisotropic submanifolds in Kähler manifolds*, Geometriae Dedicata **90** (2002), 115–125.
5. S. Helgason, *Differential Geometry, Lie groups, and Symmetric spaces*, Academic Press, New York, 1978.
6. H. Iriyeh, T. Sakai and H. Tasaki, *Lagrangian Floer homology of a pair of real forms in Hermitian symmetric spaces of compact type*, to appear in J. Math. Soc. Japan **65** (2013).
7. H. Iriyeh, T. Sakai and H. Tasaki, *On the structure of the intersection of real flag manifolds in a complex flag manifold*, to appear in Advanced Studies in Pure Mathematics.
8. D. P. S. Leung, *Reflective submanifolds. IV, Classification of real forms of Hermitian symmetric spaces*, J. Differential Geom. **14** (1979), 179–185.
9. O. Loos, *Charakterisierung symmetrischer R-Räume durch ihre Einheitsgitter*, Math. Z. **189** (1985), 211–226.
10. S. Murakami, *On the automorphisms of a real semisimple Lie algebras*, J. Math. Soc. Japan, **4** (1952), 103–133.
11. T. Nagano, *The involutions of compact symmetric spaces*, Tokyo J. Math. **11** (1988), 57–79.
12. T. Nagano, *The involutions of compact symmetric spaces, II*, Tokyo J. Math. **15** (1992), 39–82.
13. M. Takeuchi, *On the fundamental group and the group of isometries of a symmetric space*, J. Fac. Sci. Univ. Tokyo, Sect. 1 **10** (1964), 88–123.
14. M. Takeuchi, *On conjugate loci and cut loci of compact symmetric spaces I.* Tsukuba J. Math. **2** (1978), 35 – 68.
15. M. Takeuchi, *Stability of certain minimal submanifolds of compact Hermitian symmetric spaces*, Tohoku Math. J., (2) **36** (1984), 293–314.
16. M. Takeuchi, *Two-number of symmetric R-spaces*, Nagoya Math. J. **115** (1989), 43–46.
17. M. S. Tanaka and H. Tasaki, *The intersection of two real forms in Hermitian symmetric spaces of compact type*, J. Math. Soc. Japan. **64** (2012), 1297–1332.
18. M. S. Tanaka and H. Tasaki, *Correction to: "The intersection of two real forms in Hermitian symmetric spaces of compact type"*, preprint.
19. M. S. Tanaka and H. Tasaki, *Antipodal sets of symmetric R-spaces*, Osaka J. Math. **50** (2013), 161–169.
20. M. S. Tanaka and H. Tasaki, *The intersection of two real forms in Hermitian symmetric spaces of compact type II*, to appear in J. Math. Soc. Japan.

21. H. Tasaki, *The intersection of two real forms in the complex hyperquadric*, Tohoku Math. J. **62** (2010), 375–382.
22. H. Tasaki, *Antipodal sets in oriented real Grassmann manifolds*, to appear in Internat. J. Math. **24** (2013).

Received March 29, 2013.

220

Proceedings of the Workshop on
Differential Geometry of Submanifolds
and its Related Topics
Saga, August 4-6, 2012

A NOTE ON SYMMETRIC TRIAD AND
HERMANN ACTION

Osamu IKAWA*

*Department of Mathematics and Physical Sciences, Faculty of Arts and Sciences,
Kyoto Institute of Technology,
Matsugasaki, Kyoto 606-8585 Japan
E-mail: ikawa@kit.ac.jp*

Dedicated to Professor Sadahiro Maeda for his sixtieth birthday

We show that every symmetric triad is obtained from a commutative compact
symmetric triad which satisfies certain conditions.

Keywords: symmetric triad, Hermann action.

1. Introduction

In [3] we defined the notion of a symmetric triad, which is a generalization
of that of an irreducible root system, and constructed a symmetric triad
from a compact symmetric triad. A symmetric triad is a useful tool to study
properties of Hermann actions. However we did not know whether if every
symmetric triad is obtained from a compact symmetric triad or not. In
this paper we give an affirmative answer to this question. In [2] we defined
the notion of a weakly reflective submanifold, which is a kind of austere
submanifold. Although we could classify the set of all austere orbits of
Hermann actions, we do not know which orbits are weakly reflective among
them. Each point of a weakly reflective submanifold has a reflection. In this
paper we show that an austere orbit in a Euclidean space furnished with a
symmetric triad has a reflection, and austere point behaves like a weakly
reflective submanifold.

*The research was supported by Grand-in-Aid for Scientific Research (C) (No. 22540108),
Japan Society for Promotion of Science.

2. The geometry of symmetric triad

In this section we show some properties of a symmetric triad with multiplicities. We begin with recalling the definitions of a symmetric triad. Let $\mathfrak{a}$ be a finite dimensional vector space over $\mathbb{R}$ with an inner product $\langle\,,\,\rangle$. A triple $(\tilde{\Sigma}, \Sigma, W)$ is a *symmetric triad* of $\mathfrak{a}$ if it satisfies the following six conditions ([3, Def. 2.2]):

(1) $\tilde{\Sigma}$ is an irreducible root system of $\mathfrak{a}$.
(2) Σ is a root system of $\mathfrak{a}$.
(3) W is a nonempty subset of $\mathfrak{a}$, which is invariant under the multiplication by -1, and $\tilde{\Sigma} = \Sigma \cup W$.
(4) If we put $l = \max\{\|\alpha\| \mid \alpha \in \Sigma \cap W\}$, then $\Sigma \cap W = \{\alpha \in \tilde{\Sigma} \mid \|\alpha\| \leq l\}$.
(5) For $\alpha \in W$ and $\lambda \in \Sigma - W$,

$$2\frac{\langle\alpha,\lambda\rangle}{\|\alpha\|^2} \text{ is odd if and only if } s_\alpha\lambda := \lambda - 2\frac{\langle\alpha,\lambda\rangle}{\|\alpha\|^2}\alpha \in W - \Sigma.$$

(6) For $\alpha \in W$ and $\lambda \in W - \Sigma$,

$$2\frac{\langle\alpha,\lambda\rangle}{\|\alpha\|^2} \text{ is odd if and only if } s_\alpha\lambda \in \Sigma - W.$$

In the sequel, we denote by $(\tilde{\Sigma}, \Sigma, W)$ a symmetric triad of $\mathfrak{a}$. Put

$$\Gamma = \left\{X \in \mathfrak{a} \mid \langle\lambda, X\rangle \in \frac{\pi}{2}\mathbb{Z} \ (\lambda \in \tilde{\Sigma})\right\},$$

$$\Gamma_{\Sigma \cap W} = \left\{X \in \mathfrak{a} \mid \langle\alpha, X\rangle \in \frac{\pi}{2}\mathbb{Z} \ (\alpha \in \Sigma \cap W)\right\}.$$

We have $\Gamma = \Gamma_{\Sigma \cap W}$ ([3, p. 82]). A point in Γ is called a *totally geodesic point*. A point H in $\mathfrak{a}$ is *regular* if $\langle\lambda, H\rangle \notin \pi\mathbb{Z}$ and $\langle\alpha, H\rangle \notin \frac{\pi}{2} + \pi\mathbb{Z}$ for any $\lambda \in \Sigma$ and $\alpha \in W$. We denote by $\tilde{\Pi}$ and $\tilde{\Sigma}^+$ a fundamental system of $\tilde{\Sigma}$ and the set of positive roots in $\tilde{\Sigma}$ with respect to $\tilde{\Pi}$ respectively. Set $\Sigma^+ = \Sigma \cap \tilde{\Sigma}^+$ and $W^+ = W \cap \tilde{\Sigma}^+$.

Definition 2.1. [3, Def. 2.13] Let $(\tilde{\Sigma}, \Sigma, W)$ be a symmetric triad of $\mathfrak{a}$. Put $\mathbb{R}_{\geq 0} = \{x \in \mathbb{R} \mid x \geq 0\}$. Consider two mappings $m, n : \tilde{\Sigma} \to \mathbb{R}_{\geq 0}$ which satisfy the following four conditions:

(1) $m(\lambda) = m(-\lambda), \quad n(\alpha) = n(-\alpha)$ and

$$m(\lambda) > 0 \Leftrightarrow \lambda \in \Sigma, \quad n(\alpha) > 0 \Leftrightarrow \alpha \in W.$$

(2) When $\lambda \in \Sigma, \alpha \in W, s \in W(\Sigma)$ then $m(\lambda) = m(s\lambda), n(\alpha) = n(s\alpha)$, where we note that W is invariant under the action of the Weyl group $W(\Sigma)$ of Σ ([3, Prop. 2.7]).

(3) When $\sigma \in W(\tilde{\Sigma})$, the Weyl group of $\tilde{\Sigma}$, and $\lambda \in \tilde{\Sigma}$ then $n(\lambda) + m(\lambda) = n(\sigma\lambda) + m(\sigma\lambda)$.

(4) Let λ be in $\Sigma \cap W$ and α in W.

If $\frac{2\langle \alpha, \lambda \rangle}{\|\alpha\|^2}$ is even then $m(\lambda) = m(s_\alpha \lambda)$.

If $\frac{2\langle \alpha, \lambda \rangle}{\|\alpha\|^2}$ is odd then $m(\lambda) = n(s_\alpha \lambda)$.

We call $m(\lambda)$ and $n(\alpha)$ the *multiplicities* of λ and α, respectively. If multiplicities are given, we call $(\tilde{\Sigma}, \Sigma, W)$ *the symmetric triad with multiplicities.*

We denote by $(\tilde{\Sigma}, \Sigma, W)$ a symmetric triad of $\mathfrak{a}$ with multiplicities in what follows. For $H \in \mathfrak{a}$, set

$$
m_H = - \sum_{\substack{\lambda \in \Sigma^+ \\ \langle \lambda, H \rangle \notin \frac{\pi}{2}\mathbb{Z}}} m(\lambda) \cot(\langle \lambda, H \rangle)\lambda + \sum_{\substack{\alpha \in W^+ \\ \langle \alpha, H \rangle \notin \frac{\pi}{2}\mathbb{Z}}} n(\alpha) \tan(\langle \alpha, H \rangle)\alpha.
$$

We call m_H the *mean curvature vector* of H. When $m_H = 0$, we call H a *minimal point*. Set

$$
F(H) = - \sum_{\substack{\lambda \in \Sigma^+ \\ \langle \lambda, H \rangle \notin \frac{\pi}{2}\mathbb{Z}}} m(\lambda) \log|\sin(\langle \lambda, H \rangle)| - \sum_{\substack{\alpha \in W^+ \\ \langle \alpha, H \rangle \notin \frac{\pi}{2}\mathbb{Z}}} n(\alpha) \log|\cos(\langle \alpha, H \rangle)|,
$$

and $\mathrm{Vol}(H) = \exp(-F(H))(> 0)$. We call $\mathrm{Vol}(H)$ the *volume* of H.

Definition 2.2. Define a *codimension* $\mathrm{codim}(H)$ of H in $\mathfrak{a}$ by

$$
\mathrm{codim}(H) = \dim \mathfrak{a} + \sum_{\substack{\lambda \in \Sigma^+ \\ \langle \lambda, H \rangle \in \pi\mathbb{Z}}} m(\lambda) + \sum_{\substack{\alpha \in W^+ \\ \langle \alpha, H \rangle \in \frac{\pi}{2} + \pi\mathbb{Z}}} n(\alpha)(\geq \dim \mathfrak{a}).
$$

Then H is a regular point if and only if $\mathrm{codim}(H) = \dim \mathfrak{a}$. We will give a meaning of the definition of $\mathrm{codim}(H)$ in the next section.

Theorem 2.1. *[3, Thm. 2.18] A point $H \in \mathfrak{a}$ is austere (see [3, Def. 2.16] for the definition) if and only if the following three conditions hold:*

(1) $\langle \lambda, H \rangle \in \frac{\pi}{2}\mathbb{Z}$ *for any* $\lambda \in (\Sigma - W) \cup (W - \Sigma)$,

(2) $2H \in \Gamma_{\Sigma \cap W}$,

(3) $m(\lambda) = n(\lambda)$ *for any* $\lambda \in \Sigma \cap W$ *with* $\langle \lambda, H \rangle \in \frac{\pi}{4} + \frac{\pi}{2}\mathbb{Z}$.

For H in $\mathfrak{a}$, if we define a mapping f_H from $\mathfrak{a}$ onto itself by $f_H(X) = 2H - X$, then $f_H(H) = H$ and $f_H^2 = 1$. The author thinks that the following two propositions show that an austere point or an austere orbit of a Hermann action has a property like a weakly reflective submanifold (see 2 for the definition), and f_H plays a role of a reflection in the sense of weakly reflective submanifold.

Proposition 2.1. *Let H in $\mathfrak{a}$ be an austere point. Then for any X in $\mathfrak{a}$,*

$$\mathrm{codim}(f_H(X)) = \mathrm{codim}(X), \ m_{f_H(X)} = -m_X, \ \mathrm{Vol}(f_H(X)) = \mathrm{Vol}(X).$$

Proof By Definition 2.2 we have

$$\mathrm{codim}(f_H(X)) = \dim \mathfrak{a} + \sum_{\substack{\lambda \in \Sigma^+ \cap W^+ \\ \langle \lambda, f_H(X) \rangle \in \pi \mathbb{Z}}} m(\lambda) + \sum_{\substack{\lambda \in \Sigma^+ - W^+ \\ \langle \lambda, f_H(X) \rangle \in \pi \mathbb{Z}}} m(\lambda)$$

$$+ \sum_{\substack{\alpha \in W^+ - \Sigma^+ \\ \langle \alpha, f_H(X) \rangle \in \frac{\pi}{2} + \pi \mathbb{Z}}} n(\alpha) + \sum_{\substack{\lambda \in W^+ \cap \Sigma^+ \\ \langle \lambda, f_H(X) \rangle \in \frac{\pi}{2} + \pi \mathbb{Z}}} n(\lambda).$$

For $\lambda \in \Sigma^+ - W^+$ and $\alpha \in W^+ - \Sigma^+$ we have $\langle \lambda, 2H \rangle \in \pi \mathbb{Z}$ and $\langle \alpha, 2H \rangle \in \pi \mathbb{Z}$ since H is an austere point. Hence

$$\langle \lambda, f_H(X) \rangle \in \pi \mathbb{Z} \Leftrightarrow \langle \lambda, X \rangle \in \pi \mathbb{Z}, \quad \langle \alpha, f_H(X) \rangle \in \pi \mathbb{Z} \Leftrightarrow \langle \alpha, X \rangle \in \pi \mathbb{Z}.$$

Let λ be in $\Sigma^+ \cap W^+$. Then $\langle \lambda, 2H \rangle \in \pi \mathbb{Z}$ or $\langle \lambda, 2H \rangle \in \frac{\pi}{2} + \pi \mathbb{Z}$ since $2H \in \Gamma_{\Sigma \cap W}$. When $\langle \lambda, 2H \rangle \in \frac{\pi}{2} + \pi \mathbb{Z}$ then $m(\lambda) = n(\lambda)$ by (3) of Theorem 2.1. Thus

$$\sum_{\substack{\lambda \in \Sigma^+ \cap W^+ \\ \langle \lambda, f_H(X) \rangle \in \pi \mathbb{Z}}} m(\lambda) + \sum_{\substack{\lambda \in W^+ \cap \Sigma^+ \\ \langle \lambda, f_H(X) \rangle \in \frac{\pi}{2} + \pi \mathbb{Z}}} n(\lambda)$$

$$= \sum_{\substack{\lambda \in \Sigma^+ \cap W^+ \\ \langle \lambda, 2H \rangle \in \pi \mathbb{Z}, \langle \lambda, X \rangle \in \pi \mathbb{Z}}} m(\lambda) + \sum_{\substack{\lambda \in \Sigma^+ \cap W^+ \\ \langle \lambda, 2H \rangle \in \frac{\pi}{2} + \pi \mathbb{Z}, \langle \lambda, X \rangle \in \frac{\pi}{2} + \pi \mathbb{Z}}} n(\lambda)$$

$$+ \sum_{\substack{\lambda \in W^+ \cap \Sigma^+ \\ \langle \lambda, 2H \rangle \in \pi \mathbb{Z}, \langle \lambda, X \rangle \in \frac{\pi}{2} + \pi \mathbb{Z}}} n(\lambda) + \sum_{\substack{\lambda \in W^+ \cap \Sigma^+ \\ \langle \lambda, 2H \rangle \in \frac{\pi}{2} + \pi \mathbb{Z}, \langle \lambda, X \rangle \in \pi \mathbb{Z}}} m(\lambda)$$

$$= \sum_{\substack{\lambda \in \Sigma^+ \cap W^+ \\ \langle \lambda, X \rangle \in \pi \mathbb{Z}}} m(\lambda) + \sum_{\substack{\lambda \in W^+ \cap \Sigma^+ \\ \langle \lambda, X \rangle \in \frac{\pi}{2} + \pi \mathbb{Z}}} n(\lambda).$$

Hence $\mathrm{codim}(f_H(X)) = \mathrm{codim}(X)$.

We show that $m_{f_H(X)} = -m_X$ and $\mathrm{Vol}(f_H(X)) = \mathrm{Vol}(X)$. Since $2H$ is in Γ, $\langle \lambda, 2H \rangle \in \frac{\pi}{2}\mathbb{Z}$ for any $\lambda \in \tilde{\Sigma}$. Hence $\langle \lambda, f_H(X) \rangle \notin \frac{\pi}{2}\mathbb{Z}$ if and only if $\langle \lambda, X \rangle \notin \frac{\pi}{2}\mathbb{Z}$. By the definition of mean curvature vector and $\Gamma(f_H(X))$,

$$m_{f_H(X)} = - \sum_{\substack{\lambda \in \Sigma^+ - W^+ \\ \langle \lambda, X \rangle \notin \frac{\pi}{2}\mathbb{Z}}} m(\lambda) \cot(\langle \lambda, f_H(X) \rangle)\lambda - \sum_{\substack{\lambda \in \Sigma^+ \cap W^+ \\ \langle \lambda, X \rangle \notin \frac{\pi}{2}\mathbb{Z}}} m(\lambda) \cot(\langle \lambda, f_H(X) \rangle)\lambda$$

$$+ \sum_{\substack{\alpha \in W^+ - \Sigma^+ \\ \langle \alpha, X \rangle \notin \frac{\pi}{2}\mathbb{Z}}} n(\alpha) \tan(\langle \alpha, f_H(X) \rangle)\alpha + \sum_{\substack{\lambda \in W^+ \cap \Sigma^+ \\ \langle \lambda, X \rangle \notin \frac{\pi}{2}\mathbb{Z}}} n(\lambda) \tan(\langle \lambda, f_H(X) \rangle)\alpha,$$

$$F(f_H(X)) = - \sum_{\substack{\lambda \in \Sigma^+ - W^+ \\ \langle \lambda, X \rangle \notin \frac{\pi}{2}\mathbb{Z}}} m(\lambda) \log |\sin(\langle \lambda, f_H(X) \rangle)|$$

$$- \sum_{\substack{\lambda \in \Sigma^+ \cap W^+ \\ \langle \lambda, X \rangle \notin \frac{\pi}{2}\mathbb{Z}}} m(\lambda) \log |\sin(\langle \lambda, f_H(X) \rangle)|$$

$$- \sum_{\substack{\lambda \in W^+ - \Sigma^+ \\ \langle \lambda, X \rangle \notin \frac{\pi}{2}\mathbb{Z}}} n(\alpha) \log |\cos(\langle \alpha, f_H(X) \rangle)|$$

$$- \sum_{\substack{\lambda \in W^+ \cap \Sigma^+ \\ \langle \lambda, X \rangle \notin \frac{\pi}{2}\mathbb{Z}}} n(\alpha) \log |\cos(\langle \lambda, f_H(X) \rangle)|.$$

Since H is an austere point, for $\lambda \in \Sigma^+ - W^+$ and $\alpha \in W^+ - \Sigma^+$,

$$\cot(\langle \lambda, f_H(X) \rangle) = -\cot(\langle \lambda, X \rangle), \quad \tan(\langle \lambda, f_H(X) \rangle) = -\tan(\langle \lambda, X \rangle),$$
$$|\sin(\langle \lambda, f_H(X) \rangle)| = |\sin(\langle \lambda, X \rangle)|, \quad |\cos(\langle \alpha, f_H(X) \rangle)| = |\cos(\langle \alpha, X \rangle)|.$$

Let λ be in $\Sigma^+ \cap W^+$. When $\langle \lambda, 2H \rangle \in \pi\mathbb{Z}$, then

$$\cot(\langle \lambda, f_H(X) \rangle) = -\cot(\langle \lambda, X \rangle),$$
$$|\sin(\langle \lambda, f_H(X) \rangle)| = |\sin(\langle \lambda, X \rangle)|, \quad |\cos(\langle \lambda, f_H(X) \rangle)| = |\cos(\langle \lambda, X \rangle)|.$$

When $\langle \lambda, 2H \rangle \in \frac{\pi}{2} + \pi\mathbb{Z}$, then $m(\lambda) = n(\lambda)$ and

$$\cot(\langle \lambda, f_H(X) \rangle) = -\tan(\langle \lambda, X \rangle),$$
$$|\sin(\langle \lambda, f_H(X) \rangle)| = |\cos(\langle \lambda, X \rangle)|, \quad |\cos(\langle \lambda, f_H(X) \rangle)| = |\sin(\langle \lambda, X \rangle)|.$$

Thus we have $m_{f_H(X)} = -m_X$, $F(f_H(X)) = F(X)$ and $\mathrm{Vol}(f_H(X)) = \mathrm{Vol}(X)$. $\qquad \square$

Proposition 2.2. *Let H in $\mathfrak{a}$ be an austere point.*

(1) *If $X \in \mathfrak{a}$ is totally geodesic, then $f_H(X)$ is also totally geodesic.*
(2) *If $X \in \mathfrak{a}$ is an austere point, then $f_H(X)$ is also an austere point.*
(3) *If $X \in \mathfrak{a}$ is a minimal point, then $f_H(X)$ is also a minimal pint.*
(4) *If $X \in \mathfrak{a}$ is a regular point, then $f_H(X)$ is also a regular point.*

Proof (1) Since X and $2H$ are totally geodesic, for any $\alpha \in \Sigma \cap W$ we have

$$\langle \alpha, X \rangle \in \frac{\pi}{2}\mathbb{Z}, \quad \langle \alpha, 2H \rangle \in \frac{\pi}{2}\mathbb{Z},$$

which implies that for any $\alpha \in \Sigma \cap W$,

$$\langle \alpha, f_H(X) \rangle = \langle \alpha, 2H \rangle - \langle \alpha, X \rangle \in \frac{\pi}{2}\mathbb{Z}.$$

Thus $f_H(X)$ is also totally geodesic.

(2) Since it is clear that $f_H(X)$ satisfies the conditions (1) and (2) in Theorem 2.1, it is sufficient to prove the condition (3) in Theorem 2.1. For $\lambda \in \Sigma \cap W$ with $\langle \lambda, f_H(X) \rangle \in \frac{\pi}{4} + \frac{\pi}{2}\mathbb{Z}$, we have $2\langle \lambda, H \rangle - \langle \lambda, X \rangle \in \frac{\pi}{4} + \frac{\pi}{2}\mathbb{Z}$. Since $\langle \lambda, 2H \rangle \in \frac{\pi}{2}\mathbb{Z}$, we have $\langle \lambda, X \rangle \in \frac{\pi}{4} + \frac{\pi}{2}\mathbb{Z}$. Since X is austere, $m(\lambda) = n(\lambda)$. Thus $f_H(X)$ is also an austere point.

(3) and (4) follow from Proposition 2.1 immediately.   □

3. Hermann actions

Let (G, K_1) and (G, K_2) be two compact symmetric pairs: There exist two involutive automorphisms θ_1 and θ_2 on the compact connected Lie group G such that the closed subgroup K_i of G lie between G_{θ_i} and the identity component $(G_{\theta_i})_0$ of G_{θ_i}. Here we denote by G_{θ_i} $(i = 1, 2)$ the closed subgroup of G consisting of all fixed points of θ_i in G. In this case the triple (G, K_1, K_2) is called a compact symmetric triad. We assume that (G, K_1, K_2) is *commutative*, that is, θ_1 and θ_2 commute each other. We also denote by θ_i the induced involutive automorphisms of the Lie algebra $\mathfrak{g}$ of G. Denote by $\mathfrak{k}_1$ and $\mathfrak{k}_2$ the Lie algebras of K_1 and K_2, respectively, then we have two canonical decompositions of the Lie algebra $\mathfrak{g}$ of G:

$$\mathfrak{g} = \mathfrak{k}_1 \oplus \mathfrak{m}_1 = \mathfrak{k}_2 \oplus \mathfrak{m}_2,$$

where we define a subspace $\mathfrak{m}_i$ of $\mathfrak{g}$ by

$$\mathfrak{m}_i = \{X \in \mathfrak{g} \mid \theta_i(X) = -X\} \quad (i = 1, 2).$$

Take an $\mathrm{Aut}(G)$-invariant Riemannian metric $\langle \, , \, \rangle$ on G and a maximal abelian subspace $\mathfrak{a}$ of $\mathfrak{m}_1 \cap \mathfrak{m}_2$. For $\alpha \in \mathfrak{a}$, define a subspace $\mathfrak{g}(\mathfrak{a}, \alpha)$ in the complexification $\mathfrak{g}^{\mathbb{C}}$ of $\mathfrak{g}$ by

$$\mathfrak{g}(\mathfrak{a}, \alpha) = \{X \in \mathfrak{g}^{\mathbb{C}} \mid [H, X] = \sqrt{-1}\langle \alpha, H \rangle X \quad (H \in \mathfrak{a})\}.$$

and set

$$\tilde{\Sigma} = \{\alpha \in \mathfrak{a} - \{0\} \mid \mathfrak{g}(\mathfrak{a}, \alpha) \neq \{0\}\}.$$

Since $\mathfrak{g}(\mathfrak{a}, \alpha)$ is $\theta_1\theta_2$-invariant, we have $\mathfrak{g}(\mathfrak{a}, \alpha) = \mathfrak{g}(\mathfrak{a}, \alpha, 1) \oplus \mathfrak{g}(\mathfrak{a}, \alpha, -1)$, where we set

$$\mathfrak{g}(\mathfrak{a}, \alpha, \pm 1) = \{X \in \mathfrak{g}(\mathfrak{a}, \alpha) \mid \theta_1\theta_2 X = \pm X\}.$$

Define subsets Σ and W of $\tilde{\Sigma}$ by

$$\Sigma = \{\alpha \in \tilde{\Sigma} \mid \mathfrak{g}(\mathfrak{a}, \alpha, 1) \neq \{0\}\}, \quad W = \{\alpha \in \tilde{\Sigma} \mid \mathfrak{g}(\mathfrak{a}, \alpha, -1) \neq \{0\}\}.$$

Then we have $\tilde{\Sigma} = \Sigma \cup W$. For $\lambda \in \Sigma$ and $\alpha \in W$, set

$$m(\lambda) = \dim \mathfrak{g}(\mathfrak{a}, \lambda, 1), \quad n(\alpha) = \dim \mathfrak{g}(\mathfrak{a}, \alpha, -1).$$

Further we assume that (G, K_1, K_2) satisfies one of the following conditions (A) or (B):

(A) G is simple, and θ_1 and θ_2 can not transform each other by an inner automorphism of $\mathfrak{g}$.

(B) There exist a compact connected simple Lie group U and a symmetric subgroup $\bar{K}_2$ of U such that

$$G = U \times U, \quad K_1 = \Delta G = \{(u, u) \mid u \in U\}, \quad K_2 = \bar{K}_2 \times \bar{K}_2.$$

In this case $\theta_1(a, b) = (b, a)$ for $(a, b) \in G$. Denote by $\bar{\theta}_2$ the involutive automorphism of U which define $\bar{K}_2$. Then $\theta_2(a, b) = (\bar{\theta}_2(a), \bar{\theta}_2(b))$. Hence $\theta_1 \theta_2 = \theta_2 \theta_1$. Koike's paper [4] inspired the author to consider the condition (B).

Theorem 3.1. *Let (G, K_1, K_2) be a commutative compact symmetric triad which satisfies one of the conditions (A) or (B). Then the triple $(\tilde{\Sigma}, \Sigma, W)$ defined as above is a symmetric triad with multiplicities. Conversely every symmetric triad is obtained in this way.*

Proof   Since the assertion was proved in the case where (G, K_1, K_2) satisfies the condition (A) ([3, Thm 4.33]), we assume that (G, K_1, K_2) satisfies the condition (B). Take an adjoint invariant innerproduct $\langle \, , \, \rangle$ on $\mathfrak{u}$ and define an adjoint invariant innerproduct on $\mathfrak{g}$ by

$$\langle (X_1, Y_1), (X_2, Y_2) \rangle = \langle X_1, X_2 \rangle + \langle Y_1, Y_2 \rangle \quad (X_1, X_2, Y_1, Y_2 \in \mathfrak{u}).$$

Denote by $\mathfrak{u} = \bar{\mathfrak{k}}_2 \oplus \bar{\mathfrak{m}}_2$ the canonical decomposition of the Lie algebra $\mathfrak{u}$ of U by $\bar{\theta}_2$. Take a maximal abelian subspace $\bar{\mathfrak{a}}$ of $\bar{\mathfrak{m}}_2$. For λ in $\bar{\mathfrak{a}}$ define a subspace $\mathfrak{u}(\bar{\mathfrak{a}}, \lambda)$ of the complexification $\mathfrak{u}^{\mathbb{C}}$ of $\mathfrak{u}$ by

$$\mathfrak{u}(\bar{\mathfrak{a}}, \lambda) = \{X \in \mathfrak{u}^{\mathbb{C}} \mid [H, X] = \sqrt{-1} \langle \lambda, H \rangle X \quad (H \in \bar{\mathfrak{a}})\}.$$

Denote by $\bar{\Sigma}$ the restricted root system of $(U, \bar{K}_2)$ with respect to $\lambda \in \bar{\mathfrak{a}}$:

$$\bar{\Sigma} = \{\lambda \in \bar{\mathfrak{a}} - \{0\} \mid \mathfrak{u}(\bar{\mathfrak{a}}, \lambda) \neq \{0\}\}.$$

Since U is simple, $\bar{\Sigma}$ is an irreducible root system. Denote by $\bar{m}(\lambda)$ the multiplicity of $\lambda \in \bar{\Sigma}$: $\bar{m}(\lambda) = \dim \mathfrak{u}(\bar{\mathfrak{a}}, \lambda)$. If we set $\mathfrak{a} = \{(H, -H) \mid H \in \bar{\mathfrak{a}}\}$ then $\mathfrak{a}$ is a maximal abelian subspace of $\mathfrak{m}_1 \cap \mathfrak{m}_2$. Since

$$\mathfrak{g}(\mathfrak{a}, (\lambda/2, -\lambda/2)) = \mathfrak{u}(\bar{\mathfrak{a}}, \lambda) \oplus \mathfrak{u}(\bar{\mathfrak{a}}, -\lambda),$$
$$\mathfrak{g}(\mathfrak{a}, (\lambda/2, -\lambda/2), \pm 1) = \{(X, \pm\theta_2(X)) \mid X \in \mathfrak{u}(\bar{\mathfrak{a}}, \lambda)\},$$

we have

$$\tilde{\Sigma} = \Sigma = W = \{(\lambda/2, -\lambda/2) \mid \lambda \in \bar{\Sigma}\},$$
$$m(\lambda/2, -\lambda/2) = n(\lambda/2, -\lambda/2) = \bar{m}(\lambda).$$

Thus $(\tilde{\Sigma}, \Sigma, W)$ is a symmetric triad with multiplicities.

We show the converse. The set of all (equivalence class of) symmetric triad were classified ([3, Thm. 2.19]), and there are three types:

(I) $\tilde{\Sigma} = \Sigma \supset W, \Sigma \neq W$,   (II) $\tilde{\Sigma} = W \supset \Sigma, W \neq \Sigma$,   (III) $\tilde{\Sigma} = \Sigma = W$.

Since every irreducible root system is obtained as a restricted root system of a compact symmetric pair $(U, \bar{K}_2)$, where U is simple, every symmetric triad of type (III) is obtained from a compact symmetric triad which satisfies the condition (B). The table below shows that every symmetric triad of type (I) or (II) is obtained from a compact symmetric triad which satisfies the condition (A). We used [5] to make the table when G is exceptional. □

We give a meaning of the definition of $\mathrm{codim}(H)$ defined in the section 1. For λ in Σ, define a subspace $\mathfrak{m}_\lambda$ of $\mathfrak{m}_1 \cap \mathfrak{m}_2$ by

$$\mathfrak{m}_\lambda = \{X \in \mathfrak{m}_1 \cap \mathfrak{m}_2 \mid [H, [H, X]] = -\langle \lambda, H \rangle^2 X \quad (H \in \mathfrak{a})\}.$$

Then $\dim \mathfrak{m}_\lambda = m(\lambda)$. Define subspaces $V(\mathfrak{m}_1 \cap \mathfrak{k}_2)$ and $V^\perp(\mathfrak{m}_1 \cap \mathfrak{k}_2)$ by

$$V(\mathfrak{m}_1 \cap \mathfrak{k}_2) = \{X \in \mathfrak{m}_1 \cap \mathfrak{k}_2 \mid [\mathfrak{a}, X] = 0\},$$
$$V^\perp(\mathfrak{m}_1 \cap \mathfrak{k}_2) = \{X \in \mathfrak{m}_1 \cap \mathfrak{k}_2 \mid X \perp V(\mathfrak{m}_1 \cap \mathfrak{k}_2)\}.$$

For $\alpha \in W$, define a subspace $V_\alpha^\perp(\mathfrak{m}_1 \cap \mathfrak{k}_2)$ by

$$V_\alpha^\perp(\mathfrak{m}_1 \cap \mathfrak{k}_2) = \{X \in V^\perp(\mathfrak{m}_1 \cap \mathfrak{k}_2) \mid [H, [H, X]] = -\langle \alpha, H \rangle^2 X \quad (H \in \mathfrak{a})\}.$$

Then $\dim V_\alpha^\perp(\mathfrak{m}_1 \cap \mathfrak{k}_2) = n(\alpha)$. The coset manifold $M_i = G/K_i$ $(i = 1, 2)$ is a compact Riemannian symmetric space. Denote by $\pi_i : G \to M_i$ the natural projection. The isometric action of K_2 on M_1 is called a Hermann action. Consider an orbit $K_2 p \subset M_1$ for $p \in M_1$. We may assume that there exists an element $H \in \mathfrak{a}$ such that $p = \pi_1(g)$ where $g = \exp H$ (see [1]). By [3, Lemma 4.15, (2)] the normal space of $K_2 p$ is given by

$$g_*{}^1 T_{\pi_1(g)}^\perp (K_2 \pi_1(g)) = \{X \in \mathfrak{m}_1 \mid \mathrm{Ad}(\exp H)X \in \mathfrak{m}_2\}$$
$$= \mathfrak{a} \oplus \sum_{\lambda \in \Sigma, \langle \lambda, H \rangle \in \pi \mathbb{Z}} \mathfrak{m}_\lambda \oplus \sum_{\alpha \in W, \langle \alpha, H \rangle \in \frac{\pi}{2} + \pi \mathbb{Z}} V_\alpha^\perp(\mathfrak{m}_1 \cap \mathfrak{k}_2). \qquad (1)$$

Thus we get

$$\mathrm{codim}(H) = \mathrm{codim}(K_2 \pi_1(\exp H)) = \mathrm{codim}(K_1 \pi_2(\exp H)).$$

$(\tilde{\Sigma}, \Sigma, W)$	(G, K_1, K_2) which satisfies the condition (A)
(I-B_r)	$(SO(r+s+t), SO(r+s) \times SO(t), SO(r) \times SO(s+t))$ $(r < t, 1 \le s)$
(I-C_r)	$(SO(4r), S(O(2r) \times O(2r)), U(2r))$
(I'-C_r)	$(SU(2r), SO(2r), S(U(r) \times U(r)))$
(I-C_3)	$(E_7, SU(8), E_6 \cdot U(1))$
(I-BC_r-A_1^r)	$(SU(r+s+t), S(U(r+s) \times U(t)), S(U(r) \times U(s+t)))$ $(r < t, 1 \le s)$ $(Sp(r+s+t), Sp(r+s) \times Sp(t), Sp(r) \times Sp(s+t))$ $(r < t, 1 \le s)$ $(SO(4r+4), U(2r+2), U(2r+2)')$
(I-BC_r-B_r)	$(SO(2r+2s), S(O(2r) \times O(2s)), U(r+s)) \ (r < s)$
(I-BC_2-B_2)	$(E_6, SU(6) \cdot SU(2), SO(10) \cdot U(1))$ $(E_7, SO(12) \cdot SU(2), E_6 \cdot U(1))$
(I-F_4)	$(E_6, Sp(4), SU(6) \cdot SU(2))$ $(E_7, SU(8), SO(12) \cdot SU(2))$ $(E_8, SO(16), E_7 \cdot SU(2))$
(II-BC_r)	$(SU(r+s), SO(r+s), S(U(r) \times U(s))) \ (r < s)$
(II-BC_2)	$(E_6, Sp(4), SO(10) \cdot U(1))$
(III-A_r)	$(SU(2r+2), Sp(r+1), SO(2r+2))$
(III-A_2)	$(E_6, Sp(4), F_4)$
(III-C_r)	$(SU(4r), S(U(2r) \times U(2r)), Sp(2r))$ $(Sp(2r), U(2r), Sp(r) \times Sp(r))$
(III-BC_r)	$(SU(2r+2s), S(U(2r) \times U(2s)), Sp(r+s)) \ (r < s)$ $(Sp(r+s), U(r+s), Sp(r) \times Sp(s)) \ (r < s)$
(III-BC_1)	$(E_6, SU(6) \cdot SU(2), F_4)$ $(E_6, SO(10) \cdot U(1), F_4)$ $(F_4, Sp(3) \cdot Sp(1), SO(9))$

Here we used the following notations:

(I) In the case where $\Sigma \supset W, \Sigma \neq W$:

type	Σ^+	W^+
(I-B_r)	B_r^+	$\{e_i \mid 1 \le i \le r\}$
(I-C_r)	C_r^+	D_r^+
(I-BC_r-A_1^r)	BC_r^+	$\{e_i \mid 1 \le i \le r\}$
(I-BC_r-B_r)	BC_r^+	B_r^+
(I-F_4)	F_4^+	$\{\text{short roots in } F_4^+\} \cong D_4^+$

On $(SO(4r+4), U(2r+2), U(2r+2)')$, we set

$$U(2r+2) = \left\{ \begin{pmatrix} x & y \\ -y & x \end{pmatrix} \in SO(4r+4) \right\},$$

$$U(2r+2)' = \{ g \in SO(4r+4) \mid J'_{2r} g J'^{-1}_{2r} = g \},$$

$$\text{where} \quad J'_{2r} = \left(\begin{array}{c|cc} & E_{2r-1} & \\ & & -1 \\ \hline -E_{2r-1} & & \\ & 1 & \end{array} \right).$$

We note that $U(2r+2)$ and $U(2r+2)'$ can not transform each other by an inner automorphism of $SO(4r+4)$ ([3, Pro. 4.39]).

(II) In the case where $\Sigma \subset W, \Sigma \neq W$:

type	Σ^+	W^+
(II-BC_r) $(r \geq 1)$	B_r^+	BC_r^+
(I'-C_r)	D_r^+	C_r^+

We defined an equivalent relation $\sim$ on the set of all symmetric triads, and showed that (I'-C_r)$\sim$(I-C_r) ([3, Def. 2.6 and Thm. 2.19]).

(III) In the case where $\tilde{\Sigma} = \Sigma = W$: For instance (III-$A_r$) means that $\tilde{\Sigma} = \Sigma = W = A_r$.

References

1. E. Heintze, R. S. Palais, C. Terng and G. Thorbergsson, *Hyperpolar actions on symmetric spaces*, Geometry, topology, & physics, Conf. Proc. Lecture Notes Geom. Topology, IV, Int. Press, Cambridge, MA, 1995, pp. 214–245.
2. O. Ikawa, T. Sakai and H. Tasaki, Weakly reflective submanifolds and austere submanifolds, J. Math. Soc. Japan **61** (2009), 437–481.
3. O. Ikawa, *The geometry of symmetric triad and orbit spaces of Hermann actions* J. Math. Soc. Japan **63** (2011), 79–136.
4. N. Koike, *Examples of certain kind of minimal orbits of Hermann actions*, arXiv:1212.3911.
5. T. Matsuki, *Classification of two involutions on compact semisimple Lie groups and root systems*, J. Lie Theory, **12** (2002), 41–68.

Received March 29, 2013.

Proceedings of the Workshop on
Differential Geometry of Submanifolds
and its Related Topics
Saga, August 4-6, 2012

SOME TOPICS OF HOMOGENEOUS SUBMANIFOLDS IN COMPLEX HYPERBOLIC SPACES

Takahiro HASHINAGA

*Department of Mathematics, Hiroshima University,
Higashi-Hiroshima 739-8526, Japan
E-mail: hashinaga@hiroshima-u.ac.jp*

Akira KUBO

*Department of Mathematics, Hiroshima University,
Higashi-Hiroshima 739-8526, Japan
E-mail: akira-kubo@hiroshima-u.ac.jp*

Hiroshi TAMARU

*Department of Mathematics, Hiroshima University,
Higashi-Hiroshima 739-8526, Japan
E-mail: tamaru@math.sci.hiroshima-u.ac.jp*

Dedicated to Professor Sadahiro Maeda for his sixtieth birthday

Homogeneous hypersurfaces in complex hyperbolic spaces have been classified by Berndt and the third author in 2007. In this expository paper, we review some recent results on homogeneous submanifolds in complex hyperbolic
spaces, which have been obtained after that. Three topics will be mentioned.
The first one is the geometry of Lie hypersurfaces, including a classification of
Ricci soliton ones. The second topic is weakly reflective submanifolds, of which
we construct uncountably many new examples. The last one is an announce-
ment of a result on cohomogeneity two actions on the complex hyperbolic
plane.

Keywords: complex hyperbolic spaces, homogeneous submanifolds, Ricci soliton hypersurfaces, weakly reflective submanifolds, austere submanifolds, cohomogeneity two actions.

1. Introduction

Submanifold geometry in complex hyperbolic space $\mathbb{C}H^n$ has been studied very actively, and turned out to be fruitful. It is interesting from its beautiful geometry. It is also important from the fact that it gives valuable suggestions to study submanifold geometry in other symmetric spaces of noncompact type. For examples, several results on homogeneous hypersurfaces in $\mathbb{C}H^n$ ([1, 6]) act as bridgeheads for the studies on homogeneous hypersurfaces in more general symmetric spaces of noncompact type ([5, 7]). Recall that a submanifold is said to be *homogeneous* if it is an orbit of a closed subgroup of the isometry group of the ambient space.

In this expository paper, we summarize some recent results on homogeneous submanifolds in complex hyperbolic spaces $\mathbb{C}H^n$. On this topic, a classification of homogeneous hypersurfaces in $\mathbb{C}H^n$ by Berndt and the third author ([6]) is one of the milestones. This classification has already been reviewed in some survey articles, for examples, [25, 26]. In this paper, we focus on the results we have obtained after [6]. We will mention three topics in Sections 3, 4, and 5 as follows (Section 2 is a preliminary).

In Section 3, we review the extrinsic and intrinsic geometry of Lie hypersurfaces in $\mathbb{C}H^n$. Lie hypersurfaces in $\mathbb{C}H^n$ are particular homogeneous hypersurfaces. Note that the horospheres and the homogeneous minimal ruled real hypersurfaces are Lie hypersurfaces. Among our results, there is a classification of Ricci soliton Lie hypersurfaces in $\mathbb{C}H^n$.

In Section 4, we construct uncountably many new examples of weakly reflective submanifolds in $\mathbb{C}H^n$. By definition, weakly reflective submanifolds are austere submanifolds with global conditions. Several examples of homogeneous austere submanifolds in $\mathbb{C}H^n$ have been known ([1, 2, 6, 10]). We show that many of them are in fact weakly reflective. Our argument gives a very simple and unified proof for several known results.

In Section 5, we announce some results on cohomogeneity two actions on the complex hyperbolic plane $\mathbb{C}H^2$. In fact, we list all cohomogeneity two actions on $\mathbb{C}H^2$. For $\mathbb{C}H^2$, cohomogeneity two actions are closely related to homogeneous surfaces, which will be studied in the near future.

Finally in this section, the authors would like to express their gratitude to Professor Sadahiro Maeda. Thanks to his influence, the authors started to think that a complex hyperbolic space $\mathbb{C}H^n$ is not only one of noncompact symmetric spaces, but has more importance. This leads us to the works mentioned in this paper. The first author was supported in part by Grant-in-Aid for JSPS Fellows (11J05284). The third author was supported in part by KAKENHI (24654012).

2. Preliminaries

In this expository paper, we deal with homogeneous submanifolds in complex hyperbolic spaces

$$\mathbb{C}H^n := \mathrm{SU}(1,n)/\mathrm{S}(\mathrm{U}(1) \times \mathrm{U}(n)), \tag{1}$$

where $n \geq 2$. In this section, we recall the structure of the Lie groups and Lie algebras involved, and mention the solvable model of $\mathbb{C}H^n$. We refer to [17, 14] for general theory, and also [8] for more explicit descriptions.

2.1. *The Cartan decomposition*

First of all, let us recall the Cartan decomposition of the Lie algebra $\mathfrak{su}(1,n)$ of $\mathrm{SU}(1,n)$. Let $E_{i,j}$ be the usual matrix unit, and set

$$I_{1,n} := -E_{1,1} + E_{2,2} + \cdots + E_{n+1,n+1}. \tag{2}$$

The Lie algebra $\mathfrak{su}(1,n)$ is defined by

$$\mathfrak{g} := \mathfrak{su}(1,n) := \{X \in \mathfrak{sl}_{n+1}(\mathbb{C}) \mid {}^{t}X I_{1,n} + I_{1,n}\overline{X} = 0\}. \tag{3}$$

Note that ${}^{t}X$ is the transposed matrices, and $\overline{X}$ is the complex conjugate matrices. The Cartan involution is given by

$$\theta : \mathfrak{g} \to \mathfrak{g} : X \mapsto I_{1,n} X I_{1,n}. \tag{4}$$

It is obvious that $\theta^2 = \mathrm{id}$, and hence the eigenvalues are ± 1.

Proposition 2.1. *Let $\mathfrak{k}$ and $\mathfrak{p}$ be the $(+1)$ and (-1)-eigenspaces with respect to θ, respectively. Then one has*

$$\mathfrak{k} = \left\{ \left(\begin{array}{c|c} z & \\ \hline & X \end{array} \right) \in \mathfrak{sl}_{n+1}(\mathbb{C}) \mid z \in \mathfrak{u}(1),\ X \in \mathfrak{u}(n) \right\},$$

$$\mathfrak{p} = \left\{ \left(\begin{array}{c|c} & {}^{t}\overline{v} \\ \hline v & \end{array} \right) \in \mathfrak{sl}_{n+1}(\mathbb{C}) \mid v \in \mathbb{C}^n \right\}.$$

Note that $\mathfrak{k}$ coincides with the Lie algebra of $\mathrm{S}(\mathrm{U}(1) \times \mathrm{U}(n))$. The eigenspace decomposition of $\mathfrak{g}$ with respect to θ,

$$\mathfrak{g} = \mathfrak{k} \oplus \mathfrak{p}, \tag{5}$$

is called the *Cartan decomposition*. It is an easy exercise to show that

$$[\mathfrak{k}, \mathfrak{p}] \subset \mathfrak{p}, \quad [\mathfrak{p}, \mathfrak{p}] \subset \mathfrak{k}. \tag{6}$$

2.2. *The root space decomposition*

We next recall the root space decompositions. Let us take

$$A_0 := (1/2)(E_{1,2} + E_{2,1}) \in \mathfrak{p}, \quad \mathfrak{a} := \mathrm{span}\{A_0\}, \tag{7}$$

where span denotes the linear span with real coefficients. Then the standard argument shows that $\mathfrak{a}$ is a maximal abelian subspace in $\mathfrak{p}$. Let $\mathfrak{a}^*$ be the dual space of $\mathfrak{a}$, and define $\alpha \in \mathfrak{a}^*$ by

$$\alpha : \mathfrak{a} \to \mathbb{R} : cA_0 \mapsto c/2. \tag{8}$$

For each $\lambda \in \mathfrak{a}^*$, define the *root space* of λ by

$$\mathfrak{g}_\lambda := \{X \in \mathfrak{g} \mid [H, X] = \lambda(H)X \ (\forall H \in \mathfrak{a})\}. \tag{9}$$

If $\mathfrak{g}_\lambda \neq 0$ and $\lambda \neq 0$, then λ is called the *root* with respect to $\mathfrak{a}$. Since the Cartan involution θ acts on $\mathfrak{a}$ as $-\mathrm{id}$, one has

$$\theta(\mathfrak{g}_\lambda) = \mathfrak{g}_{-\lambda}. \tag{10}$$

We here see, for the case $n = 2$, an explicit matrix expression of the root spaces. Note that $\mathfrak{g}_\lambda$ easily determines $\mathfrak{g}_{-\lambda}$ from (10).

Proposition 2.2. *Let* $\mathfrak{g} := \mathfrak{su}(1,2)$, *and define*

$$T_0 := \frac{1}{6}\left(\begin{array}{cc|c} i & & \\ & i & \\ \hline & & -2i \end{array}\right), \quad X_1 := \frac{1}{2}\left(\begin{array}{cc|c} & & 1 \\ & & 1 \\ \hline 1 & -1 & \end{array}\right),$$

$$Y_1 := \frac{1}{2}\left(\begin{array}{cc|c} & & i \\ & & i \\ \hline -i & i & \end{array}\right), \quad Z_0 := \frac{1}{2}\left(\begin{array}{cc|c} -i & i & \\ -i & i & \\ \hline & & \end{array}\right).$$

Then one has

$$\mathfrak{g}_0 = \mathrm{span}\{T_0\} \oplus \mathfrak{a}, \quad \mathfrak{g}_\alpha = \mathrm{span}\{X_1, Y_1\}, \quad \mathfrak{g}_{2\alpha} = \mathrm{span}\{Z_0\}.$$

Proof. The bracket relations can be checked easily, for examples, one has

$$[A_0, X_1] = (1/2)X_1, \quad [A_0, Y_1] = (1/2)Y_1, \quad [A_0, Z_0] = Z_0. \tag{11}$$

One can complete the proof by dimension reasons. □

We can also have similar matrix expressions for the case $n > 2$. Such explicit expressions show the following.

Proposition 2.3. *For* $\mathfrak{g} := \mathfrak{su}(1,n)$, *the set of roots with respect to* $\mathfrak{a}$ *coincides with* $\{\pm\alpha, \pm 2\alpha\}$. *Furthermore, we have*

(1) *There exist bases* $\{X_1, \ldots, X_{n-1}, Y_1, \ldots, Y_{n-1}\}$ *of* $\mathfrak{g}_\alpha$ *and* $\{Z_0\}$ *of* $\mathfrak{g}_{2\alpha}$
    such that the bracket relations among them are given by

$$[X_i, Y_i] = Z_0 \quad (i = 1, \ldots, n-1).$$

(2) $A_0 \in \mathfrak{a}$ *satisfies*

$$[A_0, X_i] = (1/2)X_i, \quad [A_0, Y_i] = (1/2)Y_i, \quad [A_0, Z_0] = Z_0.$$

In fact, one has a direct sum decomposition as vector spaces

$$\mathfrak{g} = \mathfrak{g}_{-2\alpha} \oplus \mathfrak{g}_{-\alpha} \oplus \mathfrak{g}_0 \oplus \mathfrak{g}_\alpha \oplus \mathfrak{g}_{2\alpha}. \tag{12}$$

This is called the *root space decomposition.*

2.3. The solvable model

We here recall the Iwasawa decomposition and the solvable model of $\mathbb{C}H^n$, which play important roles for our study.

Proposition 2.4. *Let* $\mathfrak{g} := \mathfrak{su}(1, n)$, *and keep the notations in the previous subsections. Then one has*

(1) $\mathfrak{n} := \mathfrak{g}_\alpha \oplus \mathfrak{g}_{2\alpha}$ *is a nilpotent Lie algebra,*
(2) $\mathfrak{s} := \mathfrak{a} \oplus \mathfrak{n}$ *is a solvable Lie algebra,*
(3) $\mathfrak{g} = \mathfrak{k} \oplus \mathfrak{a} \oplus \mathfrak{n}$ *as vector spaces.*

The proof of this proposition is easy. The decomposition $\mathfrak{g} = \mathfrak{k} \oplus \mathfrak{a} \oplus \mathfrak{n}$ is called the *Iwasawa decomposition*. The Lie algebraic structures of $\mathfrak{n}$ and $\mathfrak{s}$ can be seen from Proposition 2.3. In particular, $\mathfrak{n}$ is the $(2n-1)$-dimensional Heisenberg Lie algebra.

We now turn from Lie algebras to Lie groups. Denote by A, N, and S the connected Lie subgroup of $SU(1, n)$ with Lie algebra $\mathfrak{a}$, $\mathfrak{n}$, and $\mathfrak{s}$, respectively. It is well-known that S acts simply-transitively on $\mathbb{C}H^n$, hence

$$F : S \to \mathbb{C}H^n : g \mapsto g.o \tag{13}$$

gives a diffeomorphism (in particular, S is simply-connected). More precisely, one has the following.

Proposition 2.5. *Let* $\langle, \rangle$ *be an inner product on* $\mathfrak{s}$ *such that the basis* $\{A_0, X_1, \ldots, X_{n-1}, Y_1, \ldots, Y_{n-1}, Z_0\}$ *is orthonormal, and* J *be a complex structure on* $\mathfrak{s}$ *define by*

$$J(A_0) = Z_0, \quad J(Z_0) = -A_0, \quad J(X_i) = Y_i, \quad J(Y_i) = -X_i.$$

Denote by the same symbols $\langle , \rangle$ and J the induced left-invariant metric and the complex structure on S, respectively. Then, $(S, \langle , \rangle, J)$ is holomorphically isometric to the complex hyperbolic space $\mathbb{C}H^n$ with constant holomorphic sectional curvature -1.

We call $(S, \langle , \rangle, J)$ the *solvable model* of a complex hyperbolic space $\mathbb{C}H^n$, and the basis $\{A_0, X_1, \ldots, X_{n-1}, Y_1, \ldots, Y_{n-1}, Z_0\}$ is called the *canonical basis* of $\mathfrak{s}$. We sometimes identify the solvable model $(S, \langle , \rangle, J)$ and $\mathbb{C}H^n$ in terms of the map (13).

3. Geometry of Lie hypersurfaces

In this section, we summarize the geometry of Lie hypersurfaces in complex hyperbolic spaces $\mathbb{C}H^n$. We refer to [1, 15, 16]. Especially, a classification of Ricci soliton Lie hypersurfaces in $\mathbb{C}H^n$ is contained. These results will be prototypes for further studies on homogeneous hypersurfaces in symmetric spaces of noncompact type.

3.1. *Preliminaries on Lie hypersurfaces*

In this subsection, we recall the definition and the classification of Lie hypersurfaces in complex hyperbolic spaces $\mathbb{C}H^n$. We mainly refer to [1], see also [5, 6, 15].

We start with some basic terminology for isometric actions. By *isometric action*, we mean an action of a connected closed subgroup of the isometry group.

Definition 3.1. For an isometric action on a Riemannian manifold, maximal dimensional orbits are said to be *regular*, and other orbits *singular*. The codimension of a regular orbit is called the *cohomogeneity* of an action.

By definition, a homogeneous hypersurface is nothing but a regular orbit of a cohomogeneity one action. In order to study homogeneous submanifolds up to isometric congruence, it is natural to consider isometric actions up to the following equivalence relation.

Definition 3.2. Two isometric actions are said to be *orbit equivalent* if there exists an isometry, of the ambient space, mapping the orbits of one action onto the orbits of the other action.

Here let us define the notion of Lie hypersurfaces. We state the definition in a way that the ambient space can be arbitrary.

Definition 3.3. An orbit of a cohomogeneity one action without singular orbit is called a *Lie hypersurface*.

From now on we assume that the ambient space is $\mathbb{C}H^n$. First of all, we recall the classification of cohomogeneity one actions on $\mathbb{C}H^n$ without singular orbit, up to orbit equivalence ([5], see also [6]). Let $(S, \langle , \rangle, J)$ be the solvable model of $\mathbb{C}H^n$, and take the canonical basis of $\mathfrak{s}$ defined in the previous section. Let

$$\mathfrak{s}(\theta) := \mathfrak{s} \ominus \operatorname{span}\{(\cos\theta)X_1 + (\sin\theta)A_0\}, \qquad (14)$$

where $\ominus$ means the orthogonal complement with respect to $\langle , \rangle$. It is easy to see that $\mathfrak{s}(\theta)$ is a Lie subalgebra of $\mathfrak{s}$. Denote by $S(\theta)$ the connected Lie subgroup of S with Lie algebra $\mathfrak{s}(\theta)$.

Proposition 3.1 ([5]). *An isometric action on $\mathbb{C}H^n$ is a cohomogeneity one action without singular orbit if and only if it is orbit equivalent to the action of $S(0)$ or $S(\pi/2)$.*

Originally, in [1], a Lie hypersurface is defined as an orbit of a connected codimension one subgroup S' of S. Proposition 3.1 yields that our definition of a Lie hypersurface is equivalent to the original one.

We next recall the classification of Lie hypersurfaces in $\mathbb{C}H^n$, up to isometric congruence. Denote by o the origin of $\mathbb{C}H^n$.

Proposition 3.2 ([1]). *Every Lie hypersurface in $\mathbb{C}H^n$ is isometrically congruent to the orbit $S(\theta).o$ for some $\theta \in [0, \pi/2]$.*

The action of $N = S(\pi/2)$ induces the so-called *horosphere foliation*, and each orbit is called the *horosphere*. Note that every orbit of $N = S(\pi/2)$ is isometrically congruent to each other (see [19] for an easy proof), which corresponds to $S(\pi/2).o$. The foliation induced by the action of $S(0)$ is called the *solvable foliation*. For $0 < \theta < \pi/2$, the orbit $S(\theta).o$ is isometrically congruent to some $S(0).p$, which is an equidistant hypersurface to $S(0).o$.

3.2. *Extrinsic geometry of Lie hypersurfaces*

In this subsection, we review the extrinsic geometry of Lie hypersurfaces. We refer to [1], see also [15].

Definition 3.4. Let $\bar{M}$ be a Riemannian manifold. A submanifold M is said to be *austere* if, for any $p \in M$ and for any normal vector $\xi \in \nu_p M$, the set of principal curvatures in direction ξ counted with multiplicities is invariant under the multiplication by -1.

Recall that principal curvatures in direction ξ are eigenvalues of the shape operator A_ξ. The sum of the principal curvatures counted with multiplicities is called the *mean curvature*. A submanifold is said to be *minimal* if the mean curvature is zero. Therefore, by definition, every austere submanifold is minimal.

Let $S(\theta).o$ be a Lie hypersurface. By using the identification $\mathbb{C}H^n = S$, one can also identify $T_o(\mathbb{C}H^n) = \mathfrak{s}$ and $T_o(S(\theta).o) = \mathfrak{s}(\theta)$. Let us take

$$\xi := (\cos\theta)X_1 + (\sin\theta)A_0, \tag{15}$$

a unit normal vector of $S(\theta).o$ at o. Then, the shape operator is given by

$$2\langle A_\xi(X), Y\rangle = \langle[\xi, X], Y\rangle + \langle X, [\xi, Y]\rangle \tag{16}$$

for every $X, Y \in \mathfrak{s}(\theta)$. By a direct calculation, using this formula, we have the following.

Proposition 3.3 ([1], see also [15]). *Let $S(\theta).o$ be the Lie hypersurface with $\theta \in [0, \pi/2]$. Then, the number of distinct principal curvatures is three if $\theta \neq \pi/2$, and two if $\theta = \pi/2$. The Lie hypersurface $S(\theta).o$ is minimal if and only if $\theta = 0$. Furthermore, $S(0).o$ is austere.*

The Lie hypersurface $S(0).o$ is the unique minimal one, which is called the *homogeneous minimal ruled real hypersurface*, introduced in [23]. This is isometrically congruent to a hypersurface called the *fan* (see [13]).

3.3. *Curvatures of Lie hypersurfaces*

In this subsection, we mention some intrinsic properties of Lie hypersurfaces $S(\theta).o$, which have been obtained in [15].

It is easy to see that the Lie hypersurface $S(\theta).o$ is isometric to the Lie group $S(\theta)$ equipped with the left-invariant metric corresponding to the inner product $\langle,\rangle|_{\mathfrak{s}(\theta)\times\mathfrak{s}(\theta)}$ on $\mathfrak{s}(\theta)$. For simplicity of the notation, we denote this inner product by the same symbol $\langle,\rangle$. One then can study Lie hypersurfaces in terms of the metric Lie algebras $(\mathfrak{s}(\theta), \langle,\rangle)$.

Let us denote by

$$\begin{aligned} T &:= (\cos\theta)A_0 - (\sin\theta)X_1, \\ \mathfrak{v}_0 &:= \mathrm{span}\{X_2, Y_2, \ldots, X_{n-1}, Y_{n-1}\}. \end{aligned} \tag{17}$$

One thus has an orthogonal decomposition

$$\mathfrak{s}(\theta) = \mathrm{span}\{T\} \oplus \mathrm{span}\{Y_1\} \oplus \mathfrak{v}_0 \oplus \mathrm{span}\{Z_0\}. \tag{18}$$

Consider an orthonormal basis compatible with this decomposition. In terms of this basis, the Ricci operator can be calculated directly.

Proposition 3.4 ([15]). *The Ricci operator* Ric *of the Lie hypersurface* $S(\theta).o$ *satisfies*

$$\mathrm{Ric}(T) = -(1/4)(2 + (2n-1)\cos^2\theta)T,$$
$$\mathrm{Ric}(Y_1) = -(1/4)(2 + (2n-3)\cos^2\theta)Y_1 + (n/2)(\sin\theta\cos\theta)Z_0,$$
$$\mathrm{Ric}(V) = -(1/4)(2 + (2n-1)\cos^2\theta)V \quad \text{for any } V \in \mathfrak{v}_0,$$
$$\mathrm{Ric}(Z_0) = (n/2)(\sin\theta\cos\theta)Y_1 + (1/2)((n-1) - 2n\cos^2\theta)Z_0.$$

By using this proposition, it is easy to calculate the eigenvalues of the Ricci operator. As a result, we have the following.

Proposition 3.5 ([15]). *Let* Ric *be the Ricci operator of the Lie hypersurface* $S(\theta).o$. *Then the number of distinct eigenvalues of* Ric *is three if* $\theta \neq \pi/2$, *and two if* $\theta = \pi/2$.

Furthermore, the sectional curvatures of the Lie hypersurfaces have also been calculated in [15] explicitly. In particular, we have the following.

Proposition 3.6 ([15]). *For the homogeneous minimal ruled real hypersurface* $S(0).o$, *the maximum of the sectional curvature is* $-1/4$. *Hence, if* θ *is sufficiently closed to* 0, *then* $S(\theta).o$ *has negative sectional curvature.*

3.4. *Ricci soliton Lie hypersurfaces*

In this subsection, we mention the classification of Ricci soliton Lie hypersurfaces in $\mathbb{C}H^n$ obtained in [16]. The notion of Ricci soliton manifolds is a generalization of the notion of Einstein manifolds.

Definition 3.5. A Riemannian manifold (M, g) is said to be *Ricci soliton* if the Ricci curvature ric satisfies

$$\mathrm{ric} = cg + \mathfrak{L}_X g \tag{19}$$

for some $c \in \mathbb{R}$ and a vector field $X \in \mathfrak{X}(M)$.

Note that $\mathfrak{L}_X$ denotes the Lie derivative. For a Ricci soliton manifold, a vector field X satisfying the above condition is called a *potential vector field*. A Ricci soliton manifold is said to be *gradient* if the potential vector field X can be written as $X = \mathrm{grad}(f)$ for some $f \in C^\infty(M)$.

It is easy to see that an Einstein manifold is always Ricci soliton. It is well-known that there exist no Einstein hypersurfaces in $\mathbb{C}H^n$ (see [24]). Hence the existence of Ricci soliton hypersurfaces is a next natural problem. The following theorem classifies Ricci soliton Lie hypersurfaces in $\mathbb{C}H^n$. In particular, Ricci soliton hypersurfaces in $\mathbb{C}H^n$ do exist.

Theorem 3.1 ([16]). *A Lie hypersurface in $\mathbb{C}H^n$ is Ricci soliton if and only if*

(1) *it is isometrically congruent to the horosphere $S(\pi/2).o$, or*
(2) *$n = 2$ and it is isometrically congruent to the homogeneous minimal ruled real hypersurface $S(0).o$.*

Proof. Here we mention the sketch of the proof. First of all, we recall the result by Lauret ([21]). In general, a left-invariant metric on a completely solvable Lie group G is Ricci soliton if and only if it is algebraic Ricci soliton, that is,

$$\text{Ric} = c \cdot \text{id} + D \tag{20}$$

holds for some $c \in \mathbb{R}$ and $D \in \text{Der}(\mathfrak{g})$. Here, $\mathfrak{g}$ is the Lie algebra of G and

$$\text{Der}(\mathfrak{g}) := \{D \in \mathfrak{gl}(\mathfrak{g}) \mid D[\cdot,\cdot] = [D(\cdot),\cdot] + [\cdot,D(\cdot)]\}. \tag{21}$$

Recall that the Lie hypersurface $S(\theta).o$ is identified with the Lie group $S(\theta)$ equipped with some left-invariant metric $\langle,\rangle$. Furthermore, $S(\theta)$ is completely solvable. Therefore one has only to check whether $(S(\theta),\langle,\rangle)$ is algebraic Ricci soliton or not. We already know the Ricci operator Ric (see Proposition 3.4). Thus, by studying $\text{Der}(\mathfrak{s}(\theta))$, one completes the proof. $\square$

This theorem would be interesting from the viewpoint of submanifold geometry. In particular, sometimes it happens, but it is still interesting that the geometry of submanifolds depends on whether $n = 2$ or $n > 2$.

Remark 3.1. The above theorem is relevant to the results by Cho and Kimura ([9]). Among others, for real hypersurfaces in non-flat complex space forms, they proved

(1) there do not exist compact Hopf hypersurfaces which are Ricci soliton,
(2) there do not exist ruled hypersurfaces which are gradient Ricci soliton.

Our theorem concludes that assumptions on these nonexistence results cannot be removed. Note that the horosphere $S(\pi/2).o$ is a Hopf hypersurface (see [1, 15]) which is Ricci soliton, but not compact. When $n = 2$, the homogeneous minimal ruled real hypersurface $S(0).o$ is a ruled hypersurface, which are Ricci soliton, but not gradient (see [21]).

Remark 3.2. We have to mention that Theorem 3.1 is not new intrinsically. Namely, the Ricci soliton Lie hypersurfaces, obtained above, are isometric to known examples. The horosphere $S(\pi/2).o$ is isometric to the

Heisenberg group equipped with certain left-invariant metric. When $n = 2$, the homogeneous minimal ruled real hypersurface $S(0).o$ is isometric to certain three-dimensional solvable Lie group with a left-invariant metric. They have been known to be Ricci soliton ([20, 21]).

4. Weakly reflective submanifolds

The notion of weakly reflective submanifolds has been introduced by Ikawa, Sakai, and Tasaki ([18]), as austere submanifolds with global conditions. In this section, we construct uncountably many examples of weakly reflective submanifolds in complex hyperbolic spaces $\mathbb{C}H^n$.

Definition 4.1. Let $\bar{M}$ be a Riemannian manifold. A submanifold M is said to be *weakly reflective* if, for any $p \in M$ and any normal vector $\xi \in \nu_p M$, there exists $\sigma \in \mathrm{Isom}(\bar{M})$ such that

$$\sigma(p) = p, \quad \sigma(M) = M, \quad (d\sigma)_p(\xi) = -\xi. \tag{22}$$

Recall that every austere submanifold is minimal (for austere submanifolds, see Definition 3.4). One of motivations for defining weakly reflective submanifolds is the following.

Proposition 4.1 ([18]). *Every weakly reflective submanifold is austere.*

A submanifold M is said to be *reflective* if it is a connected component of the fixed point set of an involutive isometry of the ambient space. Obviously, every reflective submanifold is weakly reflective, but the converse does not hold. Several interesting examples of weakly reflective submanifolds in spheres, which are not reflective, have been obtained in [18].

The next theorem provides a lot of examples of weakly reflective submanifolds in $\mathbb{C}H^n$. We use the solvable model $(S, \langle, \rangle, J)$. Recall that

$$\mathfrak{s} = \mathfrak{a} \oplus \mathfrak{n}, \quad \mathfrak{n} = \mathfrak{g}_\alpha \oplus \mathfrak{g}_{2\alpha}, \tag{23}$$

and $\mathfrak{g}_\alpha$ is a complex vector space with respect to J.

Theorem 4.1. *Let $\mathfrak{v}$ be any proper real linear subspace in $\mathfrak{g}_\alpha$. Then $\mathfrak{s}_\mathfrak{v} := \mathfrak{s} \ominus \mathfrak{v}$ is a Lie subalgebra of $\mathfrak{s}$, and the orbit of the corresponding connected Lie subgroup $S_\mathfrak{v}$ of S through the origin o is weakly reflective.*

Proof. We identify $\mathbb{C}H^n$ with the solvable model $(S, \langle, \rangle, J)$. Since the orbit $(S_\mathfrak{v}).o$ is homogeneous, it is enough to construct $\sigma \in \mathrm{Isom}(S)$ satisfying (22) at the identity $e \in S$. In order to do this, let us consider the linear map $\theta : \mathfrak{s} \to \mathfrak{s}$ defined by

$$\theta|_{\mathfrak{a} \oplus \mathfrak{g}_{2\alpha}} = \mathrm{id}, \quad \theta|_{\mathfrak{g}_\alpha} = -\mathrm{id}. \tag{24}$$

It is easy to see that θ is a Lie algebra isomorphism. Since S is simply-connected, there exists a Lie group automorphism $\sigma : S \to S$ such that $(d\sigma)_e = \theta$. This σ satisfies the conditions. In fact, σ is an isometry, since θ preserves the inner product $\langle , \rangle$. It is easy to see that $\sigma(e) = e$. Since θ preserves $\mathfrak{s}_\mathfrak{v}$, one has that σ preserves $(S_\mathfrak{v}).o$. Moreover, by definition, the differential $(d\sigma)_e$ maps every normal vector at e into its negative. This completes the proof of this theorem. □

Weakly reflective submanifolds obtained in Theorem 4.1 contain several interesting examples. In order to mention such examples, we need the notion of Kähler angles. For each $\xi \in \mathfrak{v}$, the angle $\theta_\xi \in [0, \pi/2]$ between $J\xi$ and $\mathfrak{v}$ is called the *Kähler angle* at ξ with respect to $\mathfrak{v}$.

Remark 4.1. Theorem 4.1 relates to the following results:

(1) If $\mathfrak{v}$ is a complex subspace in $\mathfrak{g}_\alpha$, say $\mathfrak{v} = \mathbb{C}^k$, then $(S_\mathfrak{v}).o$ is a totally geodesic $\mathbb{C}H^{n-k}$. This is known to be reflective ([22]).

(2) If $\mathfrak{v}$ is one-dimensional, then $(S_\mathfrak{v}).o$ is (isometrically congruent to) the homogeneous minimal ruled real hypersurface. It is known that this is austere (see Proposition 3.3), but the weak reflectivity seems to be a new result.

(3) If $\dim \mathfrak{v} \geq 2$ and the Kähler angle is constant for every $\xi \in \mathfrak{v}$, then $(S_\mathfrak{v}).o$ is a singular orbit of a cohomogeneity one action (see [2, 6]). In general, it is known that a singular orbit of a cohomogeneity one action is weakly reflective ([18]).

(4) For every $\mathfrak{v}$, the submanifold $(S_\mathfrak{v}).o$ has been studied in [10], in the context of (inhomogeneous) isoparametric hypersurfaces. They have proved that $(S_\mathfrak{v}).o$ is austere, but again the weak reflectivity would to be new.

We here point out that one can construct a lot of weakly reflective submanifolds in all symmetric spaces of noncompact type. The construction is exactly the same. Unfortunately, we cannot have new examples in real hyperbolic spaces $\mathbb{R}H^n$, since every homogeneous minimal submanifolds in $\mathbb{R}H^n$ is totally geodesic ([12]). But, apart from the case of $\mathbb{R}H^n$, one can easily construct new examples of weakly reflective submanifolds.

5. Cohomogeneity two actions on the complex hyperbolic plane

In this section, we announce our recent result on cohomogeneity two actions on the complex hyperbolic plane $\mathbb{C}H^2$. In particular, we list all cohomogeneity two actions on $\mathbb{C}H^2$. We also comment on a relation to polar actions, and a further problem.

First of all, let us recall the solvable model $(S, \langle, \rangle, J)$ of $\mathbb{C}H^2$, and the canonical basis $\{A_0, X_1, Y_1, Z_0\}$ of the Lie algebra $\mathfrak{s}$. As we saw in Proposition 2.2, one has

$$\mathfrak{g}_0 = \mathrm{span}\{T_0, A_0\}, \quad \mathfrak{g}_\alpha = \mathrm{span}\{X_1, Y_1\}, \quad \mathfrak{g}_{2\alpha} = \mathrm{span}\{Z_0\}. \tag{25}$$

Theorem 5.1. *An isometric action on $\mathbb{C}H^2$ is of cohomogeneity two if and only if it is orbit equivalent to the action of the connected Lie subgroup of $\mathrm{SU}(1,2)$ whose Lie algebra is one of the following:*

(1) $\mathfrak{k} \cap (\mathfrak{g}_{-2\alpha} \oplus \mathfrak{g}_0 \oplus \mathfrak{g}_{2\alpha}) \cong \mathfrak{u}(1) \oplus \mathfrak{u}(1)$. *This action has 0, 1, and 2-dimensional orbits.*

(2) $\mathfrak{g}_0 = \mathrm{span}\{T_0, A_0\}$. *This action has 1 and 2-dimensional orbits.*

(3) $\mathrm{span}\{T_0, Z_0\}$. *This action has 1 and 2-dimensional orbits.*

(4) $\mathrm{span}\{X_1, Z_0\}$. *All orbits of this action are 2-dimensional.*

(5) $\mathrm{span}\{A_0, X_1\}$. *All orbits of this action are 2-dimensional.*

(6) $\mathrm{span}\{A_0 + tT_0, Z_0\}$, *where $t \in \mathbb{R}$. All orbits of this action are 2-dimensional.*

The proof is based on a classification of certain Lie subalgebras of $\mathfrak{su}(1,2)$ up to congruence. A strategy of this classification is essentially same as the one in [6], which classifies cohomogeneity one actions on $\mathbb{C}H^n$.

We next comment on a relation to polar actions. An isometric action on a Riemannian manifold $\bar{M}$ is said to be *polar* if there exists a connected closed submanifold Σ of $\bar{M}$ which intersects all the orbits orthogonally. The submanifold Σ is called a *section* of the action. It is known that a section Σ must be totally geodesic. Polar actions on $\mathbb{C}H^2$ have been classified by Berndt and Díaz-Ramos ([4]) as follows.

Proposition 5.1 ([4]). *An isometric action on $\mathbb{C}H^2$ is polar if and only if it is of cohomogeneity one, or orbit equivalent to one of the actions (1)–(4) in Theorem 5.1.*

Other relevant results of polar actions on $\mathbb{C}H^n$ can be found in [3, 11].

Remark 5.1. The proof of Proposition 5.1 in [4] contains some geometric arguments, for example, possible candidates for sections Σ. Theorem 5.1

gives another direct proof of Proposition 5.1. In fact, cohomogeneity one actions are always polar, and cohomogeneity three actions $\mathbb{C}H^2$ cannot be polar (since there are no totally geodesic hypersurfaces). Thus, it is enough to study cohomogeneity two actions. It is not difficult to check that the action in Theorem 5.1 is polar if and only if it is one of (1)–(4).

Finally, we comment on remaining problems on cohomogeneity two actions.

Remark 5.2. We could not say that Theorem 5.1 gives a classification of cohomogeneity two actions on $\mathbb{C}H^2$, since we have not checked the orbit equivalence of the actions. In particular, the actions (6) contain a parameter $t \in \mathbb{R}$, and hence we have to check the orbit equivalence among continuously many actions. Note that the orbit of the action (6) through o is a totally geodesic $\mathbb{C}H^1$, for any $t \in \mathbb{R}$, but this does not imply that a different t gives orbit equivalent actions.

Theorem 5.1 would be useful for studying homogeneous surfaces in $\mathbb{C}H^2$, an interesting object in the surface and submanifold geometry. Note that, from dimension reasons, every homogeneous surface in $\mathbb{C}H^2$ is either a singular orbit of a cohomogeneity one action, or a regular orbit of a cohomogeneity two action. Therefore, we anyway have to study the orbit equivalence of cohomogeneity two actions.

References

1. J. Berndt, *Homogeneous hypersurfaces in hyperbolic spaces*, Math. Z. **229** (1998), 589–600.
2. J. Berndt, M. Brück, *Cohomogeneity one actions on hyperbolic spaces*, J. Reine Angew. Math. **541** (2001), 209–235.
3. J. Berndt, J. C. Díaz-Ramos, *Homogeneous polar foliations of complex hyperbolic spaces*, Comm. Anal. Geom. **20** (2012), 435–454.
4. J. Berndt, J. C. Díaz-Ramos, *Polar actions on the complex hyperbolic plane*, Ann. Global Anal. Geom. **43** (2013), 99–106.
5. J. Berndt, H. Tamaru, *Homogeneous codimension one foliations on noncompact symmetric spaces*, J. Differential Geom. **63** (2003), 1–40.
6. J. Berndt, H. Tamaru, *Cohomogeneity one actions on noncompact symmetric spaces of rank one*, Trans. Amer. Math. Soc. **359** (2007), 3425–3438.
7. J. Berndt, H. Tamaru, *Cohomogeneity one actions on symmetric spaces of noncompact type*, J. Reine Angew. Math., to appear, arXiv:1006.1980v1.
8. J. Berndt, F. Tricerri, L. Vanhecke, *Generalized Heisenberg groups and Damek-Ricci harmonic spaces*, Lecture Notes in Mathematics **1598**, Springer-Verlag, Berlin, 1995.

9. J. T. Cho, M. Kimura, *Ricci solitons of compact real hypersurfaces in Kähler manifolds*, Math. Nachricten **284** (2011), 1385–1393.
10. J. C. Díaz-Ramos, M. Domńguez-Vázquez, *Inhomogeneous isoparametric hypersurfaces in complex hyperbolic spaces*, Math. Z. **271** (2012), 1037–1042.
11. J. C. Díaz-Ramos, M. Domńguez-Vázquez, A. Kollross, *Polar actions on complex hyperbolic spaces*, preprint, arXiv:1208.2823v1.
12. A. J. Di Scala, C. Olmos, *The geometry of homogeneous submanifolds of hyperbolic space*, Math. Z. **237**, 199–209.
13. C. Gorodski, N. Gusevskii, *Complete minimal hypersurfaces in complex hyperbolic space*, Manuscripta Math. **103** (2000), 221–240.
14. A. W. Knapp, *Lie groups beyond an introduction, second edition*, Birkhäuser, Boston, 2005.
15. T. Hamada, Y. Hoshikawa, H. Tamaru, *Curvature properties of Lie hypersurfaces in the complex hyperbolic space*, J. Geom. **103** (2012), 247–261.
16. T. Hashinaga, A. Kubo, H. Tamaru, *Homogeneous Ricci soliton hypersurfaces in the complex hyperbolic spaces*, preprint, arXiv:1305.6128v1.
17. S. Helgason, *Differential geometry and symmetric spaces*, Pure and Applied Mathematics **XII**, Academic Press, New York-London, 1962.
18. O. Ikawa, T. Sakai, H. Tasaki, *Weakly reflective submanifolds and austere submanifolds*, J. Math. Soc. Japan **61** (2009), 437–481.
19. A. Kubo, H. Tamaru, *A sufficient condition for congruency of orbits of Lie groups and some applications*, Geom. Dedicata, to appear, arXiv:1212.3893v1.
20. J. Lauret, *Ricci soliton homogeneous nilmanifolds*, Math. Ann. **319** (2001), 715–733.
21. J. Lauret, *Ricci soliton solvmanifolds*, J. Reine Angew. Math. **650** (2011), 1–21.
22. D. S. P. Leung, *On the classification of reflective submanifolds of Riemannian symmetric spaces*, Indiana Univ. Math. J. **24** (1974), 327–339. Errata, ibid. **24** (1975), 1199.
23. M. Lohnherr, H. Reckziegel, *On ruled real hypersurfaces in complex space forms*, Geom. Dedicata **74** (1999), 267–286.
24. P. Ryan, *Intrinsic properties of real hypersurfaces in complex space forms*, in: Geometry and topology of submanifolds, X (Beijing/Berlin, 1999), 266–273, World Sci. Publ., River Edge, NJ, 2000.
25. H. Tamaru, *The classification of homogeneous hypersurfaces in complex hyperbolic spaces*, in: Proceedings of "Submanifolds in Yuzawa 2007", 5–15. (in Japanese)
26. H. Tamaru, *Homogeneous submanifolds in noncompact symmetric spaces*, in: Proceedings of The Fourteenth International Workshop on Diff. Geom. **14** (2010), 111–144.

Received March 31, 2013.

Proceedings of the Workshop on
Differential Geometry of Submanifolds
and its Related Topics
Saga, August 4-6, 2012

AUSTERE HYPERSURFACES IN 5-SPHERE AND REAL HYPERSURFACES IN COMPLEX PROJECTIVE PLANE

Jong Taek CHO

Department of Mathematics, Chonnam National University,
Gwangju 500-757, Korea
E-mail: jtcho@chonnam.ac.kr

Makoto KIMURA

Department of Mathematics, Ibaraki University,
Bunkyo, Mito 310-8512 Japan
E-mail: kmakoto@mx.ibaraki.ac.jp

Dedicated to Professor Sadahiro Maeda on his 60th birthday

In this paper we study an austere hypersurface M'^4 in S^5 which is invariant under the action of unit complex numbers S^1, i.e., it is the inverse image of a real hypersurface M^3 in $\mathbb{CP}^2$. We will give a characterization of a minimal isoparametric hypersurface with 4 distinct principal curvatures in S^5. Also we will construct austere hypersurfaces in S^5 which are invariant under 1-parameter subgroup of $SU(3)$. They are obtained from Levi-flat real hypersurfaces in $\mathbb{CP}^2$.

Keywords: real hypersurfaces, austere submanifolds, special Lagrangian submanifolds.

1. Introduction

Special Lagrangian submanifold in Ricci-flat Kähler manifold, i.e., Calabi-Yau manifold is defined by Harvey and Lawson [5] and now it is very important object in not only differential geometry but also symplectic geometry and mathematical physics. In the same paper they defined *austere submanifolds*, i.e., all odd degree symmetric polynomials in eigenvalues of any shape operator A_N vanish, and 2-dimensional minimal surface and complex submanifold in Kähler manifold are automatically satisfy this condition. So austere submanifolds are considered as a generalization of them. Also in [5] they showed that from austere submanifold in real Euclidean space $\mathbb{R}^n$, one can construct special Lagrangian submanifold, in complex Euclidean space $\mathbb{C}^n$. After that Karigiannis and Min-Oo [9] proved that from an austere submanifold in sphere S^n, one can construct special Lagrangian submanifold in the cotangent bundle $T^*(S^n)$ with Ricci-flat Stenzel metric [18]. Hence austere submanifolds in S^n are interesting to study. Examples are (i) focal submanifold of *any* isoparametric hypersurfaces in S^n, and (ii) minimal isoparametric hypersurface with equal multiplicities of principal curvatures. There are a lot of results concerning austere submanifolds [1, 4, 7, 8, 13, 17, 21]. Even in hypersurface case, the classification is not known. Any austere hypersurface is obtained as a tube of radius $\pi/2$ over a minimal surface in S^4. Also austere hypersurface with at most 3 distinct principal curvatures in S^5 is classified, but austere hypersurfaces with 4 *distinct* principal curvatures are not classified yet. Typical example is minimal (homogeneous) isoparametric hypersurface $SO(2) \times SO(3)/\mathbb{Z}_2$.

In this paper, we study austere hypersurfaces M^{2n} in odd-dimensional sphere S^{2n+1} such that $M^{2n} = \pi^{-1}(\Sigma)$ is the inverse image of a real hypersurface $\Sigma = \Sigma^{2n-1}$ in complex projective space $\mathbb{CP}^n$ with respect to the Hopf fibration $\pi : S^{2n+1} \to \mathbb{CP}^n$. Examples are:

(1) Let Σ^{4m-3} $(m \geq 2)$ be a tube of radius $\pi/4$ over totally geodesic $\mathbb{CP}^{m-1}$ in $\mathbb{CP}^{2m-1}$. Then $M^{4m-2} = \pi^{-1}(\Sigma^{4m-3})$ is the Riemannian product of S^{2m-1} with same radius $1/\sqrt{2}$, so M is isoparametric with 2 distinct principal curvatures.

(2) Let Σ^3 be a tube of radius $\pi/8$ over a totally geodesic $\mathbb{RP}^2$ (or equivalently a tube of radius $\pi/8$ over a complex quadric Q^1) in $\mathbb{CP}^2$. Then $M^4 = \pi^{-1}(\Sigma^3)$ is the minimal isoparametric hypersurface with 4 distinct principal curvatures.

(3) Let $\Sigma^{2n-1} \subset \mathbb{CP}^n$ be a minimal ruled real hypersurface (i.e., Levi-flat and each leaf of the foliation $\{\xi\}^\perp$ is totally geodesic complex projective hyperplane $\mathbb{CP}^{n-1}$). Then $M^{2n} \subset S^{2n+1}$ is an austere hypersurface and the type number of M is equal to 2.

In §2 and §3, we recall basic facts about austere submanifolds in sphere and real hypersurfaces in complex projective space, respectively. In §4 and §5, we study an austere hypersurface in S^5 which is invariant under the action of unit complex numbers, i.e., it is the inverse image of a real hypersurface M^3 in $\mathbb{CP}^2$ with respect to Hopf fibration $S^5 \to \mathbb{CP}^2$. We will give a characterization of a minimal isoparametric hypersurface with 4 distinct principal curvatures in S^5, which is nothing but the inverse image of minimal homogeneous real hypersurface of type (B) (Theorem 4.1). Finally we will show that for a Levi-flat real hypersurface M^3 in $\mathbb{CP}^2$ and its inverse image $M'^4 = \pi^{-1}(M)$ in S^5 with respect to the Hopf fibration $\pi : S^5 \to \mathbb{CP}^2$, if M'^4 is austere, then M^3 is minimal and foliated by a totally geodesic $\mathbb{RP}^2$. Hence M^3 (resp. M'^4) is given as an orbit of $\mathbb{RP}^2$ (resp. $\pi^{-1}(\mathbb{RP}^2)$) under a 1-parameter subgroup of $SU(3)$.

2. Austere submanifolds in sphere

In this section, we review definition and known results with respect to austere submanifolds in sphere (cf. [9]). Calibrated submanifolds (cf. [5]) are a distinguished class of submanifolds of a Riemannian manifold (M, g) which are absolutely volume minimizing in their homology class. Being minimal is a second order differential condition, but being calibrated is a first order differential condition.

Definition 2.1. A closed k-form α on M is called a *calibration* if it satisfies

$$\alpha(e_1, \ldots, e_k) \leq 1$$

for any choice of k orthonormal tangent vectors $e_1, \ldots, e_k$ at any point $p \in M$. A calibrated subspace of $T_p(M)$ is an oriented k-dimensional subspace V_p for which $\alpha(V_p) = 1$. Then a calibrated submanifold L of M is a k-dimensional oriented submanifold for which each tangent space is a calibrated subspace. Equivalently, L^k is calibrated if $\alpha|_L = vol_L$ where vol_L is the volume form of L associated to the induced Riemannian metric from M and the choice of orientation.

There are four examples of calibrated geometries:

(1) *Complex submanifolds L^{2k}* (of complex dimension k) of a Kähler manifold M where the calibration is given by $\alpha = \omega^k/k!$, and ω is the Kähler form on M. Kähler manifolds are characterized by having Riemannian holonomy contained in $U(n)$, where n is the complex dimension of M. These submanifolds have even real dimensions.

(2) *Special Lagrangian submanifolds L^n* with phase $e^{i\theta}$ of a Calabi-Yau manifold M where the calibration is given by $\mathrm{Re}(e^{i\theta})\Omega$, where Ω is the holomorphic $(n, 0)$ volume form on M. Calabi-Yau manifolds have Riemannian holonomy contained in $SU(n)$. Special Lagrangian submanifolds are always half-dimensional, but there is an S^1 family of these calibrations for each M, corresponding to the $e^{i\theta}$ freedom of choosing Ω. Note that Calabi-Yau manifolds, being Kähler, also possess the Kähler calibration.

Note that (iii) Associative submanifolds L^3 and coassociative submanifolds L^4 of a G_2 manifold M^7, and (iv) Cayley submanifolds L^4 of a $Spin(7)$ manifold M^8 are also calibrated submanifolds.

It is classically known that if X^p is a p-dimensional submanifold of $\mathbb{R}^n$, then the conormal bundle $N^*(X)$ is a Lagrangian submanifold of the symplectic manifold $T^*(\mathbb{R}^n)$ with its canonical symplectic structure. Harvey and Lawson found conditions ([5, Theorem III. 3.11]) on the immersion $X \subset \mathbb{R}^n$ that makes $N^*(X)$ a special Lagrangian submanifold of $T^*(\mathbb{R}^n) \cong \mathbb{C}^n$ if and only if X is an *austere* submanifold, i.e., for any normal vector field N, the eigenvalues of the shape operator A_N are at each point symmetrically arranged around zero in the real line (equivalently, all odd degree symmetric polynomials in these eigenvalues vanish).

Stenzel [18] showed that there exists a Ricci-flat Kähler metric on $T^*(S^n)$. Then Karigiannis and Min-Oo [9] showed that the conormal bundle L of a submanifold X in S^n is special Lagrangian in $T^*(S^n)$ equipped with Ricci-flat Stenzel metric if and only if X is austere in S^n.

Typical class of austere submanifolds are: (i) real 2-dimensional minimal surfaces, and (ii) Complex submanifolds of Kähler submanifolds. So austere submanifolds are also considered as a generalization of them. References of austere submanifolds are [1, 4, 6, 8, 17].

We recall examples of austere submanifolds in sphere.

(1) Focal submanifolds of *any* isoparametric hypersurfaces in S^n,
(2) Minimal isoparametric hypersurfaces with *same multiplicities* of (constant) principal curvatures in S^n,
(3) Some circle bundle over first order isotropic submanifold in the complex quadric Q^{n-1}, [7, 13, 21].

So *classification of all austere submanifolds or hypersurfaces* in S^n is an important problem. An austere hypersurface M^3 in 4-sphere S^4 has principal curvatures $\lambda, -\lambda, 0$ at each point of M. Hence M^3 is either totally geodesic ($\lambda = 0$), or M^3 lies on a tube of radius $\pi/2$ over a 2-dimensional minimal surface Σ^2 in S^4. Example of M^3 (resp. Σ^2) is a minimal isoparametric (Cartan) hypersurface (resp. Veronese surface $\mathbb{RP}^2$). Now we consider an austere hypersurface M^4 in 5-sphere S^5. Then there are 4 cases:

(1) When all principal curvatures are equal to 0, M^4 is totally geodesic S^4,
(2) When there are 2 distinct principal curvatures $\lambda, \lambda, -\lambda, -\lambda$ ($\lambda \neq 0$), M^4 is a Riemannian product of 2-spheres S^2 with same radius $1/\sqrt{2}$ (cf. [16]),

(3) When there are 3 distinct principal curvatures $\lambda, -\lambda, 0, 0$ ($\lambda \neq 0$), M^4 lies on a tube of radius $\pi/2$ over a 2-dimensional minimal surface in S^5,

(4) (Generic case) M^4 has 4 distinct principal curvatures $\lambda, -\lambda, \mu, -\mu$ $(\lambda\mu(\lambda^2 - \mu^2)) \neq 0$.

Example of (4) is the minimal (homogeneous) isoparametric hypersurface with 4 distinct principal curvatures $M^4 = SO(2) \times SO(3)/Z_2$ in S^5. Little seems to be known about austere hypersurface M^4 with 4 distinct principal curvatures in S^5, other than isoparametric one.

3. Real hypersurfaces in complex projective space

We first recall the Fubini-Study metric on the complex projective space $\mathbb{CP}^n$. Let $(z, w) = \sum_{k=0}^{n} z_k \bar{w}_k$ be the natural Hermitian inner product on $\mathbb{C}^{n+1}$. The Euclidean metric $\langle \, , \, \rangle$ on $\mathbb{C}^{n+1}$ is given by $\langle z, w \rangle = \mathrm{Re}(z, w)$. The unit sphere S^{2n+1} in $\mathbb{C}^{n+1}$ is the principal fibre bundle over $\mathbb{CP}^n$ with the structure group S^1 and the projection map π. The tangent space of S^{2n+1} at a point z is $T_z S^{2n+1} = \{w \in \mathbb{C}^{n+1} | \langle z, w \rangle = 0\}$. Restricting $\langle \, , \, \rangle$ to S^{2n+1}, we have a Riemannian metric whose Levi-Civita connection $\tilde{\nabla}$ satisfies

$$D_X Y = \tilde{\nabla}_X Y - \langle X, Y \rangle z$$

for X, Y tangent to S^{2n+1} at z, where D is the Levi-Civita connection of $\mathbb{C}^{n+1}$. Let $V_z = \sqrt{-1}z$ and write down the orthogonal decomposition into so-called vertical and horizontal components,

$$T_z S^{2n+1} = \mathrm{span}\{V_z\} \oplus V_z^\perp.$$

Let π be the canonical projection of S^{2n+1} to complex projective space,

$$\pi : S^{2n+1} \to \mathbb{CP}^n.$$

Note that $\pi_*(V_z) = 0$ but that π_* is an isomorphism on $V_z^\perp$. Let $z \in \mathbb{CP}^n$. For $X \in T_{\pi(z)}\mathbb{CP}^n$, let X^L be the vector in $V_z^\perp$ that projects to X. X^L is called the *horizontal lift* of X to z. Define a Riemannian metric on $\mathbb{CP}^n$ by $\langle X, Y \rangle = \langle X^L, Y^L \rangle$. It is well-defined since the metric on S^{2n+1} by the fibre S^1. Since $V_z^\perp$ is invariant under the multiplication of $\sqrt{-1}$, $\mathbb{CP}^n$ can be assigned a complex structure by $JX = \pi_*(\sqrt{-1}X^L)$. Then $\langle \, , \, \rangle$ is Hermitian on $\mathbb{CP}^n$ and that its Levi-Civita connection $\bar{\nabla}$ satisfies

$$\bar{\nabla}_X Y = \pi_*(\tilde{\nabla}_{X^L} Y^L).$$

We also note that on $\mathbb{CP}^n$

$$\bar{\nabla}_{X^L} V = \bar{\nabla}_V X^L = \sqrt{-1}X^L = (JX)^L.$$

while

$$\bar{\nabla}_V V = 0.$$

The Riemannian metric we have just constructed is nothing but the *Fubini-Study metric* on $\mathbb{CP}^n$.

Let M^{2n-1} be a real hypersurface of $\mathbb{CP}^n$. Then the Levi-Civita connection ∇ of the induced metric and the shape operator A are characterized respectively by

$$\tilde{\nabla}_X Y = \nabla_X Y + \langle AX, Y \rangle N$$

where N is a local choice of unit normal. Let J be the complex structure of $\mathbb{CP}^n$ with properties $J^2 = -1$, $\tilde{\nabla} J = 0$, and $\langle JX, JY \rangle = \langle X, Y \rangle$. Define the *structure vector*

$$\xi = -JN.$$

Clearly $\xi \in TM$, and $|\xi| = 1$. The dual 1-form η of ξ is defined by

$$\eta(X) = \langle X, \xi \rangle.$$

Define a skew-symmetric $(1,1)$-tensor ϕ from the tangent projection of J by

$$JX = \phi X + \eta(X)N.$$

Then we see that

$$\phi^2 X = -X + \eta(X)\xi, \quad \langle \phi X, \phi Y \rangle = \langle X, Y \rangle - \eta(X)\eta(Y).$$

Also we have $\phi\xi = 0$. Noting that $\phi^2 = -1$ on $\xi^\perp$, we see that ϕ has rank $2n - 2$ and that

$$\ker \phi = \operatorname{span}\{\xi\}.$$

Such a ϕ determines an almost contact metric structure and $\xi^\perp$ is caller *contact distribution*.

$\tilde{\nabla} J = 0$ implies

$$(\nabla_X \phi)Y = \eta(Y)AX - \langle AX, Y \rangle \xi, \quad \nabla_X \xi = \phi AX. \tag{1}$$

Also in the usual way, we derive the Gauss and Codazzi equations:

$$R(X,Y) = X \wedge Y + \phi X \wedge \phi Y - 2\langle \phi X, Y \rangle \phi + AX \wedge AY, \tag{2}$$

$$(\nabla_X A)Y - (\nabla_Y A)X = \eta(X)\phi Y - \eta(Y)\phi X - 2\langle \phi X, Y \rangle \xi. \tag{3}$$

On the contact distribution $\xi^\perp$, the *Levi form* is defined by $L(X,Y) = \langle [X,Y], \xi \rangle$ for $X, Y \perp \xi$. Hence $L \equiv 0$ if and only if the contact distribution

$\xi^\perp$ is involutive. Such a real hypersurface is called *Levi-flat*. Using (1), we obtain

$$L(X,Y) = \langle [X,Y],\xi \rangle = \langle \nabla_X Y - \nabla_Y X, \xi \rangle = -\langle Y, \phi AX \rangle + \langle X, \phi AY \rangle$$
$$= \langle X, (\phi A + A\phi)Y \rangle. \quad (X, Y \perp \xi) \tag{4}$$

Example of Levi-flat real hypersurfaces is a *ruled real hypersurfaces* , i.e., each leaf of the contact distribution is (an open subset of) totally geodesic complex projective hyperplane $\mathbb{CP}^{n-1}$ in $\mathbb{CP}^n$.

For a real hypersurface M in $\mathbb{CP}^n$, $M' = \pi^{-1}(M)$ is an S^1-invariant hypersurface in S^{2n+1} with respect to the Hopf fibration $\pi : S^{2n+1} \to \mathbb{CP}^n$. If N is a unit normal for M, then $N' = N^L$ is a unit normal for M'. The induced connection ∇' and the shape operator A' for M' satisfy

$$\tilde{\nabla}_X Y = \nabla'_X Y + \langle A'X, Y \rangle N', \quad \tilde{\nabla}_X N' = -A'X,$$

and the more familiar form of the Codazzi equation

$$(\nabla'_X A')Y = (\nabla'_Y A')X.$$

We also have $\xi^L = U = -\sqrt{-1}N'$.

Lemma 3.1. *([13, Lemma 1.6]) For X and Y tangent to $\mathbb{CP}^n$,*

$$(\tilde{\nabla}_X Y)^L = \tilde{\nabla}_{X^L} Y^L + \langle \sqrt{-1}X^L, Y^L \rangle V.$$

Lemma 3.2. *([13, Lemma 1.7])*

(1) $\tilde{\nabla}_V N' = \sqrt{-1}N' = -U$, so $A'V = U$.
(2) For X tangent to M, $(AX)^L = A'X^L - \langle X^L, U \rangle V$.
(3) In particular, $(A\xi)^L = A'U - V$.
(4) If $A\xi = \mu\xi$, then $A'U = \mu U + V$.

If ξ is a principal vector, M is called a *Hopf hypersurface*. Hopf hypersurfaces have several nice characterizations. The notion makes sense in any Kähler ambient space, and corresponds to the property that the integral curves of ξ are geodesics of M.

Lemma 3.3. *([13, Theorem 2.1, Corollary 2.3]) Let M^{2n-1}, where $n \geq 2$, be a Hopf hypersurface in $\mathbb{CP}^n$, and let μ be the principal curvature corresponding to ξ. Then μ must be a constant.*

(1) If $X \in \xi^\perp$ and $AX = \lambda X$, then

$$\left(\lambda - \frac{\mu}{2} \right) A\phi X = \left(\frac{\lambda\mu}{2} + 1 \right) \phi X.$$

(2) If a nonzero $X \in \xi^{\perp}$ satisfies $AX = \lambda X$ and $A\phi X = \lambda' \phi X$, then

$$\lambda \lambda' = \frac{\lambda + \lambda'}{2} \mu + 1.$$

Cecil and Ryan ([3]) obtained the following structure theorems about Hopf hypersurfaces in $\mathbb{CP}^n$:

Theorem 3.1. *([3, Corollary 3.2]) Let M be a real hypersurface in $\mathbb{CP}^n$ which lies on a tube of sufficientle small constant radius over a complex submanifold of $\mathbb{CP}^n$. Then M is a Hopf hypersurface.*

Theorem 3.2. *([3, Theorem 1, local version]) Let M be a connected, orientable Hopf hypersurface of $\mathbb{CP}^n$ satisfying $A\xi = \mu\xi$ and $\mu = 2\cot 2r$ for $0 < r < \pi/2$. Suppose the map ϕ_r from M to $\mathbb{CP}^n$, defined by $\phi_r(p) = \exp_p(rN_p)$ where N_p is a unit normal of M at p and $\exp$ is the exponential mapping of $\mathbb{CP}^n$, has constant rank q. Then q is even and every point $x_0 \in M$ has a neighborhood U such that $\phi_r(U)$ is an embedded complex $(q/2)$-dimensional submanifold of $\mathbb{CP}^n$. Moreover U lies on the tube of radius r over $\phi_r(U)$.*

Among Hopf hypersurfaces in $\mathbb{CP}^n$, most important examples *homogeneous real hypersurface*, i.e., an orbit of a subgroup of isometry group $PU(n+1)$ of $\mathbb{CP}^n$. Homogeneous real hypersurfaces in $\mathbb{CP}^n$ were determined by R. Takagi, and he showed that they are all obtained as regular orbits of isotropy representation of Hermitian symmetric spaces of rank 2. Also Takagi computed principal curvatures of homogeneous real hypersurfaces in $\mathbb{CP}^n$ and he showed that they are all Hopf hypersurfaces. After that we obtained the following:

Theorem 3.3. *Let M be a real hypersurface in $\mathbb{CP}^n$. Then M is Hopf and has constant principal curvature if and only if M is (an open subset of) homogeneous real hypersurface.*

It is known that each homogeneous real hypersurface in $\mathbb{CP}^n$ is realized as a tube of constant radius over a Kähler submanifold with *parallel second fundamental form*. According to Takagi's classification, the list is as follows.

(A1) Geodesic spheres with 2 distinct principal curvatures.

(A2) Tubes over totally geodesic complex projective space $\mathbb{CP}^k$ where $1 \leq k \leq n-2$, with 3 distinct principal curvatures.

(B) Tubes over complex quadrics Q^{n-1} and totally geodesic $\mathbb{RP}^n$, with 3 distinct principal curvatures.

(C) Tubes over the Segre embedding of $\mathbb{CP}^1 \times \mathbb{CP}^m$ where $n = 2m+1$ and $n \geq 5$, with 5 distinct principal curvatures.

(D) Tubes over the Plücker embedding of the complex Grassmann manifold $G_2(\mathbb{C}^5)$, with 5 distinct principal curvatures. Occur only for $n = 9$.

(E) Tubes over the canonical embedding of the Hermitian symmetric space $SO(10)/U(5)$, with 5 distinct principal curvatures. Occur only for $n = 15$.

4. Austere hypersurfaces in S^5 and real hypersurfaces in $\mathbb{CP}^2$

Let M^3 be a real hypersurface in $\mathbb{CP}^2$. Then we have a local field of orthonormal frame $\{\xi, \mathbf{e}_1, \mathbf{e}_2 = \phi\mathbf{e}_1\}$ of tangent space $T_x M$ at each point $x \in M^3$ such that the shape operator A is given by

$$A\xi = a\xi + b_1\mathbf{e}_1 + b_2\mathbf{e}_2, \quad A\mathbf{e}_1 = b_1\xi + c_1\mathbf{e}_1, \quad A\mathbf{e}_2 = b_2\xi + c_2\mathbf{e}_2 \qquad (5)$$

for some functions a, b_1, b_2, c_1 and c_2. By Lemma 4.2, (1), (2) and (3), the local shape operator A' of the hypersurface $M' = \pi^{-1}(M)$ of S^5 is represented by an orthonormal basis $\{V, U = \xi', \mathbf{e}_1', \mathbf{e}_2'\}$ as

$$A'V = U, \quad A'U = V + aU + b_1\mathbf{e}_1' + b_2\mathbf{e}_2',$$
$$A'\mathbf{e}_1' = b_1 U + c_1\mathbf{e}_1', \quad A'\mathbf{e}_2' = b_2 U + c_2\mathbf{e}_2'.$$

By computing coefficients of odd degree of characteristic polynomial of A', we obtain

Lemma 4.1. *Let M^3 be a real hypersurface in $\mathbb{CP}^2$ and let $M'^4 = \pi^{-1}(M^3)$ be the S^1-invariant hypersurface in S^5. Then M' is austere if and only if, with respect to a local field of orthonormal local frame $\{\xi, \mathbf{e}_1, \mathbf{e}_2\}$ of M satisfying (5),*

$$a + c_1 + c_2 = 0, \qquad (6)$$
$$(1 + b_2^2)c_1 + (1 + b_1^2)c_2 - ac_1c_2 = 0. \qquad (7)$$

Note that minimality (6) of M^3 in $\mathbb{CP}^2$ is equivalent to minimality of M'^4 in S^5.

Suppose M^3 is a *Hopf* hypersurface in $\mathbb{CP}^2$. Then the shape operator A is represented as

$$A\xi = a\xi, \quad A\mathbf{e}_1 = c_1\mathbf{e}_1, \quad A\mathbf{e}_2 = c_2\mathbf{e}_2,$$

where $\{\xi, \mathbf{e}_1, \mathbf{e}_2\}$ and c_1, c_2 are a local field of orthonormal frame and functions on M, respectively. By Lemma 4.3, a is a constant. Then Lemma 4.4

(2) implies

$$c_1 c_2 = \frac{c_1 + c_2}{2} a + 1. \tag{8}$$

Moreover we assume that M^3 is *minimal*, i.e.,

$$a + c_1 + c_2 = 0. \tag{9}$$

Then (7) and (8) yield that c_1 and c_2 are constant on M. Using Theorem 4.3, we obtain

Proposition 4.1. *Let M^3 be a minimal Hopf hypersurface in $\mathbb{CP}^2$. Then M is an open subset of homogeneous hypersurface of type (A1) or (B).*

Note that among 1-parameter family of homogeneous real hypersurfaces (regular orbits), there exists unique minimal one.

Let M^3 be a *minimal, Hopf* hypersurface in $\mathbb{CP}^2$. Then the condition (6) for which $M' = \pi^{-1}(M)$ is *austere* is

$$a(1 + c_1 c_2) = 0. \tag{10}$$

Let M^3 be a homogeneous real hypersurface of type (A1), i.e., a geodesic sphere in $\mathbb{CP}^2$. Then the principal curvatures are $\lambda, \lambda, \lambda - 1/\lambda$ where $\lambda = \pm \cot r$ and r ($0 < r < \pi/2$) denotes the radius of the geodesic sphere ([3,13,19]). If (10) hold, then $c_1 = c_2$ implies $a = \lambda - (1/\lambda) = 0$. So $\lambda = \pm 1$, but in this case (9) does not valid. Hence corresponding hypersurface M' in S^5 is never austere.

Let M^3 be a homogeneous real hypersurface of type (B), i.e., a tube over totally geodesic $\mathbb{RP}^2$ in $\mathbb{CP}^2$. Then the principal curvatures are $-1/\lambda, \lambda, 2\lambda/(1 - \lambda^2)$ where $\lambda = \pm \cot r$ and r ($0 < r < \pi/4$) denotes the radius of the tube ([3,13,19]). Then M is minimal if and only if $r = \pi/8$, and in this case (10) is also hold. So the corresponding hypersurface M' in S^5 is austere. Note that M' is nothing but the minimal homogeneous isoparametric hypersurface with 4 distinct principal curvatures in S^5. Now we have

Theorem 4.1. *Let M^3 be a Hopf hypersurface in $\mathbb{CP}^2$ and let $M'^4 = \pi^{-1}(M)$ be the hypersurface in S^5, which is the inverse image of M under the Hopf fibration $\pi : S^5 \to \mathbb{CP}^2$. Suppose M'^4 is austere. Then M^3 lies on a tube of radius $\pi/8$ over a totally geodesic $\mathbb{RP}^2$ (and a complex quadric Q^1) in $\mathbb{CP}^2$, and M'^4 is an open subset of the minimal isoparametric hypersurface with 4 distinct principal curvatures.*

5. Austere hypersurfaces in S^5 and Levi-flat real hypersurfaces in $\mathbb{CP}^2$

Next suppose M^3 is a *Levi-flat* real hypersurface in $\mathbb{CP}^2$. Then using (4), we obtain

$$c_1 + c_2 = 0, \qquad (11)$$

with respect to an orthonormal basis $\{\xi, \mathbf{e}_1, \mathbf{e}_2 = \phi\mathbf{e}_1\}$ satisfying (5). Then minimality condition (9) is equivalent to $a(= \eta(A\xi)) = 0$. Hence for Levi-flat real hypersurface M^3 in $\mathbb{CP}^2$, $M' = \pi^{-1}(M^3)$ is austere in S^5 if and only if $b_1^2 = b_2^2$. It well-known (cf. Lemma 4.3) that Hopf hypersurface is never Levi-flat, so we have $b_1^2 = b_2^2 \neq 0$. Without lost of generality, we may assume $b := b_1 = b_2 \neq 0$. Then

$$A\xi = b(\mathbf{e}_1 + \mathbf{e}_2), \quad A\mathbf{e}_1 = b\xi + c_1\mathbf{e}_1, \quad A\mathbf{e}_2 = b\xi - c_1\mathbf{e}_2. \qquad (12)$$

If we put

$$\tilde{\mathbf{e}}_1 = \frac{\mathbf{e}_1 - \mathbf{e}_2}{\sqrt{2}}, \quad \tilde{\mathbf{e}}_2 = \frac{\mathbf{e}_1 + \mathbf{e}_2}{\sqrt{2}} = \phi\tilde{\mathbf{e}}_1, \quad \tilde{b} = \sqrt{2}b, \quad \tilde{c} = \sqrt{2}c_1,$$

then (12) is

$$A\xi = \tilde{b}\tilde{\mathbf{e}}_2, \quad A\tilde{\mathbf{e}}_1 = \tilde{c}\tilde{\mathbf{e}}_2, \quad A\tilde{\mathbf{e}}_2 = \tilde{b}\xi + \tilde{c}\tilde{\mathbf{e}}_1. \qquad (13)$$

Then (1) implies

$$\nabla_\xi \xi = \phi A\xi = -\tilde{b}\tilde{\mathbf{e}}_1, \quad \nabla_{\tilde{\mathbf{e}}_1}\xi = \phi A\tilde{\mathbf{e}}_1 = -\tilde{c}\tilde{\mathbf{e}}_1. \qquad (14)$$

Using Codazzi equation (3), we have

$$(\nabla_\xi A)\tilde{\mathbf{e}}_1 - (\nabla_{\tilde{\mathbf{e}}_1} A)\xi = \phi\tilde{\mathbf{e}}_1 = \tilde{\mathbf{e}}_2. \qquad (15)$$

By taking inner product with ξ, we obtain

$$
\begin{aligned}
0 &= \langle (\nabla_\xi A)\tilde{\mathbf{e}}_1 - (\nabla_{\tilde{\mathbf{e}}_1} A)\xi, \xi \rangle \\
&= \langle \nabla_\xi(A\tilde{\mathbf{e}}_1) - A\nabla_\xi\tilde{\mathbf{e}}_1 - \nabla_{\tilde{\mathbf{e}}_1}(A\xi) + A\nabla_{\tilde{\mathbf{e}}_1}\xi, \xi \rangle \\
&= \langle \nabla_\xi(\tilde{c}\tilde{\mathbf{e}}_2), \xi \rangle - \langle \nabla_\xi\tilde{\mathbf{e}}_1, A\xi \rangle - \langle \nabla_{\mathbf{e}_1}(\tilde{b}\tilde{\mathbf{e}}_2), \xi \rangle + \langle A\phi A\tilde{\mathbf{e}}_1, \xi \rangle \\
&= -\tilde{c}\langle \tilde{\mathbf{e}}_2, \phi A\xi \rangle - \tilde{b}\langle \nabla_\xi\tilde{\mathbf{e}}_1, \tilde{\mathbf{e}}_2 \rangle + \tilde{b}\langle \tilde{\mathbf{e}}_2, \phi A\tilde{\mathbf{e}}_1 \rangle - \tilde{c}\langle A\tilde{\mathbf{e}}_1, \xi \rangle \\
&= -\tilde{b}\langle \nabla_\xi\tilde{\mathbf{e}}_1, \tilde{\mathbf{e}}_2 \rangle,
\end{aligned}
$$

which together with $\tilde{b} \neq 0$, yields

$$\langle \nabla_\xi\tilde{\mathbf{e}}_1, \tilde{\mathbf{e}}_2 \rangle = 0. \qquad (16)$$

Also by taking inner product of (15) with $\tilde{\mathbf{e}}_1$, we get

$$
\begin{aligned}
0 &= \langle (\nabla_\xi A)\tilde{\mathbf{e}}_1 - (\nabla_{\tilde{\mathbf{e}}_1} A)\xi, \tilde{\mathbf{e}}_1 \rangle \\
&= \langle \nabla_\xi(A\tilde{\mathbf{e}}_1) - A\nabla_\xi\tilde{\mathbf{e}}_1 - \nabla_{\tilde{\mathbf{e}}_1}(A\xi) + A\nabla_{\tilde{\mathbf{e}}_1}\xi, \tilde{\mathbf{e}}_1 \rangle \\
&= \langle \nabla_\xi(\tilde{c}\tilde{\mathbf{e}}_2), \tilde{\mathbf{e}}_1 \rangle - \langle \nabla_\xi\tilde{\mathbf{e}}_1, A\tilde{\mathbf{e}}_1 \rangle - \langle \nabla_{\tilde{\mathbf{e}}_1}(\tilde{b}\tilde{\mathbf{e}}_2), \tilde{\mathbf{e}}_1 \rangle + \langle A\phi A\tilde{\mathbf{e}}_1, \tilde{\mathbf{e}}_1 \rangle \\
&= -\tilde{c}\langle \tilde{\mathbf{e}}_2, \nabla_\xi\tilde{\mathbf{e}}_1 \rangle - \tilde{b}\langle \nabla_\xi\tilde{\mathbf{e}}_1, \tilde{\mathbf{e}}_2 \rangle + \tilde{b}\langle \tilde{\mathbf{e}}_2, \nabla_{\tilde{\mathbf{e}}_1}\tilde{\mathbf{e}}_1 \rangle - \tilde{c}\langle A\tilde{\mathbf{e}}_1, \tilde{\mathbf{e}}_1 \rangle \\
&= \tilde{b}\langle \tilde{\mathbf{e}}_2, \nabla_\xi\tilde{\mathbf{e}}_1 \rangle,
\end{aligned}
$$

and

$$
\langle \nabla_{\tilde{\mathbf{e}}_1}\tilde{\mathbf{e}}_1, \tilde{\mathbf{e}}_2 \rangle = 0. \tag{17}
$$

Hence (14), (16) and (17) yield that 2-dimensional distribution $D = \{\xi, \tilde{\mathbf{e}}_1\}$ on M^3 is integrable and totally geodesic. Moreover by (13), each leaf L of D is a totally geodesic Lagrangian submanifold and hence an open subset of $\mathbb{RP}^2$.

In [11], we proved that: there is a natural one-to-one correspondence between $(n+1)$-dimensional submanifold M^{n+1} foliated by totally geodesic $\mathbb{RP}^n$ in $\mathbb{CP}^n$ and a curve γ in a Riemannian symmetric space $SU(n+1)/SO(n+1)$. Then we showed that M^{n+1} is *minimal* if and only if γ is a *geodesic* of $SU(n+1)/SO(n+1)$, and M is given as an orbit of $\mathbb{RP}^n$ under a 1-parameter subgroup of $SU(n+1)$. In particular M^3 is a minimal real hypersurface foliated by $\mathbb{RP}^2$ in $\mathbb{CP}^2$ when $n = 2$. Hence we have

Theorem 5.1. *Let M^3 be a Levi-flat real hypersurface in $\mathbb{CP}^2$ and let $M'^4 = \pi^{-1}(M)$ be the hypersurface in S^5, which is the inverse image of M under the Hopf fibration $\pi : S^5 \to \mathbb{CP}^2$. Suppose M'^4 is austere. Then M^3 is minimal and foliated by totally geodesic $\mathbb{RP}^2$, and M^3 (resp. M'^4) is given as an orbit of $\mathbb{RP}^2$ (resp. $\pi^{-1}(\mathbb{RP}^2)$) under a 1-parameter subgroup of $SU(3)$.*

We will give explicitly the construction of Levi-flat M^3 in $\mathbb{CP}^2$ and austere M'^4 in S^5. Let P be a real 3×3 symmetric matrix with trace $P = 0$, and let $F_P : M' := \mathbb{R} \times S^1 \times S^2 \to S^5$ be a map defined by

$$
F_P(s, t, \mathbf{x}) = e^{it} \exp(siP)\mathbf{x}.
$$

Differential of F_p is

$$
(F_P)_*(\partial/\partial s) = ie^{it}\exp(siP)P\mathbf{x},
$$
$$
(F_P)_*(\partial/\partial t) = ie^{it}\exp(siP)\mathbf{x},
$$
$$
(F_P)_*(X) = e^{it}\exp(siP)X. \quad (X \in T_\mathbf{x}(S^2))
$$

Hence F_P is an immersion, i.e., has rank 4 at $(s, t, \mathbf{x}) \in M'$ if and only if $\mathbf{x}$ is *not* an eigenvector of P. Clearly $F_P(M')$ is S^1-invariant and gives a real hypersurface in $\mathbb{CP}^2$ with singularities.

We calculate F_p on open subset of M' such that F_p has rank 4. $v = (F_P)_*(\partial/\partial t) = iF_P$ is a unit vertical vector with respect to the Hopf fibration $\pi : S^5 \to \mathbb{CP}^2$. $\xi' = (F_P)_*((P\mathbf{x} \times \mathbf{x})/|P\mathbf{x} \times \mathbf{x}|)$ is a horizontal lift of the structure vector ξ of the real hypersurface $\pi(F_P(M'))$ in $\mathbb{CP}^2$. If we put

$$\mathbf{e}_2 := (F_P)_*((P\mathbf{x} - \langle P\mathbf{x}, \mathbf{x}\rangle\mathbf{x})/\sqrt{|P\mathbf{x}|^2 - \langle P\mathbf{x}, \mathbf{x}\rangle^2}), \quad \mathbf{e}_1 := i\mathbf{e}_2,$$

then $\{v, \xi', \mathbf{e}_1, \mathbf{e}_2\}$ is an orthonormal basis of $T_{F(s,t,\mathbf{x})}(M')$. By direct computations, the shape operator A' of M'^4 in S^5 with respect to unit normal vector field $N' = i\xi'$ is represented by the following matrix with respect to the above orthonormal basis:

$$\begin{pmatrix} 0 & 1 & 0 & 0 \\ 1 & 0 & h(\xi', \mathbf{e}_1) & 0 \\ 0 & h(\xi', \mathbf{e}_1) & 0 & h(\mathbf{e}_1, \mathbf{e}_2) \\ 0 & 0 & h(\mathbf{e}_1, \mathbf{e}_2) & 0 \end{pmatrix},$$

where

$$h(\xi', \mathbf{e}_1) = \frac{\langle P(P\mathbf{x} \times \mathbf{x}), P\mathbf{x} \times \mathbf{x}\rangle}{|P\mathbf{x} \times \mathbf{x}|^2} - \langle P\mathbf{x}, \mathbf{x}\rangle,$$

and $h(\mathbf{e}_1, \mathbf{e}_2) = \det(P^2\mathbf{x}, P\mathbf{x}, \mathbf{x})/|P\mathbf{x} \times \mathbf{x}|^2$. Hence we can see that M'^4 is an austere hypersurface in S^5. We have

$$\det A' \neq 0 \Leftrightarrow h(\mathbf{e}_1, \mathbf{e}_2) \neq 0 \Leftrightarrow \det(P^2\mathbf{x}, P\mathbf{x}, \mathbf{x}) \neq 0.$$

Consequently we obtain an austere hypersurface M'^4 in S^5 of type number 4 locally, provided $\det(P^2\mathbf{x}, P\mathbf{x}, \mathbf{x}) \neq 0$. On the other hand, when $\det(P^2\mathbf{x}, P\mathbf{x}, \mathbf{x}) = 0$ holds for any $\mathbf{x} \in S^2$, the corresponding real hypersurface M^3 in $\mathbb{CP}^2$ is nothing but the *minimal ruled* hypersurface.

We note that Levi-flat *minimal* hypersurfaces M^3 in 2-dimensional complex space form are completely classified by R. Bryant [2].

Acknowledgments

J. T. Cho was supported by Basic Science Research Program through the National Research Foundation of Korea (NRF) funded by the Ministry of Education, Science and Technology (2012R1A1B3003930). M. Kimura was partially supported by Grants-in-Aid for Scientific Research, The Ministry of Education, Science, Sports and Culture, Japan, No. 24540080.

References

1. R. L. Bryant, *Some remarks on the geometry of austere manifolds*, Bol. Soc. Brasil. Mat. (N.S.) **21** (1991) no. 2, 133-157.

2. R. L. Bryant, *Levi-flat minimal hypersurfaces in two-dimensional complex space forms*, Lie groups, geometric structures and differential equations - one hundred years after Sophus Lie (Kyoto/Nara, 1999), 1-44, Adv. Stud. Pure Math., **37**, Math. Soc. Japan, Tokyo, 2002.

3. T. E. Cecil and P. J. Ryan, *Focal sets and real hypersurfaces in complex projective space*, Trans. Amer. Math. Soc. **269** (1982), 481–499

4. M. Dajczer and L. Florit, *A class of austere submanifolds*, Illinois J. Math. **45** (2001), no. 3, 735-755.

5. R. Harvey and H. B. Lawson, *Calibrated Geometries*, Acta. Math. **148** (1982), 47-157.

6. M. Ionel and T. Ivey, *Austere submanifolds of dimension four: examples and maximal types*, Illinois J. Math. **54** (2010), no. 2, 713-746.

7. G. Ishikawa, M. Kimura and R. Miyaoka, *Submanifolds with degenerate Gauss mappings in spheres*, Lie groups, geometric structures and differential equations - one hundred years after Sophus Lie (Kyoto/Nara, 1999), 115-149, Adv. Stud. Pure Math., **37**, Math. Soc. Japan, Tokyo, 2002.

8. O. Ikawa, T. Sakai and H. Tasaki, *Weakly reflective submanifolds and austere submanifolds*, J. Math. Soc. Japan **61** (2009), no. 2, 437-481.

9. S. Karigiannis and M. Min-Oo, *Calibrated subbundles in noncompact manifolds of special holonomy*, Ann. Global Anal. Geom. **28** (2005), no. 4, 371-394.

10. M. Kimura, *Real hypersurfaces and complex submanifolds in complex projective space*, Trans. Amer. Math. Soc. **296** (1986), 137–149.

11. M. Kimura, *Sectional curvatures of holomorphic planes on a real hypersurface in $P^n(\mathbb{C})$*, Math. Ann. **276** (1987), no. 3, 487-497.

12. M. Kimura, *Curves in $SU(n+1)/SO(n+1)$ and some submanifolds in $P^n(\mathbb{C})$*, Saitama Math. J. **14** (1996), 79-89.

13. M. Kimura, *Minimal immersions of some circle bundles over holomorphic curves in complex quadric to sphere*, Osaka J. Math. **37** (2000), no. 4, 883-903.

14. R. Niebergall and P. J. Ryan, *Real hypersurfaces in complex space forms*, Tight and taut submanifolds (Berkeley, CA, 1994), Math. Sci. Res. Inst. Publ., **32**, Cambridge Univ. Press, Cambridge (1997), 233–305.

15. M. Okumura, *On some real hypersurfaces of a complex projective space*, Trans. Amer. Math. Soc., **212** (1975), 355–364

16. T. Otsuki, *Minimal hypersurfaces in a Riemannian manifolds of constant curvature*, Amer. J. Math. **92** (1970), 145–173.

17. F. Podesta, *Some remarks on austere submanifolds* Boll. Un. Mat. Ital. B (7) **11** (1997), no. 2, suppl., 157-160.

18. M. Stenzel, *Ricci-flat metrics on the complexification of a compact rank one symmetric space*, Manuscripta Math. **80** (1993), 151-163.

19. R. Takagi, *On homogeneous real hypersurfaces in a complex projective space*, Osaka J. Math. **10** (1973), 495-506.

20. R. Takagi, *Real hypersurfaces in a complex projective space with constant*

principal curvature, J. Math. Soc. Japan **27** (1975), 43-53.
21. K. Tsukada, *Isotropic Kähler immersions into a complex quadric*, Natur. Sci. Rep. Ochanomizu Univ. **57** (2007), no. 1, 1-30.

Received March 31, 2012.

Proceedings of the Workshop on
Differential Geometry of Submanifolds
and its Related Topics
Saga, August 4-6, 2012

ON THE MINIMALITY OF NORMAL BUNDLES IN THE TANGENT BUNDLES OVER THE COMPLEX SPACE FORMS

Toru KAJIGAYA

Mathematical Institute, Graduate School of Sciences,
Tohoku University, Aoba-ku, Sendai, 980-8578, Japan
E-mail: sa9m09@math.tohoku.ac.jp

Dedicated to Professor Sadahiro Maeda for his sixtieth birthday

In this note, we investigate the minimality of the normal bundle of a submanifold N of a Riemannian manifold M in the tangent bundle TM equipped with the Sasaki metric. In particular, we give details in the case M is a complex space form.

Keywords: normal bundle, tangent bundle, Sasaki metric, minimal Lagrangian submanifold, austere submanifold.

1. Introduction

Let M be a smooth manifold and N a submanifold in M. It is a classical fact that the cotangent bundle T^*M admits the standard symplectic structure, and the conormal bundle ν^*N of N becomes a Lagrangian submanifold in T^*M. If we introduce a Riemannian metric $\langle,\rangle$ on M, we have the standard identification between the conormal bundle T^*M and the tangent bundle TM. Throughout this article, we use this identification.

The Riemannian metric on M induces a Riemannian metric $\tilde{g}$ on TM, which is the so-called *Sasaki metric*. This gives an almost Kähler structure on TM (see section 2). It is natural to ask when a submanifold N in M has a *minimal* Lagrangian normal bundle νN in $(TM, \tilde{g})$. In [5], the author shows that the normal bundle over a submanifold N in the real space form $M^m(c)$ is a minimal Lagrangian submanifold in $(TM, \tilde{g})$ if and only if N is *austere* in M. This gives a generalization of the well known fact due to Harvey-Lawson in the case $M = \mathbb{R}^m$ [4].

As opposed to the situation of the real space forms, it seems more involved in the case of the complex space forms since the curvature tensor is more complicated. In this note, we derive the mean curvature formula for normal bundles over some classes of submanifolds in the complex space form $M = M^m(4c)$ of constant holomorphic sectional curvature $4c$, and investigate the minimality of the normal bundles. In particular, we show that any normal bundle over a complex submanifold in $M^m(4c)$ is minimal in $(TM, \tilde{g})$ (Theorem 3.1). Moreover, we classify totally geodesic normal bundles, and show that the totally geodesic normal bundles do not necessarily correspond to totally geodesic submanifolds in $M^m(4c)$ (Theorem 3.2). Finally, we classify Hopf hypersurfaces with constant principal curvatures in $M^m(4c)$ which give minimal normal bundles in $(TM, \tilde{g})$ (Theorem 3.3).

2. Preliminaries

2.1. *Tangent bundles and the Sasaki metric*

In this subsection, we review the geometry of tangent bundles according to [1] and [3]. Let $(M, \langle, \rangle)$ be an m-dimensional Riemannian manifold, TM the tangent bundle over M, and $\pi : TM \to M$ the natural projection. For a vector field X on M, we define two vector fields on TM, the *horizontal lift* X^h and the *vertical lift* X^v by $X^h \alpha := \nabla_X \alpha$, and $X^v \alpha = \alpha(X) \circ \pi$, respectively, where ∇ is the Levi-Civita connection with respect to the Riemannian metric $\langle, \rangle$, and α is a 1-form on M which is regarded as a function on TM. If we choose a local coordinate $(x^1, \cdots, x^m)$ of M, then we can choose a local coordinate $(p^1, \cdots, p^m, q^1, \cdots, q^m)$ of TM, where $x^i = p^i \circ \pi$, and $(q^1, \cdots, q^m)$ is the fiber coordinate. With respect to this coordinate, a local expression of X^h and X^v is given by

$$X^h = X^i \frac{\partial}{\partial p^i} - X^i q^j \Gamma_{ij}^k \frac{\partial}{\partial q^k}, \quad X^v = X^i \frac{\partial}{\partial q^i}, \qquad (1)$$

where $X = X^i \frac{\partial}{\partial x^i}$, Γ_{ij}^k's are the Christoffel symbols of ∇, and we use the Einstein convention. The *connection map* K is a bundle map $K : TTM \to TM$ defined as follows. Let $\exp_p : V' \to V$ be the local diffeomorphism of the exponential map at p, and $\tau : \pi^{-1}(V) \to T_pM$ be the smooth map which translates every $Y \in \pi^{-1}(V)$ from $q = \pi(Y)$ to p in a parallel manner along the unique geodesic curve in V. For $u \in T_pM$, the map $R_{-u} : T_pM \to T_pM$ is defined by $R_{-u} := X - u$. Then we define $K_{(p,u)} := d(\exp_p \circ R_{-u} \circ \tau)$.

Let $\tilde{c}$ be a smooth section in TM with $\tilde{c}(0) = z, \dot{\tilde{c}}(0) = \tilde{V} \in T_z TM$. Put $c := \pi \circ \tilde{c}$. Then the connection map satisfies

$$K_z \tilde{V} = \nabla_{\dot{c}} \tilde{c}|_0. \qquad (2)$$

For $z = (p, u) \in TM$, we define vector subspaces of $T_z TM$ by $\mathcal{H}_z :=$ $\mathrm{Ker} K_z$ and $\mathcal{V}_z := \mathrm{Ker} d\pi_z$. Then the tangent space of TM has a direct sum $T_z TM = \mathcal{H}_z \oplus \mathcal{V}_z$. Since the horizontal and vertical lifts are characterized by the following properties

$$\pi_*(X^h)_z = X_p, \ \pi_*(X^v)_z = 0, \ K(X^h)_z = 0, \ \text{and } K(X^v)_z = X_p$$

for any vector field X on M, we have vector space isomorphisms π_* : $\mathcal{H}_z \tilde{\to} T_p M$ and $K : \mathcal{V}_z \tilde{\to} T_p M$. Thus, every tangent vector $\tilde{V}_z \in T_z TM$ can be decomposed into $\tilde{V}_z := (X_p)^h_z + (Y_p)^v_z$, where $X_p, Y_p \in T_p M$ are uniquely determined by $X_p := \pi_*(\tilde{V}_z)$ and $Y_p := K(\tilde{V}_z)$.

The tangent bundle TM admits an almost complex structure J defined by $JX^h = X^v$ and $JX^v = -X^h$ for any vector field X on M. The *Sasaki metric* $\tilde{g}$ is a Riemannian metric on TM defined by

$$\tilde{g}(\tilde{X}, \tilde{Y})_z := \langle \pi_* \tilde{X}, \pi_* \tilde{Y} \rangle_p + \langle K\tilde{X}, K\tilde{Y} \rangle_p$$

for $\tilde{X}, \tilde{Y} \in T_z TM$. By the definition, the splitting $T_z TM = \mathcal{H}_z \oplus \mathcal{V}_z$ is orthogonal with respect to $\tilde{g}$.

The Riemannian metric $\langle , \rangle$ on M defines the standard identification between the tangent bundle TM and the cotangent bundle T^*M, namely, $\iota : TM \tilde{\to} T^*M$ via $X_p \mapsto \langle X_p, \cdot \rangle$. The Liouville form $\gamma \in \Omega^1(T^*M)$ is the 1-form defined by $\gamma_{(p,\chi)}(\tilde{V}) := \chi_p(\pi_*(\tilde{V}))$, where $\tilde{V}$ is a tangent vector of T^*M. The canonical symplectic structure on T^*M is defined by $\omega^* := -d\gamma$. Then we can induce a symplectic structure on TM by $\omega := \iota^* \omega^*$. It is easily shown that the almost complex structure J and $\tilde{g}$ are associated with each other, i.e., it gives an almost Hermitian structure on TM, and ω is the associated 2-form, i.e., $\omega = \tilde{g}(J\cdot, \cdot)$. Since ω is closed, this almost Hermitian structure defines the almost Kähler structure on TM. We remark that this almost Kähler structure is Kähler if and only if $(M, \langle , \rangle)$ is flat.

The Levi-Civita connection with respect to the Sasaki metric satisfies the following relation for the horizontal and vertical lifts:

$$(\tilde{\nabla}_{X^h} Y^h)_z = (\nabla_X Y)^h_z - \frac{1}{2}(R(X, Y)u)^v_z, \tag{3}$$

$$(\tilde{\nabla}_{X^h} Y^v)_z = \frac{1}{2}(R(u, Y)X)^h_z + (\nabla_X Y)^v_z, \tag{4}$$

$$(\tilde{\nabla}_{X^v} Y^h)_z = \frac{1}{2}(R(u, X)Y)^h_z, \tag{5}$$

$$\tilde{\nabla}_{X^v} Y^v = 0, \tag{6}$$

where $X, Y \in \Gamma(TM)$, and R denotes the curvature tensor of ∇.

2.2. Lemmata for the general setting

Let N be an n-dimensional submanifold of M. We denote the normal bundle of N by νN, i.e., $\nu N := \{z = (p, u) \in TM | \ p \in N \text{ and } u \perp T_pN\}$. Then νN is a submanifold in TM. Fix an arbitrary point $z_0 = (p_0, u_0) \in \nu N$. We can choose a local field of orthonormal frames $\{e_1, \cdots, e_n\}$ around p_0 in N such that $A^{u_0}(e_i)(p_0) = \kappa_i(z_0)e_i(p_0)$ for $i = 1, \cdots, n$, where $\kappa_i(z_0)$ is the *principal curvature* at the point p_0 with respect to the normal direction u_0. We can also choose a local field of orthonormal frames $\{\nu_{n+1}, \cdots, \nu_m\}$ of the normal space of N around p_0. The following is given in [5].

Lemma 2.1. *Let $\{e_1, \cdots, e_n\}$ and $\{\nu_{n+1}, \cdots, \nu_m\}$ be a local field of tangent frames and normal frames of N respectively as above. Then we have a local field of frames of νN as follows:*

$$E_i(z) := (e_i)_z^h - (A^u(e_i))_z^v \text{ for } i = 1, \cdots n,$$
$$E_\alpha(z) := (\nu_\alpha)_z^v \text{ for } \alpha = n+1, \cdots, m,$$

where A^u denotes the shape operator at $p \in N$ in M with respect to the normal vector u.

Proof. First, we show that $E_i(z)$ is a tangent vector of νN. Fix a point $z = (p, u) \in \nu N$. Let $c_i : I \to N$ be a smooth curve with $c_i(0) = p$ and $\dot{c}_i = e_i(p)$ where I is an interval containing 0. We can take a curve $\tilde{c}_i(t) := (c_i(t), \nu(c_i(t)))$ on $\nu N \subset TM$ such that $\nu(c_i(t))$ is a parallel transport of the normal vector u along the curve c_i with respect to the normal connection $\nabla^\perp$. Then we have

$$\pi_*(\dot{\tilde{c}}_i(0)) = \frac{d}{dt}\Big|_{t=0} \pi \circ \tilde{c}_i(t) = \dot{c}_i(0) = e_i,$$
$$K(\dot{\tilde{c}}_i(0)) = \frac{d}{dt}\Big|_{t=0} K \circ \tilde{c}_i(t) = \nabla_{e_i}\nu(c_i(t))|_{t=0} = -A^u(e_i).$$

Thus $E_i(z) = \dot{\tilde{c}}_i(0) = (e_i)_z^h - (A^u(e_i))_z^v$ is a tangent vector of νN.

Next, we show $E_\alpha(z)$ is a tangent vector of νN. We take a curve $\tilde{c}_\alpha(t) := (p, t\nu_\alpha + u)$ on $\nu N \subset TM$. Obviously, $\pi_*(\dot{\tilde{c}}_\alpha(0)) = 0$. On the other hand,

$$K(\dot{\tilde{c}}_\alpha(0)) = \frac{d}{dt}\Big|_{t=0} K \circ \tilde{c}_\alpha(t)$$
$$= \frac{d}{dt}\Big|_{t=0} \exp_p(\tau(\tilde{c}_\alpha(t)) - u)$$
$$= \frac{d}{dt}\Big|_{t=0} \exp_p(t\nu_\alpha) = \nu_\alpha.$$

Thus, $E_\alpha(z) = \dot{\tilde{c}}_\alpha(0) = (\nu_\alpha)_z^v$ is a tangent vector of νN.

For the point $z_0 = (p_0, u_0)$, we have

$$\tilde{g}(E_i, E_j)_{z_0} = \langle e_i, e_j \rangle_{p_0} + \langle A^u(e_i), A^u(e_j) \rangle_{p_0} = (1 + \kappa_i^2(z_0))\delta_{ij},$$
$$\tilde{g}(E_i, E_\alpha)_{z_0} = 0,$$
$$\tilde{g}(E_\alpha, E_\beta)_{z_0} = \langle \nu_\alpha, \nu_\beta \rangle_{p_0} = \delta_{\alpha\beta}.$$

These imply that $\{E_i, E_\alpha\}$ are linearly independent around z_0. Thus this gives a local frame of νN.   □

Remark 2.1. We should remark that the vertical parts of the vector fields E_i are *not* vertical lifts for vector fields on N (even though $(e_i)_z^h$ and $(\nu_\alpha)_z^v = E_\alpha$ are defined by the horizontal lift of e_i and the vertical lift of ν_α, respectively). This is crucial to compute the covariant derivative of the frame field $\{E_i, E_\alpha\}$ using the formula (3) through (6). In Lemma 2.1, we define these vectors at each point around z_0. The smoothness of the local frame field $\{E_i, E_\alpha\}$ is shown by using the local expression (1).

Note that by the proof of Lemma 2.1, we have an orthonormal basis of $T_{z_0}\nu N$ by

$$E_i'(z_0) := \frac{1}{\sqrt{1 + \kappa_i^2(z_0)}} E_i(z_0), \quad E_\alpha'(z_0) := E_\alpha(z_0). \tag{7}$$

Recall that the symplectic structure on $(TM, \tilde{g})$ is given by $\omega := \tilde{g}(J\cdot, \cdot)$. Then $\omega(E_i, E_j)_{z_0} = \kappa_i(z_0)\delta_{ij} - \kappa_j(z_0)\delta_{ij} = 0$, and $\omega(E_i, E_\alpha)_{z_0} = \omega(E_\alpha, E_\beta)_{z_0} = 0$ for any $i, j = 1, \cdots, n$ and $\alpha, \beta = n+1, \cdots, m$. This means that *the normal bundle νN is a Lagrangian submanifold in (TM, ω)*.

By the direct computation using the formula (3) through (6), we have the following formula at the point $z_0 = (p_0, u_0)$.

$$(\tilde{\nabla}_{E_i} E_j)(z_0) = \left\{ \nabla_{e_i} e_j - \frac{1}{2}\kappa_i(z_0)R(u_0, e_i)e_j - \frac{1}{2}\kappa_j(z_0)R(u_0, e_j)e_i \right\}_{z_0}^h \tag{8}$$
$$- \left\{ \kappa_j(z_0)\nabla_{e_i} e_j + \frac{1}{2}R(e_i, e_j)u_0 + \sum_{l=1}^n E_i \langle A^u(e_j), e_l \rangle e_l \right\}_{z_0}^v,$$

$$(\tilde{\nabla}_{E_i} E_\alpha)(z_0) = \left\{ \frac{1}{2}R(u_0, \nu_\alpha)e_i \right\}_{z_0}^h + (\nabla_{e_i}\nu_\alpha)_{z_0}^v, \tag{9}$$

$$(\tilde{\nabla}_{E_\alpha} E_i)(z_0) = \left\{ \frac{1}{2}R(u_0, \nu_\alpha)e_i \right\}_{z_0}^h - \left(\sum_{l=1}^n E_\alpha \langle A^u(e_i), e_l \rangle e_l \right)_z^v, \tag{10}$$

$$(\tilde{\nabla}_{E_\alpha} E_\beta)(z_0) = 0, \tag{11}$$

for $i, j = 1, \cdots, n$, and $\alpha, \beta = n+1, \cdots m$, where all the tensor fields and the differentials take value at p_0.

The next equality also follows from a direct computation.

$$(\tilde{\nabla}_{E_i} J) E_i(z_0) = -\frac{1}{2}(\kappa_i^2(z_0) - 1)\{R(u_0, e_i)e_i\}_{z_0}^h \qquad (12)$$
$$+\kappa_i(z_0)\{R(u_0, e_i)e_i\}_{z_0}^v.$$

We denote the second fundamental form of νN by B. Then the mean curvature vector of νN in TM is defined by $H := \mathrm{tr}B$.

Lemma 2.2. *For the local frame* $\{E_i, E_\alpha\}$, *we have the following formula at the point* $z_0 = (p_0, u_0)$,

$$\tilde{g}(JB(E_i, E_i), E_j)(z_0) = E_j\langle A^u(e_i), e_i\rangle(z_0)$$
$$+(\kappa_i(z_0)\kappa_j(z_0) - 1)\langle R(u_0, e_i)e_i, e_j\rangle,$$
$$\tilde{g}(JB(E_i, E_i), E_\alpha)(z_0) = E_\alpha\langle A^u(e_i), e_i\rangle(z_0) - \kappa_i(z_0)\langle R(u_0, e_i)e_i, \nu_\alpha\rangle,$$

for $i, j = 1, \cdots, n$ *and* $\alpha = n+1, \cdots, m$.

Proof. Since νN is Lagrangian, we have

$$\tilde{g}(JB(E_i, E_i), E_\lambda) - \tilde{g}(JB(E_\lambda, E_i), E_i) = \tilde{g}(J(\tilde{\nabla}_{E_i}E_i) - \tilde{\nabla}_{E_i}(JE_i), E_\lambda)$$

for $\lambda = j, \alpha$. It follows from this equation and (12) that

$$\tilde{g}(JB(E_i, E_i), E_j) = \tilde{g}(JB(E_j, E_i), E_i) \qquad (13)$$
$$+ \left\{\frac{1}{2}(\kappa_i^2(z_0) - 1) + \kappa_i(z_0)\kappa_j(z_0)\right\}\langle R(u_0, e_i)e_i, e_j\rangle,$$
$$\tilde{g}(JB(E_i, E_i), E_\alpha) = \tilde{g}(JB(E_\alpha, E_i), E_i) - \kappa_i(z_0)\langle R(u_0, e_i)e_i, \nu_\alpha\rangle. \qquad (14)$$

On the other hand, by a direct computation using (8), (10) and the Lagrangian condition, we have in (13) and (14),

$$\tilde{g}(JB(E_j, E_i), E_i) = E_j\langle A^u(e_i), e_i\rangle - \frac{1}{2}\left\{(1 + \kappa_i^2(z_0))\langle R(u_0, e_i)e_i, e_j\rangle\right\},$$
$$\tilde{g}(JB(E_\alpha, E_i), E_i) = E_\alpha\langle A^u(e_i), e_i\rangle.$$

Thus, we get the desired formula. $\square$

3. On the minimality of normal bundles in the tangent bundle over the complex space form

In what follows, we assume that the ambient space is the complex space form $M = M^m(4c)$ of the holomorphic sectional curvature $4c$ with complex dimension m. Then the curvature tensor of the complex space form $M(4c)$ is given by

$$R(X,Y)Z = c\{\langle Y, Z\rangle X - \langle X, Z\rangle Y + \langle J_0Y, Z\rangle J_0 X$$
$$- \langle J_0 X, Z\rangle J_0 Y + 2\langle X, J_0 Y\rangle J_0 Z\} \tag{15}$$

for $X, Y, Z \in \Gamma(TM)$, where J_0 denotes the complex structure of M. In this section, we give some examples of minimal normal bundles in $(TM, \tilde{g})$. Define the mean curvature form on νN by $\alpha_H := \tilde{g}(JH, \cdot)|_{\nu N}$.

3.1. Complex submanifolds

Lemma 3.1. *Let N^k be a complex submanifold with complex dimension k in $M = M^m(4c)$. Then the mean curvature form of the normal bundle νN in $(TM, \tilde{g})$ is given by*

$$\alpha_H = d\Big(\sum_{i=1}^{2k} \tan^{-1}\kappa_i\Big) - cU\Big(\sum_{i=1}^{2k} \tan^{-1}\kappa_i\Big)U^\flat,$$

where the functions κ_i $(i = 1, \cdots, 2k)$ on νN are the eigenvalues of the shape operator A, and U is the canonical vertical vector field defined by $U_z := u_z^v$, and $U^\flat := \tilde{g}(U, \cdot)|_{\nu N}$.

Proof. We choose a local frame $\{E_i, E_\alpha\}$ around $z_0 \in \nu N$ given in section 2. Recall that this frame gives an orthonormal basis $\{E_i', E_\alpha'\}$ of $T_{z_0}\nu N$ (cf. (7)). We calculate the mean curvature at z_0. First, for a complex submanifold N in the complex space form, we have

$$\langle R(u_0, e_i)e_i, e_j\rangle = 0, \quad \langle R(u_0, e_i)e_i, \nu_\alpha\rangle = c\langle u_0, \nu_\alpha\rangle, \tag{16}$$

since the tangent space of N is J_0-invariant, and M satisfies the curvature condition (15). Since

$$H = \sum_{i=1}^{n} B(E_i', E_i') + \sum_{\alpha=n+1}^{2m} B(E_\alpha', E_\alpha') = \sum_{i=1}^{n} \frac{1}{1 + \kappa_i^2(z_0)} B(E_i, E_i),$$

where we denote $n = 2k$, we find by using Lemma 2.2 and (16)

$$\alpha_H(E_j) = \tilde{g}(JH, E_j) = \sum_{i=1}^{n} \frac{1}{1 + \kappa_i^2(z_0)} \tilde{g}(JB(E_i, E_i), E_j)$$

$$= \sum_{i=1}^{n} \frac{1}{1 + \kappa_i^2(z_0)} E_j\langle A^u(e_i), e_i\rangle = E_j\Big(\sum_{i=1}^{n} \tan^{-1}\langle A^u(e_i), e_i\rangle\Big).$$

One can show that the sum of arctangents is written by the elementary symmetric polynomials with respect to $\{\langle A^u(e_i), e_i \rangle\}_i$ (cf. [5]). Thus the value of the (local) function $\sum_{i=1}^{n} \tan^{-1}\langle A^u(e_i), e_i \rangle$ is independent of the choice of the local orthonomal frame $\{e_i\}$ (up to constant factor). In particular, we have,

$$\alpha_H(E_j) = E_j\Big(\sum_{i=1}^{n} \tan^{-1}\langle A^u(e_i), e_i \rangle \Big) = E_j\Big(\sum_{i=1}^{n} \tan^{-1}\kappa_i \Big).$$

By a similar calculation, we have

$$\alpha_H(E_\alpha) = E_\alpha\Big(\sum_{i=1}^{n} \tan^{-1}\kappa_i \Big) - c\Big(\sum_{i=1}^{n} \frac{\kappa_i(z_0)}{1 + \kappa_i^2(z_0)} \Big)\langle u_0, \nu_\alpha \rangle.$$

Thus, we obtain

$$\alpha_H = d\Big(\sum_{i=1}^{n} \tan^{-1}\kappa_i \Big) - c\Big(\sum_{i=1}^{n} \frac{\kappa_i(z_0)}{1 + \kappa_i^2(z_0)} \Big)\Big(\sum_{\alpha=n+1}^{n} \langle u_0, \nu_\alpha \rangle dE_\alpha \Big).$$

We note that $\langle u_0, \nu_\alpha \rangle = \tilde{g}(U, E_\alpha)$, and hence

$$\sum_{\alpha=n+1}^{2m} \langle u_0, \nu_\alpha \rangle dE_\alpha = U^\flat.$$

On the other hand, we have

$$U\Big(\sum_{i=1}^{n} \tan^{-1}\kappa_i \Big)(z_0) = \sum_{i=1}^{n} \frac{d}{dt}\Big|_{t=0} \tan^{-1}\kappa_i(p_0, (t+1)u_0)$$

$$= \sum_{i=1}^{n} \frac{\kappa_i(z_0)}{1 + \kappa_i^2(z_0)}, \tag{17}$$

where we use $\kappa_i(p, tu) = t\kappa_i(p, u)$ for $(p, u) \in \nu N$ and $t \in \mathbb{R}$. $\square$

By virtue of this Lemma, we have the following:

Theorem 3.1. *Let N be a complex submanifold in the complex space form $M = M^m(4c)$. Then the normal bundle νN is a minimal Lagrangian submanifold in $(TM, \tilde{g})$.*

Proof. Since any complex submanifold is austere (i.e., for any normal vector u, the set of eigenvalues of the shape operator A^u is invariant under multiplication of -1), we have $\sum_{i=1}^{2k} \tan^{-1}\kappa_i = 0$. Thus, by Lemma 3.1, $\alpha_H = 0$ follows. $\square$

Remark 3.1. In the case $c = 0$, the statement of Theorem 3.1 has already appeared in [4].

3.2. *Totally geodesic submanifolds*

In this subsection, we investigate the simplest case, namely, totally geodesic case. First, we discuss it in a general setting.

Let N^n be a totally geodesic submanifold in a Riemannian manifold $(M^m, \langle,\rangle)$. Then, by using the formula (8) through (11), we can easily compute the second fundamental form of the normal bundle νN in $(TM, \tilde{g})$. For simplicity, we set $S_{\lambda\mu\zeta} := \tilde{g}(JB(E_\lambda, E_\mu), E_\zeta)$ for $\lambda, \mu, \zeta = 1, \cdots, m$. For $i, j, k = 1, \cdots, n$, we see

$$S_{ijk}(z_0) = \frac{1}{2}\langle R(e_i, e_j)u_0, e_k\rangle = -\frac{1}{2}\langle R(e_i, e_j)e_k, u_0\rangle$$
$$= -\frac{1}{2}\langle (\nabla^\perp_{e_i} B')(e_j, e_k) - (\nabla^\perp_{e_j} B')(e_i, e_k), u_0\rangle = 0,$$

where $B'(\equiv 0)$ is the second fundamental form of N in M, and we use the Codazzi equation. For $i = 1, \cdots, n$ and $\alpha, \beta = n+1, \cdots, m$, we get

$$S_{i\alpha\beta}(z_0) = \frac{1}{2}\langle R(u_0, \nu_\alpha)e_i, \nu_\beta\rangle. \tag{18}$$

Otherwise, we obtain $S_{\lambda\mu\zeta} = 0$. In particular, $S_{iij} = S_{iia} = S_{\alpha\alpha i} = S_{\alpha\alpha\beta} = 0$ holds at arbitrary point z_0, and hence, the following proposition follows:

Proposition 3.1. *Let N^n be a totally geodesic submanifold in a Riemannian manifold $(M^m, \langle,\rangle)$. Then the normal bundle νN is a minimal Lagrangian submanifold in $(TM, \tilde{g})$.*

In the case of real space forms, if N is totally geodesic, then νN is also totally geodesic, and the converse is true (cf. [5], we remark that this includes the case $\mathbb{C}^n \simeq \mathbb{R}^{2n}$). However, this is not the case in general.

Lemma 3.2. *Let N^n be a submanifold in a Riemannian manifold $(M^m, \langle,\rangle)$. If the normal bundle νN is totally geodesic in $(TM, \tilde{g})$, then for each $z_0 = (p_0, u_0) \in \nu N$, and each principal curvature $\kappa_i(z_0)$ $(i = 1, \cdots, n)$, one of the following holds:*

(1) $\kappa_i(z_0) = 0$, *or*
(2) $\kappa_i(z_0) \neq 0$ *and* $R(u_0, e_i)e_i(p_0) = 0$,

where e_i is the principal direction with respect to $\kappa_i(z_0)$.

Proof. Assume νN is totally geodesic. Then, by (13) and (14), we have

$$\left\{\frac{1}{2}(\kappa_i^2(z_0) - 1) + \kappa_i(z_0)\kappa_j(z_0)\right\}\langle R(u_0, e_i)e_i, e_j\rangle = 0, \tag{19}$$
$$\kappa_i(z_0)\langle R(u_0, e_i)e_i, \nu_\alpha\rangle = 0, \tag{20}$$

for $i, j = 1, \cdots, n$ and $\alpha = n + 1, \cdots, m$. If $u_0 = 0$, then we have $\kappa_i(z_0) = 0$. Thus, we assume $u_0 \neq 0$. Then, for any $t \in \mathbb{R}$, every e_i is the principal direction with respect to the point (p_0, tu_0), i.e., $A^{tu_0}(e_i) = t\kappa_i(z_0)e_i$ for $i = 1, \cdots, n$. Hence, from (17), we have $t\{(t^2\kappa_i^2(z_0) - 1) + 2t^2\kappa_i(z_0)\kappa_j(z_0)\}\langle R(u_0, e_i)e_i, e_j \rangle = 0$. By this identity, we obtain

$$\langle R(u_0, e_i)e_i, e_j \rangle = 0 \tag{21}$$

for $i, j = 1, \cdots, n$. From (20) and (21), the statement follows. $\qquad\square$

We return to the case of the complex space forms.

Theorem 3.2. *Let N^n be a submanifold in $M = \mathbb{C}P^m(4c)$ (resp. $\mathbb{C}H^m(4c)$). Then the normal bundle νN is totally geodesic in $(TM, \tilde{g})$ if and only if N is locally congruent to the totally geodesic complex submanifold $\mathbb{C}P^n$ (resp. $\mathbb{C}H^n$) where $n = 1, \cdots, m - 1$, or the totally geodesic Lagrangian submanifold $\mathbb{R}P^m$ (resp. $\mathbb{R}H^m$).*

Proof. We prove the case $M = \mathbb{C}P^m(4c)$ (the case where $\mathbb{C}H^m(4c)$ is similar). Assume that νN is totally geodesic. Then, we shall show that N must be totally geodesic in M. By the curvature condition (15), we have

$$R(u_0, e_i)e_i = c(u_0 + 3\langle u_0, J_0 e_i \rangle J_0 e_i).$$

If $\kappa_i(z_0) \neq 0$, we have $u_0 + 3\langle u_0, J_0 e_i \rangle J_0 e_i = 0$ by Lemma 3.2. However, this occurs only in the case $u_0 = 0$, and this is a contradiction. Thus we have $\kappa_i(z_0) = 0$ for $i = 1, \cdots, n$ and $z_0 \in \nu N$, and hence, N is totally geodesic in M.

Conversely, assume that N is a totally geodesic submanifold in M. It is well-known that N is (locally) congruent to the complex submanifold $\mathbb{C}P^n$ ($n = 1, \cdots, m-1$) or the totally real submanifold $\mathbb{R}P^n$ ($n = 1, \cdots, m$). If N is a complex submanifold or a Lagrangian submanifold in $\mathbb{C}P^m$, we have $\langle R(u_0, \nu_\alpha)e_i, \nu_\beta \rangle = 0$ in (18). Thus, for $N = \mathbb{C}P^n$ or $\mathbb{R}P^m$, νN is totally geodesic. However, if N is the totally real submanifold $\mathbb{R}P^n$ ($n = 1, \cdots, m - 1$), then we can take the normal orthonormal basis at p_0 so that $\nu_{n+i} := J_0 e_i$ for $i = 1, \cdots, n$ and other ν_α are orthogonal to $\{J_0 e_i\}_i$. Then we have $\langle R(u_0, \nu_{n+i})e_i, \nu_\beta \rangle = c\langle u_0, J\nu_\beta \rangle$ for $i = 1, \cdots, n$, and $\beta = n + 1, \cdots, m$, and we can choose ν_β so that $\langle u_0, J\nu_\beta \rangle \neq 0$. Therefore, we have $S_{i\alpha\beta} \neq 0$ for some points, and hence, νN is not totally geodesic. $\qquad\square$

3.3. *Hopf hypersurfaces*

Let N^{2m-1} be a Hopf hypersurface in $M = M^m(4c)$, i.e., a real hypersurface of M such that the characteristic vector field $\xi := -J_0\nu$ is a principal direction of N (where J_0 is the complex structure of M and ν is a local normal vector of N in M). A real hypersurface in the complex space form M admits a contact structure induced from the Kähler structure of M. More precisely, we define the contact 1-form η by $\eta := \langle \xi, \cdot \rangle|_N$. Then $\mathrm{Ker}\eta$ is a $2m - 2$ dimensional distribution.

For the Hopf hypersurface N, we can take the local orthonormal tangent frame $\{e_1, \cdots, e_{2m-1}\}$ in such a way that $e_{2m-1} = \xi$, $\mathrm{Ker}\eta$ is spanned by $\{e_1, \cdots, e_{2m-2}\}$ around p_0 and $A(e_i)(p_0) = \kappa_i(z_0)e_i(p_0)$ for $i = 1, \cdots, 2m - 1$. Then, by the curvature condition (15), we find

$$\langle R(u_0, e_i)e_i, e_j \rangle = 0, \text{ for } i, j = 1, \cdots, 2m - 1,$$
$$\langle R(u_0, e_i)e_i, \nu \rangle = c\langle u_0, \nu \rangle, \text{ for } i = 1, \cdots, 2m - 2,$$
$$\langle R(u_0, \xi)\xi, \nu \rangle = 4c\langle u_0, \nu \rangle.$$

From these equalities, we can derive the mean curvature formula for the normal bundle over the Hopf hypersurface as follows. The proof is similar to that of Lemma 3.1, and we omit it.

Proposition 3.2. *Let N^{2m-1} be a Hopf hypersurface in the complex space form $M = M^m(4c)$. Then the mean curvature form of νN in $(TM, \tilde{g})$ is given by*

$$\alpha_H = d\left(\sum_{i=1}^{2m-1} \tan^{-1}\kappa_i \right) - c\left\{ U\left(\sum_{i=1}^{2m-1} \tan^{-1}\kappa_i \right) + \frac{3\kappa_{2m-1}}{1 + \kappa_{2m-1}^2} \right\} U^\flat,$$

where the functions κ_i ($i = 1, \cdots, 2m-1$) on νN are the eigenvalues of the shape operator A, κ_{2m-1} is the eigenvalue with respect to the direction ξ, and U is the canonical vertical vector field defined by $U_z := u_z^v$, and $U^\flat := \tilde{g}(U, \cdot)|_{\nu N}$.

For Hopf hypersurfaces with constant principal curvatures, we obtain the following result.

Theorem 3.3. *Let N^{2m-1} be a Hopf hypersurface with constant principal curvatures in the non-flat complex space form $M = M^m(4c)$. Then*

(1) if $c > 0$, νN is a minimal Lagrangian submanifold in $(TM, \tilde{g})$ if and only if m is odd and N is the tube of radius $\pi/(4\sqrt{c})$ over totally geodesic $\mathbb{C}P^k$ in $\mathbb{C}P^{2k+1}(4c)$.

(2) if $c < 0$, there exist no Hopf hypersurface with constant principal curvatures whose normal bundle is minimal in $(TM, \tilde{g})$.

Proof. (1) *The case $c > 0$.* By Proposition 3.2, the normal bundle νN is minimal in $(TM, \tilde{g})$ if and only if the following equality holds:

$$d\theta = c\{U(\theta) + 3\alpha\}U^\flat, \tag{22}$$

where $\theta := \sum_{i=1}^{2m-1} \tan^{-1} \kappa_i$, and $\alpha := \kappa_{2m-1}/(1 + \kappa_{2m-1}^2)$. If N is the tube of radius $\pi/(4\sqrt{c})$ over totally geodesic $\mathbb{C}P^k$ in $\mathbb{C}P^{2k+1}(4c)$, then we have $\theta = \kappa_{2m-1} = 0$ on νN, and hence, the equality (22) is satisfied. Conversely, assume the equality (22) holds. Substituting the canonical vertical vector U to (22), we have

$$(1 - c|u|^2)U(\theta) = 3c\alpha|u|^2, \tag{23}$$

for any normal vector u. Since $c > 0$, we can take a normal vector u_0 so that $|u_0|^2 = 1/c$. Then we have $\alpha(u_0) = 0$, and hence $\kappa_{2m-1}(p_0, u_0) = 0$. This implies that $\kappa_{2m-1}(p_0, u) = 0$ for any $u \in \nu_{p_0} N$. Since p_0 is an arbitrary point, we have $\kappa_{2m-1} = 0$ on νN. By [2], such a Hopf hypersurface is locally congruent to the tube of radius $\pi/(4\sqrt{c})$ over a complex submanifold in $M = \mathbb{C}P^m(4c)$. From this investigation, using the classification theorem of Hopf hypersurfaces with constant principal curvatures in $\mathbb{C}P^m$ (cf. [6]), one can easily check that the tube of radius $\pi/(4\sqrt{c})$ over $\mathbb{C}P^k$ in $\mathbb{C}P^{2k+1}$ only satisfies the equality (22).

(2) *The case $c < 0$.* Assume the equality (22) holds. Then by the equalities (17) and (23), we have,

$$\sum_{i=1}^{2m-1} \frac{\kappa_i(z_0)}{1 + \kappa_i^2(z_0)} = \frac{3c}{1 - c|u|^2} \cdot \frac{\kappa_{2m-1}(z_0)}{1 + \kappa_{2m-1}^2(z_0)}. \tag{24}$$

By the classification theorem (cf. [6]), we can show that all principal curvatures have the same sign. However, since $c < 0$, the equality (24) holds if and only if both sides of (24) vanish. Then one can easily check that this dose not occur. $\qquad\square$

Remark 3.2. By Theorem 3.3, we see that the minimality of the submanifold N in M does not imply the minimality of the normal bundle in general. However, one can show that the minimality of the normal bundle over a Hopf hypersurface is equivalent to the austere condition of the hypersurface if it satisfies $\kappa_{2m-1} = 0$ (cf. [5]). In fact, the tube of radius $\pi/(4\sqrt{c})$ over $\mathbb{C}P^k$ in $\mathbb{C}P^{2k+1}(4c)$ is the only austere Hopf hypersurface with constant principal curvatures and $\kappa_{2m-1} = 0$.

By virtue of the above results, it is natural to ask the following question: *Which submanifolds in the non-flat complex space form $M = M^m(4c)$ have* *minimal normal bundles? We note that all examples obtained in this article are austere.* By using lemmas in the section 2, one can derive the local expression of the mean curvature form for any submanifold N in M.

Acknowledgment

This work has been done while the author was staying at the King's College London by the "Strategic Young Researcher Overseas Visits Program for Accelerating Brain Circulation by JSPS". He expresses his sincere thanks to the hospitality of the college. He also thanks to Professors Jürgen Berndt and Reiko Miyaoka for valuable discussion and helpful comments. The author is partially supported by Grant-in-Aid for JSPS Fellows.

References

1. D. E. BLAIR, *Riemannian geometry of contact and symplectic manifolds*, Progress in Math. 203, Birkhäuser, Inc., Boston, MA, 2002.
2. T. E. CECIL AND P. J. RYAN, *Focal sets and real hypersurfaces in complex space projective space*, Trans. Am. Math. Soc. 269, 481–498 (1982).
3. S. GUDMUNDSSON AND E. KAPPOS, *On the Geometry of Tangent bundles*, Expo. Math. 20 (2002): 1–41.
4. R. HARVEY AND H. B. LAWSON, *Calibrated geometry*, Acta Math. **148** (1982), 47–157.
5. T. KAJIGAYA, *On the minimality of normal bundles and austere submanifolds*, in preparation.
6. R. NIEBERGALL AND P. J. RYAN, *Real hypersurfaces in complex space forms*, Tight and Taut submanifolds, Vol. 32, 1997, 233–305.

Received March 31, 2013.

273

Proceedings of the Workshop on
Differential Geometry of Submanifolds
and its Related Topics
Saga, August 4-6, 2012

OVER-DETERMINED SYSTEMS ON SURFACES

Naoya ANDO

*Graduate School of Science and Technology, Kumamoto University,
2-39-1 Kurokami, Kumamoto 860-8555 Japan
E-mail: ando@sci.kumamoto-u.ac.jp*

Dedicated to Professor Sadahiro Maeda for his sixtieth birthday

From the equations of Gauss and Codazzi-Mainardi, we can obtain an over-determined system on a surface in E^3 with no umbilical point and nowhere zero Gaussian curvature. The set of the solutions of this system is determined by the first fundamental form and principal distributions. In this paper, we will survey results with respect to over-determined systems on surfaces.

Keywords: semisurface, canonical pre-divergence, Codazzi-Mainardi polynomial, principal distribution, equation of Gauss, equations of Codazzi-Mainardi, over-determined system, compatibility condition, sinh-Gordon equation.

1. Introduction

The equations of Gauss and Codazzi-Mainardi give relations among the first fundamental form, principal distributions and principal curvatures of a surface, and these relations determine the shape of each surface in the space. Principal curvatures are almost determined by the first fundamental form and principal distributions: for a surface S in E^3 with no umbilical point and nowhere zero Gaussian curvature, a pair (k_1, k_2) of two principal curvatures of S is a zero point of the Codazzi-Mainardi polynomial; if S is oriented, then this polynomial is determined by the first fundamental form and principal distributions at each point of S ([3, 6]). Based on these, *we consider S to be a two-dimensional Riemannian manifold (M, g) with nowhere zero curvature and an orthogonal pair of two one-dimensional distributions $\mathcal{D}_1$, $\mathcal{D}_2$ on M such that $(\mathcal{D}_1, \mathcal{D}_2)$ is connected with the metric g by some good relation.*

In [6], the author studied the geodesic curvatures of lines of curvature on a surface in a 3-dimensional space form, in order to discuss the relation between the induced metric and a pair of principal distributions. The author described the equations of Gauss and Codazzi-Mainardi for a surface without any umbilical point, as relations between a pair of two principal curvatures and the canonical pre-divergence, which is a vector field on the surface defined by the sum of the geodesic curvature vectors of the two lines of curvature through each point and determined by the first fundamental form and principal distributions. The Codazzi-Mainardi polynomial can be represented by the geodesic curvatures of lines of curvature. The author characterized surfaces with constant mean curvature and surfaces with constant curvature, in terms of the geodesic curvatures of lines of curvature.

From the equations of Gauss and Codazzi-Mainardi, we can obtain over-determined systems on surfaces in E^3 with no umbilical point and nowhere zero Gaussian curvature. The purpose of the present paper is to survey results with respect to over-determined systems on surfaces. Whether an over-determined system on a surface as above has a solution is determined by the first fundamental form g and principal distributions $\mathcal{D}_1$, $\mathcal{D}_2$. In addition, the set of the solutions of the system is determined by g and $\mathcal{D}_1$, $\mathcal{D}_2$. We can find an over-determined system of the same type on a two-dimensional Riemannian manifold (M, g) with nowhere zero curvature and an orthogonal pair of two one-dimensional distributions $\mathcal{D}_1$, $\mathcal{D}_2$ on M. Then the system does not necessarily have any solution. *The condition that the system has a solution corresponds to a good relation between g and $(\mathcal{D}_1, \mathcal{D}_2)$. Each solution of the system corresponds to a surface such that g and $(\mathcal{D}_1, \mathcal{D}_2)$ are the first fundamental form and a pair of principal distributions, respectively.* We can obtain a necessary and sufficient condition for the existence of a solution of the system and this condition is represented by quantities in relation to g and $(\mathcal{D}_1, \mathcal{D}_2)$ ([8]).

If the system has a unique solution, then g and $(\mathcal{D}_1, \mathcal{D}_2)$ determine a unique surface up to an isometry of E^3. If the system satisfies the compatibility condition, then the system has a unique solution for an arbitrarily given initial value and the surface corresponding to each solution is said to be *molding*. Any molding surface has a family of lines of curvature which consists of geodesics ([12, pp. 152–153], [11, pp. 277–281], [7]). Therefore referring to [3], we see that some neighborhood of each point of a molding surface S is a canonical parallel curved surface and this means that we can grasp the shape of S in detail (see [1, 2]). If the system does not satisfy the compatibility condition, then the system has at most two solutions. We can represent a necessary and sufficient condition for the existence of just two solutions by the sinh-Gordon equation ([8]).

Several properties of an over-determined system on a surface S in E^3 with no umbilical point and nowhere zero Gaussian curvature are determined by principal distributions $\mathcal{D}_1$, $\mathcal{D}_2$ and some 1-form ω on S. We can represent the system on S in terms of $(\omega, \mathcal{D}_1, \mathcal{D}_2)$. Then referring to [8], we see that the system has just two solutions if and only if the system appears on a surface in E^3 with nonzero constant mean curvature which is not part of any surface of revolution. We can characterize an over-determined system on a minimal surface in E^3 which is not part of any catenoid ([9]).

This work was supported by Grant-in-Aid for Scientific Research (No. 24740048), Japan Society for the Promotion of Science.

2. Semisurfaces

Let (M, g) be a two-dimensional Riemannian manifold and $\mathcal{D}_1$, $\mathcal{D}_2$ one-dimensional distributions on M. We call $(M, g, \mathcal{D}_1, \mathcal{D}_2)$ a *semisurface* if $\mathcal{D}_1$ is orthogonal to $\mathcal{D}_2$ with respect to g at any point of M. Let $(M, g, \mathcal{D}_1, \mathcal{D}_2)$ be a semisurface. Then a triplet $(g, \mathcal{D}_1, \mathcal{D}_2)$ is called a *semisurface structure* of M. The *canonical pre-divergence* V_K of $(M, g, \mathcal{D}_1, \mathcal{D}_2)$ is a vector field on M defined by the sum of the geodesic curvature vectors of the two integral curves of $\mathcal{D}_1$ and $\mathcal{D}_2$ through each point of M. For each point p of M, there exist smooth unit vector fields $\boldsymbol{U}_1$, $\boldsymbol{U}_2$ on a neighborhood of p satisfying $\boldsymbol{U}_i \in \mathcal{D}_i$ for $i = 1, 2$. Then V_K is given by

$$V_K = \nabla_{\boldsymbol{U}_1} \boldsymbol{U}_1 + \nabla_{\boldsymbol{U}_2} \boldsymbol{U}_2,$$

where ∇ is the covariant differentiation with respect to the Levi-Civita connection of a Riemannian manifold (M, g). For a vector field V on M, let $\mathrm{div}(V)$ denote the divergence of V with respect to ∇: $\mathrm{div}(V)$ is the trace of a tensor field τ_V on M of type $(1, 1)$ defined by $\tau_V(w) := \nabla_w V$ for each tangent vector w at each point of M. The following holds ([4, 6]):

Proposition 2.1. *The divergence* $\mathrm{div}(V_K)$ *of* V_K *with respect to* ∇ *is equal to the curvature* K *of* (M, g).

Let (u, v) be local coordinates compatible with $(\mathcal{D}_1, \mathcal{D}_2)$, that is, suppose that (u, v) satisfy $\partial/\partial u \in \mathcal{D}_1$, $\partial/\partial v \in \mathcal{D}_2$. Then g is locally represented as $g = A^2 du^2 + B^2 dv^2$, where A, B are positive-valued functions. Suppose that there exists a number L_0 such that the curvature K of (M, g) satisfies

$K \neq L_0$ on M. We set

$$c_{20} := \frac{1}{AB}\left\{\left(\log|K-L_0|A^2\right)_v(\log B)_u - (\log B)_{uv}\right\},$$

$$c_{11} := \frac{1}{AB}\left\{(\log|K-L_0|AB)_{uv} - 4(\log A)_v(\log B)_u\right\}, \qquad (1)$$

$$c_{02} := \frac{1}{AB}\left\{\left(\log|K-L_0|B^2\right)_u(\log A)_v - (\log A)_{uv}\right\}.$$

If M is oriented, then c_{20}, c_{11}, c_{02} are determined by the semisurface structure $(g, \mathcal{D}_1, \mathcal{D}_2)$ and do not depend on the choice of (u, v) which give the positive orientation of M. We set

$$P_{\mathrm{CM}}(X_1, X_2) := c_{20}X_1^2 + c_{11}X_1X_2 + c_{02}X_2^2.$$

We call P_{CM} the *Codazzi-Mainardi polynomial* of $(M, g, \mathcal{D}_1, \mathcal{D}_2)$ for the number L_0.

Remark 2.1. We can represent c_{20}, c_{11}, c_{02} by the geodesic curvatures of integral curves of $\mathcal{D}_1$, $\mathcal{D}_2$. Let $\boldsymbol{U}_1$, $\boldsymbol{U}_2$ be smooth unit vector fields on a neighborhood of each point of M satisfying $\boldsymbol{U}_i \in \mathcal{D}_i$. Then there exist functions l_1, l_2 satisfying

$$\nabla_{\boldsymbol{U}_1}\boldsymbol{U}_1 = l_1\boldsymbol{U}_2, \qquad \nabla_{\boldsymbol{U}_2}\boldsymbol{U}_2 = -l_2\boldsymbol{U}_1,$$

respectively. Then l_i is the geodesic curvature of each integral curve of $\mathcal{D}_i$. The following hold ([6]):

$$c_{20} = l_2\boldsymbol{U}_2(\log|K-L_0|) - \boldsymbol{U}_2(l_2) - l_1l_2,$$

$$c_{11} = \frac{1}{2}\{\boldsymbol{U}_1\boldsymbol{U}_2(\log|K-L_0|) + \boldsymbol{U}_2\boldsymbol{U}_1(\log|K-L_0|)$$

$$- l_1\boldsymbol{U}_1(\log|K-L_0|) + l_2\boldsymbol{U}_2(\log|K-L_0|)\}$$

$$- \boldsymbol{U}_1(l_1) + \boldsymbol{U}_2(l_2) + 2l_1l_2,$$

$$c_{02} = -l_1\boldsymbol{U}_1(\log|K-L_0|) + \boldsymbol{U}_1(l_1) - l_1l_2.$$

3. The fundamental equations of surfaces

Let M be a two-dimensional manifold and $\iota : M \longrightarrow N$ an immersion of M into a 3-dimensional space form N with constant sectional curvature L_0. Suppose that M has no umbilical point with respect to ι. Let g be the metric on M induced by ι. Let $\mathcal{D}_1$, $\mathcal{D}_2$ be principal distributions on M with respect to ι which give the two principal directions of M at each point. Then g, $\mathcal{D}_1$, $\mathcal{D}_2$ form a semisurface structure of M. Let k_1, k_2 be principal curvatures of M with respect to ι corresponding to $\mathcal{D}_1$, $\mathcal{D}_2$, respectively. Then we obtain the following ([6]):

Theorem 3.1. *The equation of Gauss is represented as*

$$\operatorname{div}(V_K) = L_0 + k_1 k_2; \tag{2}$$

the equations of Codazzi-Mainardi are represented as

$$-\boldsymbol{U}_1(k_2)\boldsymbol{U}_1 + \boldsymbol{U}_2(k_1)\boldsymbol{U}_2 = (k_1 - k_2)V_K, \tag{3}$$

where V_K is the canonical pre-divergence of $(M, g, \mathcal{D}_1, \mathcal{D}_2)$ and $\boldsymbol{U}_1, \boldsymbol{U}_2$ are smooth unit vector fields on a neighborhood of each point of M satisfying $\boldsymbol{U}_i \in \mathcal{D}_i$ for $i = 1, 2$.

Let (u, v) be local coordinates compatible with $(\mathcal{D}_1, \mathcal{D}_2)$. Then g is locally represented as $g = A^2 du^2 + B^2 dv^2$. We can rewrite (3) into

$$\begin{aligned}
(k_2)_u &= (\log B)_u(k_1 - k_2), \\
(k_1)_v &= -(\log A)_v(k_1 - k_2).
\end{aligned} \tag{4}$$

Suppose that the curvature K of (M, g) is nowhere equal to L_0 and set

$$\begin{aligned}
\alpha &:= 2(\log |K - L_0|B)_u, & \beta &:= -2\frac{(\log B)_u}{K - L_0}, \\
\gamma &:= -2(\log A)_v, & \delta &:= 2(K - L_0)(\log A)_v.
\end{aligned} \tag{5}$$

Then from $\operatorname{div}(V_K) = K$, (2) and (4), we obtain

Proposition 3.1. *A function $F := \log k_1^2$ is a solution of the following over-determined system*

$$F_u = \alpha + \beta e^F, \qquad F_v = \gamma + \delta e^{-F}, \tag{6}$$

where α, β, γ, δ are as in (5).

Let $(M, g, \mathcal{D}_1, \mathcal{D}_2)$ be a semisurface and K the curvature of (M, g). Suppose that a number L_0 satisfies $K \neq L_0$ on M. Let (u, v) be local coordinates compatible with $(\mathcal{D}_1, \mathcal{D}_2)$ on a neighborhood of each point of M. We locally represent g as $g = A^2 du^2 + B^2 dv^2$. Let α, β, γ, δ be as in (5). Then whether an over-determined system (6) with these α, β, γ, δ has a solution is determined by the semisurface structure $(g, \mathcal{D}_1, \mathcal{D}_2)$ and does not depend on the choice of local coordinates (u, v). In addition, the set of solutions of (6) with (5) is determined by $(g, \mathcal{D}_1, \mathcal{D}_2)$ and does not depend on the choice of (u, v). Suppose that (6) with (5) has a solution F. Then if we set

$$k_1 := e^{F/2}, \qquad k_2 := \frac{K - L_0}{k_1},$$

then k_1, k_2 satisfy (2) and (3). Therefore from the fundamental theorem of the theory of surfaces, we obtain

Theorem 3.2. *Let* $(M, g, \mathcal{D}_1, \mathcal{D}_2)$ *be a semisurface and* K *the curvature of* (M, g). *Suppose that a number* L_0 *satisfies* $K \neq L_0$ *on* M *and that* (6) *with* (5) *has a solution* F *on a neighborhood of each point* p *of* M. *Then there exists an isometric immersion of a neighborhood of* p *into* N, *unique up to an isometry of* N, *satisfying*

(i) $\mathcal{D}_1$, $\mathcal{D}_2$ *give principal distributions;*
(ii) $k_1 := e^{F/2}$ *and* $k_2 := (K - L_0)/k_1$ *are principal curvatures corresponding to* $\mathcal{D}_1$, $\mathcal{D}_2$, *respectively.*

In addition, we can obtain the following ([8]):

Theorem 3.3. *Let* $(M, g, \mathcal{D}_1, \mathcal{D}_2)$ *be a semisurface and* K *the curvature of* (M, g). *Suppose that a number* L_0 *satisfies* $K \neq L_0$ *on* M. *Then for each* $p \in M$, *the following are mutually equivalent:*

(i) *there exists an isometric immersion of a neighborhood of* p *into* N *such that* $\mathcal{D}_1$, $\mathcal{D}_2$ *give principal distributions;*
(ii) *an over-determined system* (6) *with* (5) *has a solution on a neighborhood* U *of* p;
(iii) *we can choose local coordinates* (u, v) *compatible with* $(\mathcal{D}_1, \mathcal{D}_2)$ *on* U *satisfying*

$$(K - L_0)^2 A^2 B^2$$
$$= \left(\int_{v_0}^{v} (K - L_0)(A^2)_v dv + 1 \right) \tag{7}$$
$$\times \left(\int_{u_0}^{u} (K - L_0)(B^2)_u du + 1 \right),$$

where a point (u_0, v_0) *corresponds to* p.

Remark 3.1. From discussions in the present section, we see that condition (i) in Theorem 3.3 is equivalent to condition (ii) in Theorem 3.3. We see that (7) is represented by quantities in relation to $(g, \mathcal{D}_1, \mathcal{D}_2)$.

Let F be a solution of (6) with (5). Then by $F_{uv} = F_{vu}$, we obtain the following ([3, 6]):

Theorem 3.4. *Let* M, ι, g, $\mathcal{D}_i$, k_i *be as in the beginning of the present section. Then*

$$P_{\mathrm{CM}}(k_1, k_2) = 0,$$

where P_{CM} *is the Codazzi-Mainardi polynomial of* $(M, g, \mathcal{D}_1, \mathcal{D}_2)$ *for the number* L_0.

4. The compatibility condition

Let Φ, Ψ be functions of u, v, w on a domain D of $\mathbf{R}^3$. Let F be a function of u, v. Suppose that F is a solution of an over-determined system

$$F_u = \Phi(u, v, F), \qquad F_v = \Psi(u, v, F). \tag{8}$$

Then by $F_{uv} = F_{vu}$, we obtain

$$\begin{aligned} \Phi_v(u, v, F) + \Phi_w(u, v, F)\Psi(u, v, F) \\ = \Psi_u(u, v, F) + \Psi_w(u, v, F)\Phi(u, v, F). \end{aligned} \tag{9}$$

The *compatibility condition* of (8) is given by

$$\Phi_v + \Phi_w\Psi \equiv \Psi_u + \Psi_w\Phi$$

on D. The following theorem is known (for example, see [14, p. 393]):

Theorem 4.1. *If (8) satisfies the compatibility condition, then for each* $(u_0, v_0, w_0) \in D$, *there exists a unique solution* F *of (8) on a neighborhood of* (u_0, v_0) *in* $\mathbf{R}^2$ *satisfying* $F(u_0, v_0) = w_0$.

Let $(M, g, \mathcal{D}_1, \mathcal{D}_2)$ be a semisurface such that the curvature K of (M, g) satisfies $K \neq L_0$ on M for some number L_0, and α, β, γ, δ functions of u, v as in (5). We set

$$\Phi(u, v, w) := \alpha(u, v) + \beta(u, v)e^w,$$
$$\Psi(u, v, w) := \gamma(u, v) + \delta(u, v)e^{-w}.$$

Then (8) is represented as (6) and the compatibility condition $\Phi_v + \Phi_w\Psi \equiv \Psi_u + \Psi_w\Phi$ is represented as

$$Xe^w + Y + Ze^{-w} \equiv 0,$$

where

$$X := \beta_v + \beta\gamma, \quad Y := \alpha_v - \gamma_u + 2\beta\delta, \quad Z := -\delta_u + \alpha\delta. \tag{10}$$

Since

$$c_{20} = \frac{K - L_0}{2AB}X, \quad c_{11} = \frac{1}{2AB}Y, \quad c_{02} = \frac{1}{2AB(K - L_0)}Z,$$

where c_{20}, c_{11}, c_{02} are as in (1), we obtain

Proposition 4.1. *The following are mutually equivalent:*

(i) *(6) satisfies the compatibility condition;*
(ii) *all of* $X \equiv 0$, $Y \equiv 0$ *and* $Z \equiv 0$ *hold;*
(iii) *the Codazzi-Mainardi polynomial of* $(M, g, \mathcal{D}_1, \mathcal{D}_2)$ *vanishes.*

Remark 4.1. From Proposition 4.1, we see that whether (6) satisfies the compatibility condition is determined by the semisurface structure $(g, \mathcal{D}_1, \mathcal{D}_2)$ and does not depend on the choice of local coordinates (u, v).

Since (9) with $\Phi = \alpha + \beta e^w$, $\Psi = \gamma + \delta e^{-w}$ is represented as $Xe^F + Y + Ze^{-F} = 0$, we obtain

Proposition 4.2. *If* (6) *does not satisfy the compatibility condition, then* (6) *has at most two solutions.*

Let S be a surface in E^3 with no umbilical point and nowhere zero Gaussian curvature. Then S is said to be *molding* if an over-determined system (6) on a neighborhood of each point of S satisfies the compatibility condition. Therefore S is molding if and only if S belongs to a continuous family of surfaces which are connected by isometries preserving principal distributions but not congruent in E^3 with one another. The following holds ([7]):

Theorem 4.2. *Any molding surface satisfies* $\beta\delta \equiv 0$.

Remark 4.2. The condition $\beta\delta \equiv 0$ is equivalent to the condition that some family of lines of curvature consists of geodesics. Originally, this remarkable property of molding surfaces was stated in [12, pp. 152–153] and provided with a careful proof in [11, pp. 277–281].

Remark 4.3. Although S has a family of lines of curvature which consists of geodesics, S is not necessarily molding. In [5], we can find a characterization of the semisurface structure of such a surface.

Molding surfaces are closely related to parallel curved surfaces, which were studied in [1, 2, 3]. A surface S in E^3 is said to be *parallel curved* if there exists a plane P such that at each point of S, at least one principal direction is parallel to P. A *canonical* parallel curved surface is represented as a disjoint union of plane curves which are congruent in E^3 with one another and tangent to principal directions. These curves are lines of curvature and geodesics. Noticing Theorem 4.2 and referring to [3], we obtain

Theorem 4.3. *For a surface S in E^3 with no umbilical point and nowhere zero Gaussian curvature, S is molding if and only if some neighborhood of each point of S is a canonical parallel curved surface.*

5. The existence of just two solutions

Let $(M, g, \mathcal{D}_1, \mathcal{D}_2)$ be a semisurface satisfying $K \neq L_0$ on M for some number L_0. Although (6) with (5) has a solution and does not satisfy the compatibility condition, $(g, \mathcal{D}_1, \mathcal{D}_2)$ does not necessarily determine a unique surface: according to Proposition 4.2, it is possible that (6) has distinct two solutions although (6) does not satisfy the compatibility condition. Noticing Theorem 3.2, we see that (6) with (5) has just two solutions on a neighborhood of each $p \in M$ if and only if there exist two isometric immersions ι_1, ι_2 of a neighborhood U of p into N satisfying

(i) $\mathcal{D}_1$, $\mathcal{D}_2$ give principal distributions;
(ii) there does not exist any isometry φ of N satisfying $\varphi \circ \iota_1 = \iota_2$;
(iii) for an isometric immersion ι of U into N satisfying the above (i), there exists an isometry φ of N satisfying $\varphi \circ \iota = \iota_1$ or ι_2.

The following holds ([8]):

Theorem 5.1. *Let $(M, g, \mathcal{D}_1, \mathcal{D}_2)$ be a semisurface with $K \neq 0$ on M. On a neighborhood of each point of M, an over-determined system (6) with (5) and $L_0 = 0$ has just two solutions if and only if we can choose local coordinates (u, v) compatible with $(\mathcal{D}_1, \mathcal{D}_2)$ so that g is locally represented as $g = A^2 du^2 + B^2 dv^2$, where A, B satisfy*

$$\frac{B_u}{A} = w_u, \qquad \frac{A_v}{B} = w_v,$$

and w is a nowhere zero function satisfying

$$(\log |\tanh w|)_{uv} \neq 0$$

and the sinh-Gordon equation

$$w_{uu} + w_{vv} = -\sinh 2w.$$

Remark 5.1. The condition $(\log |\tanh w|)_{uv} \neq 0$ means that the compatibility condition does not hold. Noticing Theorem 4.2 and referring to [3], we see that an over-determined system (6) on an isothermic surface S in E^3 with nowhere zero Gaussian curvature satisfies the compatibility condition if and only if S is part of a surface of revolution. Therefore (6) with (5) and $L_0 = 0$ has just two solutions so that we can choose isothermal coordinates as (u, v) in Theorem 5.1 if and only if the system has a solution such that the corresponding surface in E^3 is with nonzero constant mean curvature and not part of any surface of revolution.

6. Over-determined systems on minimal surfaces in E^3

Noticing the previous section, we naturally have interest in over-determined systems on minimal surfaces in E^3.

Let $(M, g, \mathcal{D}_1, \mathcal{D}_2)$ be a semisurface with $K \neq 0$ on M. Let X, Y, Z be as in (10) with $L_0 = 0$. Then whether $Y^2 - 4XZ$ is positive, zero or negative is determined by the semisurface structure $(g, \mathcal{D}_1, \mathcal{D}_2)$ and does not depend on the choice of local coordinates (u, v). The following holds ([9]):

Theorem 6.1. *Let $(M, g, \mathcal{D}_1, \mathcal{D}_2)$ be a semisurface with $K \neq 0$ on M. On a neighborhood of each point of M, an over-determined system (6) with (5) and $L_0 = 0$ has a unique solution so that $Y^2 - 4XZ$ vanishes if and only if we can choose local coordinates (u, v) compatible with $(\mathcal{D}_1, \mathcal{D}_2)$ so that g is locally represented as $g = A^2 du^2 + B^2 dv^2$, where A, B satisfy*

$$\frac{B_u}{A} = w_u, \qquad \frac{A_v}{B} = w_v,$$

and w is a function satisfying

$$w_{uv} + 2w_u w_v \neq 0, \qquad w_{uu} + w_{vv} = e^{-2w}.$$

Remark 6.1. The condition $w_{uv} + 2w_u w_v \neq 0$ means that the compatibility condition does not hold. We see that (6) with (5) and $L_0 = 0$ has a unique solution and satisfies $Y^2 - 4XZ = 0$ and the condition that we can choose isothermal coordinates as (u, v) in Theorem 6.1 if and only if the system has a solution such that the corresponding surface in E^3 is minimal and not part of any catenoid.

7. A two-dimensional manifold equipped with a 1-form and two one-dimensional distributions

Let $(M, g, \mathcal{D}_1, \mathcal{D}_2)$ be a semisurface with $K \neq 0$ on M. The following holds:

Proposition 7.1. *A function F is a solution of (6) with (5) and $L_0 = 0$ if and only if $F' := F + \log A^2$ is a solution of the following over-determined system*

$$F'_u = \alpha' + \beta' e^{F'}, \qquad F'_v - \delta' e^{-F'}, \tag{11}$$

where

$$\alpha' := (\log(I_u + J_v)^2)_u,$$
$$\beta' := \frac{2I}{I_u + J_v}, \tag{12}$$
$$\delta' := -2(I_u + J_v)J$$

and

$$I := \frac{B_u}{A}, \qquad J := \frac{A_v}{B}.$$

Let V_K be the canonical pre-divergence of $(M, g, \mathcal{D}_1, \mathcal{D}_2)$ and θ_K a 1-form on M defined by

$$\theta_K(w) := g(V_K, w)$$

for each tangent vector w at each point of M. Let M be oriented and set $\omega := - * \theta_K$, where $*$ is Hodge's $*$-operator on an oriented Riemannian manifold (M, g). Then for local coordinates (u, v) which are compatible with $(\mathcal{D}_1, \mathcal{D}_2)$ and give the positive orientation of M, ω is represented as

$$\omega = -J du + I dv.$$

By $K \neq 0$ together with

$$d\omega = (I_u + J_v) du \wedge dv,$$

we obtain $d\omega \neq 0$.

Let M be a two-dimensional manifold. Let $\mathcal{D}_1$, $\mathcal{D}_2$ be two one-dimensional distributions on M satisfying $\mathcal{D}_1(p) \neq \mathcal{D}_2(p)$ for any $p \in M$. Let ω be a 1-form on M satisfying $d\omega \neq 0$ at any point of M. Let (u, v) be local coordinates compatible with $(\mathcal{D}_1, \mathcal{D}_2)$ on a neighborhood U of each point p of M. We locally represent ω as $\omega = -J du + I dv$. Then we obtain $I_u + J_v \neq 0$. Let α', β', δ' be as in (12). Referring to the proof of Theorem 3.3, we see that there exist local coordinates (s, t) on U compatible with $(\mathcal{D}_1, \mathcal{D}_2)$ satisfying

$$\begin{aligned}
(I_u + J_v)^2 &= \left(-2 \int_{u_0}^{u} (I_u + J_v) I du + \left(dt \left(\frac{\partial}{\partial v} \right) \right)^2 \right) \\
&\quad \times \left(-2 \int_{v_0}^{v} (I_u + J_v) J dv + \left(ds \left(\frac{\partial}{\partial u} \right) \right)^2 \right),
\end{aligned} \tag{13}$$

where (u_0, v_0) corresponds to p, if and only if (11) with (12) has a solution. Whether there exist (s, t) as in (13) is determined by a triplet $(\omega, \mathcal{D}_1, \mathcal{D}_2)$ and does not depend on the choice of local coordinates (u, v). Therefore whether (11) with (12) has a solution is determined by $(\omega, \mathcal{D}_1, \mathcal{D}_2)$ and does not depend on the choice of (u, v). Referring to Theorem 3.2, we see that if there exists a solution F' of (11) for $(\omega, \mathcal{D}_1, \mathcal{D}_2)$, then for each pair (A, B) of positive-valued functions A, B satisfying $B_u/A = I$ and $A_v/B = J$, there exists an immersion of a neighborhood of each point of M into E^3, unique up to an isometry of E^3, satisfying

(i) the induced metric is given by $g = A^2 du^2 + B^2 dv^2$;
(ii) $\mathcal{D}_1$, $\mathcal{D}_2$ give principal distributions;
(iii) $k_1 := e^{F'/2}/A$, $k_2 := K/k_1$ are principal curvatures corresponding to $\mathcal{D}_1$, $\mathcal{D}_2$, respectively, where K is the curvature of (M, g).

As is explained in [9], there exist many pairs as the above (A, B), because of the existence and the uniqueness of a solution of some semilinear hyperbolic equation with a given initial data (see Chapter 5 of [13]) and therefore in the case where an over-determined system (11) for $(\omega, \mathcal{D}_1, \mathcal{D}_2)$ has a solution, each solution of (11) gives a family of surfaces in E^3. Then for a surface S in E^3 with no umbilical point and nowhere zero Gaussian curvature, an over-determined system (11) on S appears on other surfaces in the family which contains S. From the above discussion, we obtain the following ([9]):

Theorem 7.1. *Let M be an orientable two-dimensional manifold. Let $\mathcal{D}_1$, $\mathcal{D}_2$ be two one-dimensional distributions on M satisfying $\mathcal{D}_1(p) \neq \mathcal{D}_2(p)$ for any $p \in M$. Let ω be a 1-form on M satisfying $d\omega \neq 0$ at any point of M. Then for each $p \in M$, the following are mutually equivalent:*

(i) *there exists an immersion of a neighborhood of p into E^3 satisfying*

    (a) *$\mathcal{D}_1$, $\mathcal{D}_2$ give principal distributions,*
    (b) *$\omega = - * \theta_K$, where $*$ is Hodge's $*$-operator with respect to the induced metric g and some orientation of M, and θ_K is a 1-form defined by $\theta_K(\,\cdot\,) = g(V_K, \cdot)$ for the canonical pre-divergence V_K of $(M, g, \mathcal{D}_1, \mathcal{D}_2)$;*

(ii) *an over-determined system (11) with (12) has a solution on a neighborhood U of p;*

(iii) *a triplet $(\omega, \mathcal{D}_1, \mathcal{D}_2)$ satisfies (13) on U.*

Remark 7.1. We set

$$X' := \beta'_v, \quad Y' := \alpha'_v + 2\beta'\delta', \quad Z' := -\delta'_u + \alpha'\delta'. \tag{14}$$

Then the compatibility condition of (11) is given by $X' \equiv 0$, $Y' \equiv 0$ and $Z' \equiv 0$. Whether (11) with (12) satisfies the compatibility condition is determined by a triplet $(\omega, \mathcal{D}_1, \mathcal{D}_2)$ and does not depend on the choice of local coordinates (u, v). Referring to the proof of Theorem 4.2, we see that if (11) with (12) satisfies the compatibility condition, then $IJ \equiv 0$.

Remark 7.2. In the case where (11) with (12) does not satisfy the compatibility condition, the number of solutions of (11) with (12) is at most two and determined by $(\omega, \mathcal{D}_1, \mathcal{D}_2)$. Referring to the proof of Theorem 5.1, we can obtain a necessary and sufficient condition for the existence of just two solutions of (11) with (12). We see that (11) with (12) has just two solutions if and only if the system has a solution such that the corresponding family of surfaces contains a surface with nonzero constant mean curvature which is not part of any surface of revolution.

Remark 7.3. By (10) with (5) and $L_0 = 0$, we see that if X', Y', Z' are as in (14) with (12), $I = B_u/A$ and $J = A_v/B$, then

$$(Y')^2 - 4X'Z' = Y^2 - 4XZ.$$

For X', Y', Z' as in (14) with (12), whether $(Y')^2 - 4X'Z'$ is positive, zero or negative is determined by $(\omega, \mathcal{D}_1, \mathcal{D}_2)$ and does not depend on the choice of (u, v). Referring to the proof of Theorem 6.1, we can obtain a necessary and sufficient condition for the existence of a unique solution of (11) with (12) and $(Y')^2 - 4X'Z' = 0$. In [9], Theorem 6.1 is stated in terms of $(\omega, \mathcal{D}_1, \mathcal{D}_2)$. We see that (11) with (12) has a unique solution so that $(Y')^2 - 4X'Z'$ vanishes if and only if the system has a solution such that the corresponding family of surfaces contains a minimal surface which is not part of any catenoid.

8. A generalization of an over-determined system on a surface

Let M be a two-dimensional manifold. Let $\mathcal{D}_1$, $\mathcal{D}_2$ be two one-dimensional distributions on M satisfying $\mathcal{D}_1(p) \neq \mathcal{D}_2(p)$ for any $p \in M$. Let ω be a 1-form on M and Ω a 2-form on M which never vanishes at any point of M. Let (u, v) be local coordinates compatible with $(\mathcal{D}_1, \mathcal{D}_2)$ on a neighborhood of each point p of M. We locally represent ω and Ω as

$$\omega = -J du + I dv, \qquad \Omega = \rho du \wedge dv,$$

respectively. Then an over-determined system (11) with

$$\alpha' := (\log \rho^2)_u, \qquad \beta' := \frac{2I}{\rho}, \qquad \delta' := -2J\rho \tag{15}$$

is a generalization of (11) with (12). We see that

(i) whether (11) with (15) has a solution,
(ii) whether (11) with (15) satisfies the compatibility condition,
(iii) the number of solutions of (11) with (15) in the case where (11) with (15) does not satisfy the compatibility condition (this number is at most two),

(iv) whether $(Y')^2 - 4X'Z'$ is positive, zero or negative, where X', Y', Z' are defined as in (14) with (15)

are determined by ω, Ω, $\mathcal{D}_1$ and $\mathcal{D}_2$, and do not depend on the choice of (u, v). Referring to the proofs of Theorem 3.3, Theorem 4.2, Theorem 5.1 and Theorem 6.1, we can obtain the following ([10]):

Theorem 8.1. *An over-determined system* (11) *with* (15) *has a solution on a neighborhood U of each $p \in M$ if and only if there exist local coordinates (u, v) on U compatible with $(\mathcal{D}_1, \mathcal{D}_2)$ satisfying*

$$\rho^2 = \left(-2\int_{u_0}^u \rho I du + 1\right)\left(-2\int_{v_0}^v \rho J dv + 1\right),$$

where (u_0, v_0) corresponds to a point p.

Theorem 8.2. *An over-determined system* (11) *with* (15) *satisfies the compatibility condition on a neighborhood U of each $p \in M$ if and only if there exist local coordinates (u, v) on U compatible with $(\mathcal{D}_1, \mathcal{D}_2)$ such that ω and Ω are represented as in one of the following:*

(*i*) $\omega = f(u)dv$ *or* $g(v)du$, *and* $\Omega = du \wedge dv$, *where f, g are functions of one variable;*

(*ii*) $\omega = \rho(-du + dv)$ *and* $\Omega = \rho du \wedge dv$, *where ρ is a nowhere zero function such that $\psi := \log 8\rho^2$ is a solution of Liouville's equation $\psi_{uv} = e^{\psi}$.*

Theorem 8.3. *An over-determined system* (11) *with* (15) *has just two solutions on a neighborhood U of each $p \in M$ if and only if there exist local coordinates (u, v) on U compatible with $(\mathcal{D}_1, \mathcal{D}_2)$ such that*

$$f := -\frac{1}{2}\log\left(\sqrt{\rho^2 + 1} + |\rho|\right)$$

satisfies

$$(\log|\tanh f|)_{uv} \neq 0, \qquad \omega(w) = \Omega(\nabla f, w)$$

for any tangent vector w at any point of U, where ∇f is the gradient vector field of f with respect to a metric $|\rho|(du^2 + dv^2)$.

Theorem 8.4. *An over-determined system* (11) *with* (15) *has a unique solution so that $(Y')^2 - 4X'Z'$ vanishes on a neighborhood U of each $p \in M$ if and only if there exist local coordinates (u, v) on U compatible with $(\mathcal{D}_1, \mathcal{D}_2)$ such that*

$$f := -\frac{1}{2}\log|\rho|$$

satisfies

$$f_{uv} + 2f_u f_v \neq 0, \qquad \omega(w) = \Omega(\nabla f, w),$$

where w and ∇f are as in Theorem 8.3.

Remark 8.1. From Theorem 8.2, we see that although (11) with (15) satisfies the compatibility condition, IJ does not necessarily vanish. In the case where $\Omega = d\omega$, referring to [3] or [7], we can show that the compatibility condition of (11) with (15) is equivalent to condition (i) in Theorem 8.2.

References

1. N. Ando, A class of real-analytic surfaces in the 3-Euclidean space, Tsukuba J. Math. **26** (2002) 251–267.
2. N. Ando, Parallel curved surfaces, Tsukuba J. Math. **28** (2004) 223–243.
3. N. Ando, A two-dimensional Riemannian manifold with two one-dimensional distributions, Kyushu J. Math. **59** (2005) 285–299.
4. N. Ando, Semisurfaces and the equations of Codazzi-Mainardi, Tsukuba J. Math. **30** (2006) 1–30.
5. N. Ando, A surface which has a family of geodesics of curvature, Beiträge zur Algebra und Geometrie **48** (2007) 237–250.
6. N. Ando, The geodesic curvatures of lines of curvature, preprint.
7. N. Ando, Molding surfaces and Liouville's equation, preprint.
8. N. Ando, Over-determined systems in relation to principal curvatures, preprint.
9. N. Ando, A family of surfaces in E^3 given by an over-determined system, preprint.
10. N. Ando, Two generalizations of an over-determined system on a surface, preprint.
11. R. L. Bryant, S. S. Chern and P. A. Griffiths, Exterior differential systems, Proceedings of the 1980 Beijing Symposium on Differential Geometry and Differential Equations, Vol. 1 (1980) 219–338, Science Press, Beijing, 1982.
12. E. Cartan, Les systèmes différentiels extérieurs et leurs applications géométriques, 2nd ed., Hermann, 1971.
13. R. Courant and D. Hilbert, Methods of mathematical physics, Vol. II, John Wiley & Sons, 1962.
14. J. J. Stoker, Differential geometry, John Wiley & Sons, 1969.

Received July 30, 2012.

AUTHOR INDEX

Adachi, T., 44
Ando, N., 273

Boumuki, N., 119

Cheng, Q.-M., 147
Cho, J. T., 87, 245

Enomoto, K., 190

Fujimori, S., 19
Fujioka, A., 180

Hamada, T., 82
Hashimoto, K., 135
Hashinaga, T., 230

Ichiyama, T., 32
Ikawa, O., 220
Inoguchi, J., 87
Itoh, J., 190

Kajigaya, T., 260
Kim, D.-S., 128

Kim, Y. H., 128
Kimura, M., 245
Kobayashi, O., 16
Kubo, A., 230

Maeda, S., 1, 113

Noda, T., 119

Ohnita, Y., 60
Okumura, K., 98

Peng, Y., 147

Shoda, T., 19

Tamaru, H., 230
Tanabe, H., 113
Tanaka, M. S., 205

Udagawa, S., ix, 32

Zeng, L., 164